T0138358

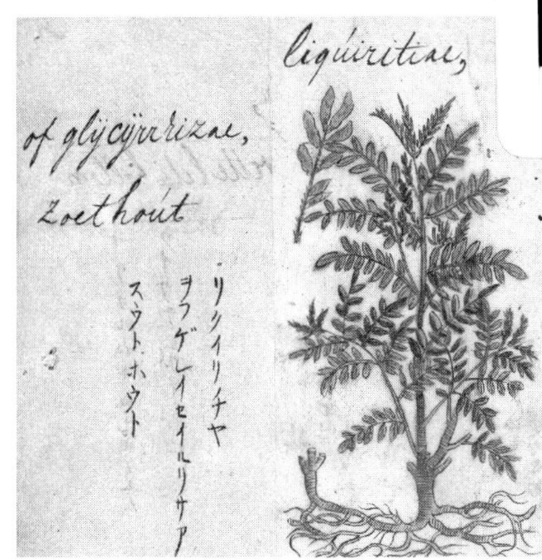

TRANSLATING MEDICINE ACROSS PREMODERN WORLDS

EDITED BY

Tara Alberts, Sietske Fransen, and Elaine Leong

OSIRIS | 37

A Research Journal Devoted to the
History of Science and Its Cultural Influences

Osiris

Series Editors, 2018–2022

W. PATRICK MCCRAY, *University of California Santa Barbara*

SUMAN SETH, *Cornell University*

Volumes 33 to 37 aim to connect the history of science with other areas of historical scholarship. Volumes of the journal are designed to explore how, where, and why science draws upon and contributes to society, culture, and politics. The journal's editors and board members strongly encourage proposals that engage with and examine broad themes while aiming for diversity across time and space. The journal is also very interested in receiving proposals that assess the state of the history of science as a field, broadly construed, in both established and emerging areas of scholarship.

33 LUKAS RIEPPEL, EUGENIA LEAN, & WILLIAM DERINGER, EDS., *Science and Capitalism: Entangled Histories*

34 AMANDA REES & IWAN RHYS MORUS, EDS., *Presenting Futures Past: Science Fiction and the History of Science*

35 E. C. SPARY & ANYA ZILBERSTEIN, EDS., *Food Matters: Critical Histories of Food and the Sciences*

36 HELEN TILLEY, ED., *Therapeutic Properties: Global Medical Cultures, Knowledge, and Law*

37 TARA ALBERTS, SIETSKE FRANSEN, & ELAINE LEONG, EDS., *Translating Medicine Across Premodern Worlds*

Series editor, 2013–2017

ANDREA RUSNOCK, *University of Rhode Island*

Volumes 28 to 32 in this series are designed to connect the history of science to broader cultural developments, and to place scientific ideas, institutions, practices, and practitioners within international and global contexts. Some volumes address new themes in the history of science and explore new categories of analysis, while others assess the "state of the field" in various established and emerging areas of the history of science.

28 ALEXANDRA HUI, JULIA KURSELL, & MYLES W. JACKSON, EDS., *Music, Sound, and the Laboratory from 1750 to 1980*

29 MATTHEW DANIEL EDDY, SEYMOUR H. MAUSKOPF, & WILLIAM R. NEWMAN, EDS., *Chemical Knowledge in the Early Modern World*

30 ERIKA LORRAINE MILAM & ROBERT A. NYE, EDS., *Scientific Masculinities*

31 OTNIEL E. DROR, BETTINA HITZER, ANJA LAUKÖTTER, & PILAR LEÓN-SANZ, EDS., *History of Science and the Emotions*

32 ELENA ARONOVA, CHRISTINE VON OERTZEN, & DAVID SEPKOSKI, EDS., *Data Histories*

On the Cover:

Japanese Herbal, 17th century. Contains various drawings of European plants with names/text in Old Dutch, Chinese, Japanese & Latin. Illustration 52. Attribution 4.0 International (CC BY 4.0)

Acknowledgments

As with many collaborative intellectual endeavors, *Translating Medicine across Premodern Worlds* is a project of long duration, and we have incurred many debts on our journey. This volume began as a "working group" project at the Max Planck Institute for the History of Science, and we are grateful to the participants of the two exploratory workshops—held in June 2017 in Berlin and London—many of whom have contributed essays to this collection. We are extremely grateful to Elma Brenner and Sandra Cavallo for their support and advice in organizing these workshops. Authors of this volume gathered again in Berlin and Rome for intense and lively discussions about the themes relating to histories of medicine and science, translation, and global histories. We thank especially Ronit Yoeli-Tlalim for her participation and input in the workshops, as well as commentators Maria Conforti and Sven Dupré, who have also helped us at various stages of the editing process of this *Osiris* volume. We are grateful to our funders: the Max Planck Society, Wellcome Collections, the Centre for the Study of the Body and Material Culture (Royal Holloway, University of London), the Culture and Communication Research Priming Fund (University of York), and the Society for the Social History of Medicine for their generous support. Finally, we thank all of the contributors for taking the time to think with us and for making editorial work such an enjoyable and rewarding experience.

The volume editors would like to thank *Osiris* editors Patrick McCray and Suman Seth and the *Osiris* editorial board for their support for and encouragement with this project. We are particularly grateful to the two anonymous reviewers for their richly detailed and immensely constructive comments on the volume. And we want to give special thanks to Beth Ina for copyediting the entire volume so meticulously.

Tara Alberts would also like to thank Mark Jenner, Simon Ditchfield, and Caroline Edwards for their advice and support with various funding applications, and Adam Perry for his unwavering support and for uncomplainingly reading endless drafts.

Sietske Fransen would like to thank Lorraine Daston, Sachiko Kusukawa, and Katherine Reinhart for their ongoing support and advice regarding this project, as well as Elisabetta Pastore, Anna Paulinyi, Ornella Rodengo, and Charlott Böhm for organizing the Rome author's workshop at the Bibliotheca Hertziana.

Elaine Leong would like to extend her sincere thanks to Lorraine Daston, Christine von Oertzen, members of former Department II (Daston), and the librarians at the Max Planck Institute for the History of Science for their unfailing support of this project.

Translating Medicine, ca. 800–1900:
Articulations and Disarticulations

by Tara Alberts, Sietske Fransen,† and Elaine Leong‡*

ABSTRACT

Research located at the nexus of medicine and translation deals with some of the fundamentals of human experience: the basic drive to survive and flourish and the urge to gather and to share information that might assist in this. Using a series of case studies ranging from ninth-century Baghdad, to fourteenth-century Aragon, to seventeenth-century Cartagena, to nineteenth-century Bengal, this volume weaves together an interconnected, long-view history of the translation of medicine. The geographically and temporally diverse contexts of our case studies explore common themes and divergent experiences, connected by our historical actors' varied endeavors to "translate" knowledge about health and the body across languages, practices, and media. Collectively, we offer a new approach to histories of (medical) knowledge, relocalizing and deconstructing traditional narratives, and de-emphasizing well-worn dichotomies.

The therapy of moxibustion, which involves the burning of the herb mugwort (moxa) on parts of the body, was widespread in various parts of premodern Asia. Chinese narratives connected the development of the therapy to legendary figures such as Fu Xi or the Yellow Emperor.[1] In Korea, legend attributed the invention of the therapy to Dangum, the legendary founder of the first Korean kingdom.[2] Japanese tradition held that the practice was introduced in 642 by the Buddhist monk Kiga Hotorike no Nanba, who had learned the technique in Korea.[3] Understandings of the theoretical underpinnings, the techniques, the material culture, and the expertise of practitioners, which

* Department of History, University of York, York, YO10 5DD, UK; tara.alberts@york.ac.uk.
† Bibliotheca Hertziana—Max Planck Institute for Art History, Via Gregoriana 28, 00187, Rome; fransen@biblhertz.it.
‡ Department of History, University College London, Gower Street, London, WC1E 6BT, UK; e.leong@ucl.ac.uk.

[1] Wang Shumin and Gabriel Fuentes, "A Survey of Images from the Chinese Medical Classics," in *Imagining Chinese Medicine*, ed. Vivienne Lo and Penelope Barrett (Leiden: Brill, 2018), 29–50; see 38.

[2] Lu Gwei-Djen and Joseph Needham, *Celestial Lancets: A History and Rationale of Acupuncture and Moxa* (Cambridge, UK: Cambridge Univ. Press, 1980), 262.

[3] Giovanni Borriello, "The Introduction of Moxibustion and Acupuncture in Europe from the Early Modern Period to the Nineteenth Century," in *New Directions in Literature and Medicine Studies*, ed. Stephanie M. Hilger (London: Palgrave Macmillan, 2017), 305–16, at 306.

Osiris, volume 37, 2022. © 2022 History of Science Society. All rights reserved. Published by The University of Chicago Press for the History of Science Society. https://doi.org/10.1086/719045.

together made up the practice, were similarly diverse, and reflected the wider cultural contexts in which the therapy was adopted.[4]

In the seventeenth century the practice came to the attention of European publics, first through travelers' tales and then through enterprising physicians who offered the novel therapy to their clients.[5] Commentators such as the Portuguese Jesuit Luis Frois (1532–97) and the Dutch minister Hermann Buschoff (1620–74) wrote of their observations of the practice in Asia, using terminology familiar to European audiences to explain the therapy.[6] Within decades, physicians in Europe developed new tools to try out the technique and developed new understandings of its efficacy.[7]

Therapies such as moxibustion traveled across the globe as texts, as material objects in the form of specimens, as images in herbals and diagrams illustrating practices, and as embodied practice. The construction and transmission of knowledge about health and the body required constant and multiple forms of translation.[8] It was linguistic and textual—with concepts framed in new languages as they traveled across cultural boundaries and medical traditions and between oral, manuscript, and print cultures. It was material, as items passed into new zones of understanding and were reinterpreted and refashioned. It was bodily, with the experiences of those who were healed bearing witness to new epistemologies of sickness and cure. Moreover, categories such as language, materiality, and body were never discrete and separate but rather co-constitutive. Translation occurred across and between them: as textual knowledge produced gesture, motion, and action; as materials were redefined in texts and images, and through practical use; as physical objects were reinterpreted as usable tools or ingestible curatives.[9] By analyzing these complexities of translation, the essays in this volume reimagine cultures of sickness and health.[10]

[4] See especially Lu and Needham, *Celestial Lancets* (cit. n. 2), and Vivienne Lo and Ronit Yoeli-Tlalim, "Travelling Light: Sino-Tibetan Moxa-Cautery from Dunhuang," in Lo and Barrett, *Imagining Chinese Medicine* (cit. n. 1), 271–90.

[5] For an overview see Michel Wolfgang, "Japanese Acupuncture and Moxibustion in 16–18th-century Europe," *Journal of the Japan Society of Acupuncture and Moxibustion* 61 (2011): 150–63.

[6] See Luis Frois, *The First European Description of Japan, 1585: A Critical English-language Edition of* Striking Contrasts in the Customs of Europe and Japan *by Luis Frois, SJ*, trans. and ed. Richard K. Danford, Robin D. Gill, and Daniel T. Reff (Abingdon, UK: Routledge, 2014); and Hermann Buschoff, Het pogagra *nader als oyt nagevorst en uytgevonden, midsgaders des selfs sekere genesingh of ontlastend hulp-middel* (Amsterdam: Jacobus de Jonge, 1675).

[7] See, for example, Margaret D. Garber, "Domesticating Moxa: The Reception of Moxibustion in a Late Seventeenth-Century German Medical Journal," in *Translation at Work: Chinese Medicine in the First Global Age*, ed. Harold J. Cook (Leiden: Brill, 2020), 134–56.

[8] On medicine and translation important recent interventions have been offered by, for example, Shigehisa Kuriyama, *The Expressiveness of the Body and the Divergence of Greek and Chinese Medicine* (New York: Zone Books, 1999); Andrew E. Goble, *Confluences of Medicine in Medieval Japan: Buddhist Healing, Chinese Knowledge, Islamic Formulas, and Wounds of War* (Honolulu: Univ. of Hawaii Press, 2011); C. Pierce Salguero, *Translating Buddhist Medicine in Medieval China* (Philadelphia: Univ. of Pennsylvania Press, 2014); Hans Pols, C. Michele Thompson, and John Harley Warner, eds., *Translating the Body: Medical Education in Southeast Asia* (Singapore: National Univ. of Singapore Press, 2017); Cook, *Translation at Work* (cit. n. 7); and Ronit Yoeli-Tlalim, *ReOrienting Histories of Medicine: Encounters along the Silk Roads* (London: Bloomsbury Academic, 2021).

[9] See, for example, Elisabeth Hu, "Towards a Science of Touch, Part I: Chinese Pulse Diagnostics in Early Modern Europe," *Anthropology and Medicine* 7.2 (2000): 251–68; and Hu, "Towards a Science of Touch, Part II: Representations of the Tactile Experience of the Seven Chinese Pulses indicating Danger of Death in Early Modern Europe," *Anthropology and Medicine* 7.3 (2000): 3–16.

[10] In doing so, we join and build on a rich field of studies. See, for example, the valuable interventions made by contributors to the Isis Focus section "Global Histories of Science," ed. Sujit Sivasundaram, *Isis* 101 (2010); and the recent special issue of *History and Technology* 34 (2018), ed. Gabriela Soto Laveaga and Pablo F. Gómez.

Our articles approach translation practices through a series of temporally and chronologically diverse case studies. The period between 800 and 1900 saw a dramatic increase in travel and trade across the globe, which came with intensified exchange of knowledge, goods, and practices. In the early part of the period overland trade routes were expanded, consolidated, or redrawn as Muslim empires extended their reach across Eurasia and into Africa, and as "silk road" trade developed and changed. Maritime trade—across the Indian Ocean, into the Pacific, and later into the Atlantic—also expanded, as did the colonial ambition of various imperial powers. At the end of the period, the rise of modern colonialism forced new connections, mobility, and exchange.

Merchants and other travelers to new regions strove to maintain their health in novel environments with unfamiliar tools, materia medica, and foodstuffs. They were bombarded with unaccustomed ways of thinking about the human body and nature and, crucially for them, a range of wondrous new drugs. As travel increased, so too did the shared experience of epidemic disease, which drove interest in new remedies and approaches. New medical theories were constructed under the influence of encounters with other cultures. As longer-distance travel and trade increased, so did the complexities of these processes of translation.

Translating Medicine is framed by the metaphor of articulation/disarticulation, through which we argue for a new approach to these complex histories of translation and exchange. We examine in tandem the constructive and destructive processes inherent in translation practices and draw attention to that which is lost, destroyed, omitted, and erased. We explore how these processes play out in multiple spheres and contexts. First, we reassess translation as a textual practice, arguing for the need to see translation as a form of "archive making" and part of a wider, interconnected array of epistemic practices. Second, we show how analysis of textual and linguistic translation practices must be firmly grounded in the broader contexts of the material, visual, oral, and sociocultural worlds of actors. Third, we turn to analyzing the agency, identities, and expertise of our historical actors, demonstrating how translators, healers, and "translator-practitioners" articulated their authority and expertise in these complex spaces of exchange. Synergies and connections join narratives across the volume, illustrating the interconnectedness of the processes explored in each section. Examining the collected case studies comparatively and connectively proposes a range of techniques for analyzing processes of translation and for uncovering voices muffled by historical practices of translation. We offer a new paradigm for approaching histories of knowledge creation.

TRANSLATION, MEDICINE, AND THE HISTORY OF SCIENCE

Translation has come under intense scrutiny by historians of science, and recent years have witnessed a flourishing of publications.[11] The cultural turn in history, literary, and translation studies opened new horizons, prompting scholars to examine "cultures

[11] For an overview, see Marwa Elshakry and Carla Nappi, "Translations," in *A Companion to the History of Science*, ed. Bernard Lightman (Chichester, West Sussex, UK: Wiley-Blackwell, 2016), chap. 26; Bettina Dietz, ed., "Translating and Translations in the History of Science," special issue, *Annals of Science* 73, no. 2 (2016); Michael Gordon, ed., "Linguistic Hegemony and the History of Science," Focus section, *Isis* 108, no. 3 (2017): 606–50; Sven Dupré, ed., "Translating Science over Time," Focus section, *Isis* 109, no. 2 (2018): 302–45; and H. Floris Cohen, ed., "Historians of Science Translating the History of Science," Focus section, *Isis* 109, no. 4 (2018): 774–95. For the premodern context, see, for example, Scott L. Montgomery, *Science in Translation: Movements of Knowledge*

of translation" and the "translation of cultures," variously defined.[12] Taking a cue from these approaches, historians of science also look beyond notions of the fidelity of the translator or the translation, and while not adopting the term directly, often align with the view of translations as "transformissions," emphasizing that each literary act, practice, and text has its own story.[13]

Until recently, histories of translation in science often excluded medicine from their purview. This volume reaffirms not only that medicine is "the most universal and oldest form of translation" but also that studies of medical translation can have wider applications for other fields of science.[14] Analysis of scholarly traditions and the writings of learned practitioners, for example, have begun to reorient our picture of the language regimes and intellectual networks that facilitated the exchange of medical knowledge.[15] Work on the gradual, iterative, and multi-actor processes of translation in Byzantine, Abbasid, Mughal, Ottoman, and Persian contexts, for example, has offered a polycentric and dynamic picture of multiple, interactive networks of translation between a bewildering variety of languages.[16] Such work has also brought into focus multilingual textual production activities at intellectual hubs—from silk road centers, including Duanhuang, Turfan, and Gandahara, to the maritime polities of the Malay archipelago—complicating existing narratives of knowledge exchange.[17]

---

through Cultures and Time (Chicago: Univ. of Chicago Press, 2000); Michèle Goyens, Pieter de Leemans, and An Smets, eds., *Science Translated: Latin and Vernacular Translations of Scientific Treatises in Medieval Europe* (Leuven: Leuven Univ. Press, 2008); Sietske Fransen, Niall Hudson, and Karl E. Enenkel, eds., *Translating Early Modern Science* (Leiden: Brill, 2017); Jaime Marroquin Arredondo and Ralph Bauer, eds., *Translating Nature: Cross-Cultural Histories of Early Modern Science* (Philadelphia: Univ. of Pennsylvania Press, 2019); and Rocio G. Sumillera, Jan Surman, and Katharina Kühn, eds., *Translation in Knowledge, Knowledge in Translation* (Amsterdam: John Benjamins, 2020).

[12] Jeanette Beer and Kenneth Lloyd-Jones, eds., *Translation and the Transmission of Culture between 1300 and 1600* (Kalamazoo, MI: Western Michigan Univ. Medieval Institute, 1995); Peter Burke and R. Po-chia Hsia, eds., *Cultural Translation in Early Modern Europe* (Cambridge, UK: Cambridge Univ. Press, 2007); Sara Barker and Brenda M. Hosington, eds., *Renaissance Cultural Crossroads: Translation, Print and Culture in Britain, 1473–1640* (Boston, MA: Brill, 2013); Francesca Orsini and Neelam Srivastava, "Translations and the Postcolonial," *Interventions* 15 (2013): 323–31; Karen Newman and Jane Tylus, *Early Modern Cultures of Translation* (Philadelphia: Univ. of Pennsylvania Press, 2015).

[13] On the idea of translation as "transformissions," see Marie-Alice Belle and Brenda M. Hosington, "Introduction: Translation as 'Transformission' in Early Modern England and France," *Canadian Review of Comparative Literature / Revue Canadienne de Littérature Comparée* 46 (2019): 201–4.

[14] Henry Fischbach, *Translation and Medicine* (Amsterdam: Benjamins, 1998), 1.

[15] See, for example, Nancy Siraisi, *History, Medicine, and the Traditions of Renaissance Learning* (Ann Arbor: Univ. of Michigan Press, 2007), chap. 2. Scholars of medieval medicine such as Monica Green and Michael McVaugh have been particularly active in this area; see also Montserrat Cabré, "Female Authority in Translation," in this volume (*Osiris* 37), for further references. See also Marta Hanson and Gianna Pomata, "Medicinal Formulas and Experiential Knowledge in the Seventeenth-Century Epistemic Exchange between China and Europe," *Isis* 108 (2017): 1–25.

[16] See, for example, David Bennett, "Medical Practice and Manuscripts in Byzantium," *Soc. Hist. Med.* 13 (2000): 279–91; Sheldon Pollack, ed., *Forms of Knowledge in Early Modern Asia* (Durham, NC: Duke Univ. Press, 2011); and Ahmed Ragab, "'In a Clear Arabic Tongue': Arabic and the Making of a Science-Language Regime," *Isis* 108 (2017): 612–20.

[17] See, for example, Thomas T. Allsen, *Culture and Conquest in Mongol Eurasia* (Cambridge, UK: Cambridge Univ. Press, 2001); Vivienne Lo and Christopher Cullen, eds., *Medieval Chinese Medicine: The Dunhuang Medical Manuscripts* (London: Routledge, 2004); Ronit Ricci, *Islam Translated: Literature, Conversion, and the Arabic Cosmopolis of South and Southeast Asia* (Chicago: Univ. of Chicago Press, 2010); and Ronit Yoeli-Tlalim, "The Silk Roads as a Model for Exploring Eurasian Transmissions of Medical Knowledge: Views from the Tibetan Medical Manuscripts of Dunhuang," in *Entangled Itineraries: Materials, Practices, and Knowledges across Eurasia*, ed. Pamela Smith (Pittsburgh, PA: Univ. of Pittsburgh Press, 2019), 47–62.

Within the history of medicine, the material turn has also brought a range of fresh perspectives that can be fruitfully applied to the study of translation.[18] Historians have charted how materia medica traveled across the premodern world and how various processes of "translation" were required to render substances used in one context to fit the needs of another.[19] For instance, recent work has illuminated how the curative properties and cultural meanings of substances such as cinchona bark, asafoetida, musk, and china root changed as they circulated.[20] Attention to layered meanings and shifting connotations has helped to elucidate some of the ways in which crossing cultural and linguistic borders re-entangled materials in new webs of significance.[21] Similarly, historians who introduced new focuses—from practice, to bodily experience, to emotion—have offered new approaches to these circulations and exchanges.[22]

This volume bring this rich historiography to current conversations about translation and the history of science, answering the call for a closer engagement between historians of science and technology and historians of medicine.[23] As John Pickstone and Michael Worboys have argued, not only are the histories of science, technology, and medicine closely interlinked, but the historiography of medicine, strong on social and cultural analysis, brings new insight to studies of scientists as professional and public figures and highlights the study of vernacular knowledge and everyday practices, thereby shining light on "synchronic assemblages of practices and hybrid knowledges."[24] This emphasis is particularly relevant for our case studies, which examine a broad set of activities located across wide swathes of time and space where cognates of concepts such as "medicine," "science," and "technology" might connote diverse practices. As Marwa Elshakry has argued, the history of science to a great extent has escaped the

[18] See especially Arjun Appadurai, ed., *The Social Life of Things: Commodities in Cultural Perspective* (Cambridge, UK: Cambridge Univ. Press, 1986); Finbarr B. Flood, *Objects of Translation: Material Culture and Medieval "Hindu–Muslim" Encounter* (Princeton, NJ: Princeton Univ. Press, 2009); Daniela Bleichmar and Meredith Martin, eds., "Objects in Motion in the Early Modern World," special issue, *Art History* 38, no. 4 (2015); Craig Clunas, "Connected Material Histories: A Response," *Modern Asian Studies* 50 (2016): 61–74; and Anne Gerritsen and Giorgio Riello, eds., *The Global Lives of Things: The Material Culture of Connections in the Early Modern World* (London: Routledge, 2016).

[19] David Arnold, ed., *Warm Climates and Western Medicine: The Emergence of Tropical Medicine* (Amsterdam: Rodolphi Press, 1996); Kuriyama, *Expressiveness of the Body* (cit. n. 8); Harold J. Cook, *Matters of Exchange: Commerce, Medicine, and Science in the Dutch Golden Age* (New Haven, CT: Yale Univ. Press, 2007); Pratik Chakrabarti, *Materials and Medicine: Trade, Conquest and Therapeutics in the Eighteenth Century* (Manchester: Univ. of Manchester Press, 2010); Goble, *Confluences of Medicine in Medieval Japan* (cit. n. 8); Samir Boumediene, *La colonisation du savoir* (Vaulx-en-Velin: Édition des Mondes à faire, 2016).

[20] Anna E. Winterbottom, "Of the China Root: A Case Study of the Early Modern Circulation of Materia Medica," *Soc. Hist. Med.* 28 (2015): 22–44; Matthew James Crawford, *The Andean Wonder Drug: Cinchona Bark and Imperial Science in the Spanish Atlantic, 1630–1800* (Pittsburgh, PA: Univ. of Pittsburgh Press, 2016); Anya H. King, *Scent from the Garden of Paradise: Musk and the Medieval Islamic World* (Leiden: Brill, 2017); Angela Ki Che Leung and Ming Chen, "The Itinerary of Hing/Awei/Asafetida across Eurasia, 400–1800," in Smith, *Entangled Itineraries* (cit. n. 17), 141–64; Yoeli-Tlalim, *ReOrienting Histories of Medicine* (cit. n. 8), chap. 3.

[21] See, for example, Linda L. Barnes, *Needles, Herbs, Gods, and Ghosts: China, Healing, and the West to 1848* (Cambridge, MA: Harvard Univ. Press, 2005), esp. 162–211; and Tara Alberts, "Curative Commodities in Southeast Asia," in Smith, *Entangled Itineraries* (cit. n. 17), 79—98.

[22] Kuriyama, *Expressiveness of the Body* (cit. n. 8); Pols, Thompson, and Warner, *Translating the Body* (cit. n. 8).

[23] John Pickstone and Michael Worboys, eds., "Between and Beyond 'Histories of Science' and 'Histories of Medicine,'" Focus section, *Isis* 102, no. 1 (2011): 97–133.

[24] Ibid., 98–99, on 99.

postmodern disruption of epistemological categories that has challenged heuristic certainties in other disciplines.[25] We see throughout our essays how the boundaries of what counted as "medicine" could shift for our actors as a result of moments of translation and exchange.

Marwa Elshakry and Carla Nappi have argued that a focus on translation can also help us rethink conventional periodizations—ancient, medieval or early modern, and modern—and geographical distinctions. They contend that translation can "show us how traditional modes of shaping historical time have been forged or broken . . . enabl[ing] the creation of new dialogues and relationships across time."[26] Our twin analytical lenses of translation and medicine enable us to converse across chronologically and geographically diverse case studies. We thereby offer new perspectives on the movement of knowledge, skills, material objects, and people around the globe, and suggest how these stories might disrupt traditional narratives.

TRANSLATION AS AN ANALYTIC: ARTICULATION AND DISARTICULATION

To interrogate a number of related practices across varied cultural contexts, we argue for the need to explore "translation" from a number of perspectives. In translation studies and cultural studies, the concept of translation underwent transformation owing to the application of theories from the philosophy of language, linguistics, semiotics, and sociology.[27] This has helped to problematize the processes of translation and the figure of the translator.[28]

As many have noted, the English term *translation* itself could be seen as untranslatable.[29] While many societies have developed practices and norms concerning the transmission of written or oral texts from one language, script, or medium to another, the semantic field of descriptors used to describe these practices carry continually shifting connotations.[30] To elucidate knowledge transfer across time and space, we examine a wide range of practices that share certain features. For our authors, "translation" involves, first, the movement of a subject across a boundary (linguistic, cultural, material, real, or imagined). Second, coterminous with this movement, there is a change or alteration in the subject. Third, the change or alteration is *intentionally*

[25] Marwa Elshakry, "When Science Became Western: Historiographical Reflections," *Isis* 101 (2010): 98–109, on 99.

[26] Elshakry and Nappi, "Translations" (cit. n. 11), 381–2.

[27] See especially Walter Benjamin, "The Task of the Translator," in *Selected Writings*, ed. Marcus Bullock and Michael W. Jennings (Cambridge, MA: Belknap Press, 2002), 1:253–63; Roman Jakobson, "On Linguistic Aspects of Translation," in *On Translation*, ed. Reuben A. Brower (Cambridge, MA: Harvard Univ. Press, 1959), 232–9; Lydia H. Liu, "The Question of Meaning-Value in the Political Economy of the Sign," in *Tokens of Exchange: The Problem of Translation in Global Circulations*, ed. Liu (Durham, NC: Duke Univ. Press, 1999), 13–42; and Maria Tymoczko and Edwin Gentzler, eds., *Translation and Power* (Amherst, MA: Univ. of Massachusetts Press, 2002).

[28] Helpful overviews of developments in these fields are provided by Susan Bassnett and André Lefevere, eds., *Translation, History, and Culture* (London: Pinter Publishers, 1990); George Steiner, *After Babel: Aspects of Language and Translation*, 3rd ed. (Oxford: Oxford Univ. Press, 1995); Umberto Eco, *Experiences in Translation* (Toronto: Univ. of Toronto Press, 2001); Lawrence Venuti, *The Translator's Invisibility: A History of Translation* (London: Routledge, 1995); and Mark Polizzotti, *Sympathy for the Traitor: A Translation Manifesto* (London: MIT Press, 2018).

[29] Ricci, *Islam Translated* (cit. n. 17), chap. 2, "On 'Translation' and Its Untranslatability," 31–65.

[30] Ibid., 31–3; Martha P. Y. Cheung, "Chinese Discourse on Translation as Intercultural Communication: The Story of *jihe* (幾何)," in *Translation: A Multidisciplinary Approach*, ed. Juliane House (Basingstoke: Palgrave Macmillan, 2014), 56–72.

brought about by actors who are intent on making the subject utilizable for a new audience. Our broad analytic approach provides historians with a methodological armamentarium for use in approaching histories of knowledge creation.

Our analysis throughout is animated by the twin themes of articulation and disarticulation. These metaphors, which are at once literary and medical, capture the spectrum of constructive and destructive aspects of translation processes that concern putting things into words, joining things together, and also creating divisions to render things comprehensible.[31] In their multiple meanings, each term evokes a number of fruitful debates that have animated various scholarly disciplines. Cultural theorists, sociologists, and human geographers have made extensive use of the concept of "articulation," engaging especially with Stuart Hall's exegesis of the term.[32] Hall emphasized the duality of the term: on the one hand it evoked "two parts . . . connected to each other, but through a specific linkage, that can be broken," and on the other, "languageing, of expressing."[33] Building on this work, and on Antonio Gramsci's theories of language and "translatability," scholars have drawn attention to the linguistic dimensions of these processes, examining the role of language in bringing together disparate things in the construction of various "unities" in the social world and in its discourses.[34] Similarly, Bruno Latour suggests using the metaphor of "articulation" to examine the connections established between all sorts of entities, from words to all aspects of the world they describe, including "gestures, papers, settings, instruments, sites, trials."[35] He explores how in a scientist's laboratory, these disparate things brought into conjunction are both articulated and become "more articulate" themselves, indeed create new "articulate" entities.[36]

For our purposes, the most important insight from this type of analysis is the focus placed on the contingency and context-dependence of all forms of "articulation," from the connections between social groups, to the semiotic connections between words and the things they represent.[37] Inherent in all these notions of articulation is the possibility of disarticulation, as linkages are dissolved or no longer sustained. In this way translation creates contingent connections between texts, materials, languages, and meaning in varied contexts.

[31] A few other scholars have noted the connections between linguistic and bodily disarticulation. See, for example, Marjorie Garber, "Out of Joint," and Stephen Greenblatt, "Mutilation and Meaning," in *The Body in Parts: Fantasies of Corporeality in Early Modern Europe*, ed. David Hillman and Carla Mazzio (Madison, NY: Routledge, 1997), 23–52 and 221–42 respectively; and Kylee-Anne Hingston, *Articulating Bodies: The Narrative Form of Disability and Illness in Victorian Fiction* (Liverpool: Liverpool Univ. Press, 2019).

[32] Useful overviews of the application of this concept in various disciplines are given by Jennifer Daryl Slack, "The Theory and Method of Articulation in Cultural Studies," in *Stuart Hall: Critical Dialogues*, ed. David Morley and Kuan-Hsing Chen (London: Routledge, 1996), 124–40.

[33] Stuart Hall and Lawrence Grossman, "On Postmodernism and Articulation: An Interview with Stuart Hall," *Journal of Communication Inquiry* 10 (1986): 45–60, on 53. See also John Clarke, "Stuart Hall and the Theory and Practice of Articulation," *Discourse: Studies in the Cultural Politics of Education* 36 (2015): 275–86.

[34] See especially Kevin DeLuca, "Articulation Theory: A Discursive Grounding for Rhetorical Practice," *Philosophy and Rhetoric* 32 (1999): 334–48; Peter Ives and Rocco Lacorte, eds., *Gramsci, Language, and Translation* (Lanham, MD: Lexington Books, 2010); and Michael Ekers, Stefan Kipfer, and Alex Loftus, "On Articulation, Translation, and Populism: Gillian Hart's Postcolonial Marxism," *Annals of the American Association of Geographers* (2020): 1–17.

[35] Bruno Latour, *Pandora's Hope: Essays on the Reality of Science Studies* (Cambridge, MA: Harvard Univ. Press, 1999), 142–3.

[36] Ibid., 144.

[37] Hall and Grossman, "On Postmodernism and Articulation" (cit. n. 33).

Disarticulation could be practiced intentionally with the intent of preserving and improving, but it could also be a form of erasure, as concepts and connotations are left fragmented or unspoken. The theme of disarticulation thus speaks to broader issues of the operation of power in translation.[38] In a Gramscian sense translation is then always understood as political: investigating translations provides the analytical means to understand sociocultural-political circumstances.[39] Our articles address crucial structural themes, including the consolidation of social and political hierarchies and the making and maintenance of state and colonial power. The processes of making legible, demarking boundaries, and ordering necessarily involve erasure and destruction as these processes impose their own intellectual hegemonies, often resulting in what Rolando Vázquez has termed "epistemicide."[40] For example, concepts of universalism or the search for universal languages and shared meanings result in erasure, as incommensurate understandings and ontologies are misunderstood, disregarded. or reinterpreted.[41] Translation into and between languages such as Latin, Greek, Arabic, and Classical Chinese could have a silencing effect on other languages, which were considered unequal to carry the burden of weighty knowledge concepts.[42] The disarticulation or violence of translation goes beyond that which is exemplified in studies framed around ideas about colonialism, empire, or commercial history.[43]

In her recent article in *History and Technology*, Gabriela Soto Laveaga challenged us to adopt the "largo dislocare" approach of connecting microhistories as a way to "dislocate histories not just geographically but also chronologically to better understand the motion of people, ideas and objects."[44] The study of translation is a key part of this approach and, indeed, many of our contributions adopt microhistory as a methodology to interrogate global exchange.[45] Our focus on the articulations and disarticulations of translation raises a set of common questions about situated epistemic practices and the various frames—institutional, political, economic, cultural, social—that shape knowledge production. Putting our stories in concert, we seek out resonances such as cognate practices or points of resistance. Examining translation through texts, beyond texts, and through experiences of historical actors, we suggest a new approach to exploring "knowledge in transit."[46]

[38] See especially Mona Baker, *Translation and Conflict: A Narrative Account* (London: Routledge, 2006); Edwin Gentzler and Maria Tymoczko, "Introduction," in *Translation and Power* (cit. n. 27), xi–xxviii; and Vicente L. Rafael, *Contracting Colonialism: Translation and Christian Conversion in Tagalog Society under Early Spanish Rule* (Durham, NC: Duke Univ. Press, 1992).

[39] Ives and Lacorte, *Gramsci, Language, and Translation* (cit. n. 34), 9.

[40] See Rolando Vázquez, "Translation as Erasure: Thoughts on Modernity's Epistemic Violence," *Historical Sociology* 24 (2011): 27–44, on 29. On legibility see also James C. Scott, *Seeing Like a State: How Certain Schemes to Improve the Human Condition Have Failed* (New Haven, CT: Yale Univ. Press, 1999).

[41] Vázquez, "Translation as Erasure" (cit. n. 40), 36–40.

[42] See, for example, Rafael, *Contracting Colonialism* (cit. n. 38).

[43] See Rosalind C. Morris, "Introduction," in *Can the Subaltern Speak? Reflections on the History of an Idea*, ed. Morris (New York: Columbia Univ. Press, 2010), 1–20.

[44] Gabriela Soto Laveaga, "Largo Dislocare: Connecting Microhistories to Remap and Recenter Histories of Science," *Hist. & Tech.* 34 (2018): 21–30.

[45] On global history and microhistory, see John-Paul A. Ghobrial, ed., "Global History and Microhistory," issue supplement 14, *Past and Present* 242 (2019).

[46] James A. Secord, "Knowledge in Transit," *Isis* 95 (2004): 654–72; Kapil Raj, *Relocating Modern Science: Circulation and Construction of Knowledge in South Asia and Europe, 1650–1900* (Basingstoke: Palgrave Macmillan, 2007); Neil Safier, "Global Knowledge on the Move: Itineraries, Amerindian Narratives, and Deep Histories of Science," *Isis* 101 (2010): 133–45; Patrick Manning and Abigail Owen, eds., *Knowledge in Translation: Global Patterns of Scientific Exchange, 1000–1800 CE* (Pittsburgh, PA: Univ. of Pittsburgh Press, 2018).

THE VOLUME

Section 1: Archives and the Authority of Practice

In recent years, archives have received considerable attention from historians of science and medicine.[47] As others have noted, the material turn drove us to consider pen and paper–based practices as technologies of information management, and as we shone light on those practices, archives have emerged as central to the production of knowledge. Thus far, these conversations have largely concentrated on two historical subfields: archival practices and learned practices of note taking, excerpting, and compiling.[48] Somewhat surprisingly, practices of translation have not been centrally featured. Yet, it is clear that as cognate processes, translation, note taking, and archive making often went hand in hand, and as historians of science have argued, these also functioned in conjunction with other epistemic practices, such as observation in the making of natural knowledge.[49] Putting histories of translation in conversation with these recent analytical discussions, the four essays in this section contend that acts of translation need to be interrogated alongside other paper-based knowledge practices. In studying these acts of articulation, and particularly disarticulation, we pay heed to the hegemonic tendencies inherent in archive creation.

Our case studies concern knowledge communities situated across diverse contexts and yet connected by a similar set of textual practices: translating, compiling, *and* archiving. Using the notion of an archive as a heuristic device, we take a macro-level view and attend to the changing power relations underlying knowledge practices.[50]

[47] For an overview, see Elizabeth Yale, "The History of Archives: The State of the Discipline," *Book History* 18 (2015): 332–59. Recent publications include Kathryn Burns, *Into the Archive: Writing and Power in Colonial Peru* (Durham, NC: Duke Univ. Press, 2010); Markus Friedrich, *Die Geburt des Archivs, Eine Wissensgeschichte* (Berlin: De Gruyter, 2013); and Kate Peters, Alexandra Walsham, and Liesbeth Corens, eds., *Archives and Information in the Early Modern World* (Oxford: Oxford Univ. Press, 2018). Key works by historians of medicine, science, and technology include Michael Hunter, ed., *Archives of the Scientific Revolution: The Formation and Exchange of Ideas in Seventeenth-Century Europe* (Rochester, NY: Boydell Press, 1998); Warwick Anderson, "The Case of the Archive," *Critical Inquiry* 39 (2013): 532–47; Lorraine Daston, *Science in the Archives: Pasts, Presents, Futures* (Univ. of Chicago Press, 2017); and Vera Keller, Anna Marie Roos, and Yale, *Archival Afterlives: Life, Death, and Knowledge-Making in Early Modern British Scientific and Medical Archives* (Leiden: Brill, 2018).

[48] Keller, Roos, and Yale, *Archival Afterlives* (cit. n. 47), 8. On learned practices see, for example, Ann Blair, *Too Much to Know. Managing Scholarly Information before the Modern Age* (New Haven, CT: Yale Univ. Press, 2010); Anthony Grafton and Glenn W. Most, *Canonical Texts and Scholarly Practices: A Global Comparative Approach* (New York: Cambridge Univ. Press, 2016); Anthony Grafton, *Inky Fingers: The Making of Books in Early Modern Europe* (Cambridge, MA: Harvard Univ. Press, 2020); and Blair, Paul Duguid, Anja-Silvia Goeing, and Grafton, eds., *Information: A Historical Companion* (Princeton, NJ: Princeton Univ. Press, 2021).

[49] Lorraine Daston, "The Sciences of the Archive," *Osiris* 27 (2012): 156–87. Recent overviews of early modern note-taking and the history of science include Dana Jalobeanu, "The Toolbox of the Early Modern Natural Historian: Notebooks, Commonplace Books, and the Emergence of Laboratory Records," *Journal of Early Modern Studies* 4 (2015): 107–23; Boris Jardine, "State of the Field: Paper Tools," *Studies in History and Philosophy of Science, Part A* 64 (2017): 53–63; and Elaine Leong, "Read. Do. Observe. Take Note!," *Centaurus* 60 (2018): 87–103. Anke te Heesen uses the term *paper technologies* in "The Notebook: A Paper-Technology," in *Making Things Public: Atmospheres of Democracy*, ed. Bruno Latour and Peter Weibel (Cambridge, MA: MIT Press, 2005), 582–9.

[50] Ann Laura Stoler, *Along the Archival Grain* (Princeton, NJ: Princeton Univ. Press, 2008); Burns, *Into the Archive* (cit. n. 47); Keller, Roos, and Yale, *Archival Afterlives* (cit. n. 47).

This focus on viewing translation processes within the larger schemes of archive making leads us to craft longer, more expansive narratives of knowledge production. If past studies tend to shine light on single instances of linguistic translations, we have chosen to recover epistemic acts occurring *before*, *during*, and *after* points of translation.

The archives featured in this volume vary considerably. Some, such as the library created by Liu Zhi (1660–1730), the Chinese literatus in Dror Weil's essay, are physical and paper-filled, consisting of a plethora of books and scrolls. Others are abstract ideas, such as Alisha Rankin's notion of an "archive of practice." In this case, the "archive" is a corpus of indigenous Amerindian knowledge based on practices rather than texts which was articulated or disarticulated by European vernacular translators as they sought to convey information about New World drugs to home audiences. We are concerned with both large-scale institutional repositories, such as the Abbasid court at the center of Ahmed Ragab's study, and personal household archives, such as the eighteenth-century Englishwoman Rebecca Tallamy's (fl. 1730s) recipe-filled copy of a printed distillation manual as outlined by Elaine Leong. By scrutinizing a wide range of archives under the same analytical lens, we bring into conversation practices situated across time and space and track the epistemic impact of translation and archiving practices across public and private spheres and communication media. Moreover, where past studies of medical archives tend to feature cases and observations, our broader remit brings to light the myriad ways in which health practitioners constructed fonds of knowledge as part of their everyday medical practices.

For many historical actors in our stories, translation efforts began with a search for textual and oral knowledge. For instance, in his essay on Chinese translations of Arabo-Persian natural philosophical texts in seventeenth-century Nanjing, Weil paints a vivid picture of Liu Zhi scouring the Chinese empire for manuscripts. Visiting bookstores and private libraries and relying on local literati for linguistic help, Liu assembled a treasure trove of Arabic-Persian texts that served as the core of his scholarly practices.[51] While a need for completion is often the driving force behind these initial hunts and collation of texts, the parameters of what counted as "complete" were shaped by social and political contexts, and continually negotiated. In examining these negotiations, we especially attend to the agency of our historical actors in determining the shape of the archive, paying heed to the influence of courtly patrons, scholarly communities, health practitioners, readers, users, and consumers of books and medicine.

Once assembled, the archive served as a dynamic space for knowledge making, and as we demonstrate, translation was only one component of a range of connected epistemic processes. In his revisionist account of the Islamic "Translation Movement," Ragab urges us to view translation not as processes of "encapsulating texts into a new language" but rather as practices of knowledge making built on existing scholarly traditions. The construction of an emerging archive was part and parcel of this work. Ragab argues that for figures such as Ḥunayn ibn Isḥāq (d. 873), translation involved more than simply moving texts faithfully from one language to another; it also included concerted efforts to create a body of knowledge. Ḥunayn built up a collection of Greek texts by many authors and sources, including several copies of the same text, offering

[51] Dror Weil, "Unveiling Nature," in this volume (*Osiris* 37).

possibilities for comparison and correction. A similar case is made by Weil, who demonstrates that the translation of Arabic-Persian natural philosophy into Chinese required not only linguistic alternations but also careful reconstructions of concepts and theories. Textual collation, extraction, and validation, based on an archive of amassed texts, was at the heart of these activities. In both Ragab and Weil's stories, the translation of medical ideas and theories was achieved through a variety of complex processes of articulation: textual collation, excerption, and validation, which, intentionally or not, all resulted in the amplification or silencing of particular voices.[52]

While Ragab and Weil situate these textual practices within large-scale archives, other authors in this section explore their impact on a smaller scale, in many cases in single books or textual objects. Much like archive-building, the creation of compendia via practices of reading and note-taking required the deliberate selection and linking of different kinds of knowledge. In many of our stories, what first appear as single titles or books turn out to have convoluted production histories arising from an "archive" of interconnected texts. Translation played a key role in these sorts of epistemic activities. The Tallamy family's annotated copy of *The Art of Distillation*, as analyzed in Leong's essay, for example, demonstrates how a single book served as an archive recording multiple layers of epistemic work conducted by men and women living decades and hundreds of miles apart. Leong shows how the physician and translator John French (1616–57) and members of the Tallamy family translated and gathered together textual excerpts, recipes, and other know-how to create a general guide for distillation and household medicine. Adopted into the household context, French's printed book transformed from a manual of artisanal how-to into a family archive, recording instances of social interactions, reading practices, and, crucially, the family's firsthand experiences with recipe testing.[53]

Textual juxtaposition, compilation, and translation also stand at the heart of Alisha Rankin's essay. Tracking the sixteenth-century Spanish physician Nicolás Monardes's treatises on New World drugs across Europe, Rankin shows how the modularity of his work allowed translators to choose portions most relevant to their linguistic communities, rather than producing only complete translations of the work. Key to these, Rankin argues, is inclusion or omission of indigenous knowledge of New World informants, Monardes and other physicians, and trustworthy contemporaries. By selectively including or excluding observations from the "archive of practice," translators were able to create new texts tailored to local interests, reflecting broader trends in Renaissance Europe.[54]

Moreover, while textual practices—in manuscript and in print—take center stage in all our case studies, our focus on health concerns means that *medical practices* often initiated these endeavors. Scholars have pointed to the common practice of recording cases and observations encountered in quotidian medical practice, and practitioners' subsequent efforts to organize and categorize these records.[55] In connection to translation practices, Harold Cook has recently pointed out how hope for medical

[52] Ahmed Ragab, "Translation and the Making of a Medical Archive," in this volume (*Osiris* 37); Weil, "Unveiling Nature" (cit. n. 51).

[53] Elaine Leong, "Translating, Printing, and Reading," in this volume (*Osiris* 37).

[54] Alisha Rankin, "New World Drugs and the Archive of Practice," in this volume (*Osiris* 37).

[55] On cases and observations in early modern medicine, see, for example, Lauren Kassell, "Casebooks in Early Modern England: Medicine, Astrology, and Written Records," *Bull. Hist. Med.* 88

innovations and advances in patient care prompted physicians to explore medical ideas from other cultures.[56] This optimism is particularly valent in Leong and Rankin's stories, where the translation of know-how concerning new processes and materia medica was seen as a path to improved health provision.

Furthermore, our historical actors' focus on medical necessity often delineated what was translated. In Ragab's study, translation was driven by the quotidian needs of medical practice and communication between patients and practitioners. As such, the translation of concepts, symptoms, diagnostic categories, and names of materia medica were prioritized over the translation of specific texts or corpora. Everyday realities also deeply shaped the diverse choices made by translators of Monardes. For example, to convince German readers to use sassafras, the physician Johann Wittich (1537–96) not only offered anecdotes on how the herbs were used locally but also extended the text by including practical recipes to help readers incorporate the drug into their own practices.[57] Finally, the focus on medical practice also highlights that the movement of knowledge was more often than not accompanied by the transfer of tacit skills, as in the case of the translation of Johann Rudolf Glauber's (1604–70) *Furni nove philosophici*. In these instances, a mere linguistic translation, whether textual or oral, simply did not suffice.[58]

Perhaps owing to the focus on practice, moreover, verification and validation occupied central roles. In the case of Ḥunayn and Liu Zhi, the continual hunt for and comparison of manuscripts was part of a larger scheme of textual refinement. As Ragab points out, in these cases, translation worked alongside other scholarly practices to weed out repetitious and inferior knowledge, resolve contradictions, and cross-reference between works. In some instances, the needs of medical practitioners to offer assurances of the safety and reliability of drugs and interventions lent a new edge to these processes, pushing practitioners to look beyond considerations of linguistic accuracy and the establishment of urtexts.[59] Other kinds of testing and trying play a key role in our stories. Experiential knowledge shines brightly in Rankin's study, where Monardes and his translators argued for the importance of firsthand experience as a means to verify and authenticate New World drugs described in various works. Observations of successful cures were a key component of this "archive of practice." In other words, the practice of translating medicine was dynamic, requiring continual refinement, assessment, and reassessment.

Finally, as noted above, by placing translation and medical concerns front and center, these essays encourage us to attend to the "afterlives" of translated texts and archives. In "Archives of the Sciences," Lorraine Daston noted how early modern scientific archives were often described as "granaries," "warehouses," or "treasuries" and served as "sites of discovery and serendipity" as well as "provisions laid up for future inquirers."[60] Similarly, the editors and authors of the collective volume *Archival*

(2014): 595–625; Volker Hess and J. Andrew Mendelsohn, "Case and Series: Medical Knowledge and Paper Technology, 1600–1900," *History of Science* 48 (2010): 288–314; and Gianna Pomata, "Observation Rising: Birth of an Epistemic Genre, 1500–1650," in *Histories of Scientific Observation*, ed. Lorraine Daston and Elizabeth Lunbeck (Chicago: Univ. of Chicago Press, 2011), 45–80.

[56] Cook, "Introduction," in *Translation at Work* (cit. n. 7), 17.

[57] Rankin, "New World Drugs" (cit. n. 54).

[58] Leong, "Translating, Printing, and Reading" (cit. n. 53).

[59] Ragab, "Translation and the Making of a Medical Archive" (cit. n. 52).

[60] Daston, "Sciences of the Archive" (cit. n. 49), 171.

Afterlives have offered nuanced readings of how early modern scientists engaged with paper archives both in terms of how they "attended to the material record of the scientific past" and "their efforts to preserve, transmit and make use of that record."[61] By adopting the analytics of translation and archive making, our case studies extend these explorations in a number of ways.

Rankin's notion of the "archive of practice" amplifies the voices of indigenous Amerindian actors and illustrates how their experiential knowledge was often obscured in sixteenth-century European translations of Monardes. Crucially, though the knowledge of indigenous actors and European translators/authors often traveled together, they were viewed in a vastly different light in terms of authority and validity.[62] Focusing on the notion of "knowledge itineraries," Leong's essay impresses upon us the utility of reconstructing the backstories and afterlives of early modern printed medical works and encourages us to view skill and knowledge acquisition, translation, printing, reading, and compilation in a continuous spectrum—each as part of the same journey—suggesting that there is much to be gained by attending to multilingual, multisited long-view histories of book production and use.[63] This emphasis is echoed in Weil's contribution, which concludes with brief illustrations of how three subsequent scholars expanded upon Liu Zhi's *Human Nature and Cosmic Principles in Islam* (*Tianfang xingli* 天方性理) in different ways, each reflecting their own interests. For Weil, the rich afterlife of Liu's text attests to the open-endedness of philosophies of nature and views of the human body.[64] Finally, Ragab takes altogether a more expansive view, challenging historians of science and medicine to reflect on the historiographical legacies of our narratives. *How* we view and value translation as an epistemic practice, he argues, has profound impact on how we impose value and hierarchy in past knowledge systems, and especially in the context of archive making.[65]

In sum, by merging the analytical frameworks of the history of archives and of translation, the four essays in this section demonstrate that stories such as ours are about much more than just dissemination of knowledge. By attending to the moments before and after translation, rather than just the act itself, we place translation within webs of interconnected practices and outline their role in constructions of cultural and linguistic hegemony. Our essays open new conversations about the authority of practice, complicate existing ideas about archives and textual scholarship, and bring to the fore how every moment of translation needs to be read with knowledge of its various contexts.

Section 2: Translation beyond the Textual

The central focus on practice and material culture in the history of medicine pushes us to examine nontextual sources, and in this section we consider how spoken language, images, objects, and practices passed into new zones of understanding and were reinterpreted or reinvented.[66] These essays illuminate the ways translation occurred

[61] Keller, Roos, and Yale, *Archival Afterlives* (cit. n. 47), 7.
[62] Rankin, "New World Drugs" (cit. n. 54).
[63] Leong, "Translating, Printing and Reading" (cit. n. 53).
[64] Weil, "Unveiling Nature" (cit. n. 51).
[65] Ragab, "Translation and the Making of a Medical Archive" (cit. n. 52).
[66] On medicine and practice, see Mary E. Fissell, "Making Meaning from the Margins: The New Cultural History of Medicine," in *Locating Medical History: The Stories and Their Meanings*, ed. Frank

across and between media: how oral knowledge of materia medica became verified in written form in glossaries (Hamza); how material objects, such as pipes, were described, used, and redefined in texts and through practical use, gaining new meanings in the process (Breen); how images were copied and reinterpreted to affirm or modify new medical theories (Trambaiolo); and how experiments and their conceptual implications were translated into various medical genres (Mukharji). By exploring how medicine was translated beyond the text, the articles in this section recover the work of invisible actors and their hitherto unheard voices, recalibrate ideas of time and place in the histories of medicine, science, and technology, and suggest new approaches to the complex interactions of knowledge traditions. From the blending of learned medicine with wider oral worlds of healing detailed in Hamza's essay to the repurposing of European anatomical images in a Sino-Japanese text on acupuncture discussed by Trambaiolo, translation practices can be seen as "braided sciences," as Projit Mukharji terms it, in which old and new traditions, or concepts from disparate medical systems, interweave to form a new pattern.[67] By considering these braided patterns of articulation (the visible, top strands) and disarticulation (the strands covered by others), we trace how some practices, theories, and concepts were successfully encoded and decoded, while others were left by the wayside.

Focusing on a fourteenth-century learned Persian medical text, the *Ṭibb-i Shihābī*, Shireen Hamza uses a study of vernacular glossaries of plant and disease names to recover the "lost" oral histories of medical practice.[68] She details how medical practitioners translated learned medicine to languages spoken by patients, suppliers of materia medica, and pharmacists, emphasizing processes of verification and legitimation. Hamza argues that translation between textual and oral sources stood at the core of medical practices and involved hitherto hidden local actors. Alongside Pablo Gómez's contribution, Hamza's study makes clear that uncovering these rarely heard voices in a vernacular, nonhegemonic language decenters the focus of medical activities from learned, bookish practices to foreground vernacular medicine, thereby complicating traditional narratives.[69]

The focus on the oral can also extend our understanding of medical practice in other ways, such as the identification of new "origin stories." In his study of "pyric technologies,"

Huisman and John Harley Warner (Baltimore, MD: Johns Hopkins Univ. Press, 2004), 364–90; Pickstone and Worboys, "Between and Beyond" (cit. n. 23); and Claudia Stein, "Introduction: The Early Modern Cultural History of Medicine," in *A Cultural History of Medicine in the Renaissance*, ed. Stein and Elaine Leong (London: Bloomsbury Academic, 2021), 1–22. Other recent publications in which the definition of translation has been broadened to analyze translation across media in the history of science and medicine include Arredondo and Bauer, eds., *Translating Nature* (cit. n. 11); Sumillera, Surman, and Kühn, *Translation in Knowledge, Knowledge in Translation* (cit. n. 11); and Cook, *Translation at Work* (cit. n. 7).

[67] See Shireen Hamza, "Vernacular Languages and Invisible Labor in Ṭibb"; Benjamin Breen, "Where There's Smoke, There's Fire"; Daniel Trambaiolo, "Translating the Inner Landscape"; and Projit Bihari Mukharji, "Casting Blood Circulations," all in this volume (*Osiris* 37); as well as Mukharji, *Doctoring Traditions: Ayurveda, Small Technologies, and Braided Sciences* (Chicago: Univ. of Chicago Press, 2016), 25–7.

[68] On orality and translation, see Walter J. Ong, *Orality and Literacy: The Technologizing of the Word* (London: Routledge, 1982); Montgomery, *Science in Translation* (cit. n. 11); Gordon Brotherston, "Contact Situations and Barriers to Intercultural Communication: Orality, Non-alphabetic Writing Systems and Translation," in *Übersetzung—Translation—Traduction: 1. Teilband*, ed. Harald Kittel et al. (Berlin: De Gruyter, 2008), 30–7; and Paul F. Bandia, ed., "Orality and Translation," special issue, *Translation Studies* 8.2 (2015).

[69] Hamza, "Vernacular Languages" (cit. n. 67); Pablo Gómez, "'[Un]Muffled Histories," in this volume (*Osiris* 37).

Benjamin Breen argues that terminology for the pipe in European vernaculars contributes to understandings of the object's origin within particular linguistic regions. He shows how *cachimbo*, the Portuguese term for pipe, is etymologically linked to *kixima*, the object's name in a West Central African Bantu language, which refers to a "water well." Breen suggests that the West African word was picked up by Portuguese traders, through whom the word and the object became part of Portuguese society.

Adopting methodology from geography and archeology to analyze pipes or *cachimbos*, Breen also attends to material histories.[70] His revisionist "origin story" argues that pipe smoking was already present in the Old World via routes coming from sub-Saharan Africa and South Asia, albeit without the presence of tobacco. Thus, he challenges conventional histories which depict pipe smoking as a habit and technology that came to Europe from the New World. The new focus on non-European locality, space, and materiality foregrounds an understanding of the usage of pipes in the African context. Furthermore, Breen argues that, as with new materia medica, the introduction of new medical technologies required not only new terminology but also a translation into local medical theory. He offers potential routes for assimilation: either an adjustment and reinterpretation of local medical theories to fit the technology, or an adaptation of the technology to fit local theories. Breen's study offers an example for understanding complicated translational processes in different medical contexts between material objects and across immaterial subjectivities.

Examining translation through images and visual culture can also offer new perspectives. Vivienne Lo and Ronit Yoeli-Tlalim have demonstrated, for example, that a style of medical illustration depicting cautery therapies like moxibustion "seems to have emerged simultaneously in 9th–10th century Europe, Tibet and China," complicating traditional narratives of transmission and dissemination.[71] Similarly, Daniel Trambaiolo's essay in this section takes us to nineteenth-century Japan, showcasing translation across visual media and turning our lens to the archival afterlives of translated anatomical texts and images, indicating how similar images can change purpose and meaning over time.[72] Trambaiolo argues that while the much studied first *rangaku*

[70] Breen, "Where There's Smoke, There's Fire" (cit. n. 67). On objects as carriers of knowledge and the translatability of materials, see Sven Dupré and Christoph Lüthy, eds., *Silent Messengers: The Circulation of Material Objects of Knowledge in the Early Modern Low Countries* (Münster: Lit, 2011); Ursula Klein and Emma C. Spary, eds., *Materials and Expertise in Early Modern Europe: Between Market and Laboratory* (Chicago: Univ. of Chicago Press, 2010); Beate Fricke and Finbarr Barry Flood, *Object Lessons: Artifacts as Archives of Pre-Modern Globalism* (Princeton, NJ: Princeton Univ. Press, forthcoming); Lorraine Daston, ed., *Things That Talk: Object Lessons from Art and Science* (New York: Zone, 2004); Paula Findlen, ed., *Early Modern Things: Objects and Their Histories, 1500–1800* (London: Routledge, 2013); Domenico Bertoloni Meli, *Thinking with Objects: The Transformation of Mechanics in the Seventeenth Century* (Baltimore, MD: Johns Hopkins Univ. Press, 2006); and Chakrabarti, *Materials and Medicine* (cit. n. 19).

[71] Lo and Yoeli-Tlalim, 'Travelling Light" (cit. n. 4), 271.

[72] On images as carriers of translatable knowledge and information, see Walter Benjamin, "The Work of Art in the Age of Mechanical Reproduction," in *Illuminations*, ed. Hannah Arendt (London: Fontana, 1992), 211–44; Suzanne Kathleen Karr Schmidt and Edward H. Wouk, eds., *Prints in Translation, 1450–1750: Image, Materiality, Space* (London: Routledge, 2017); and Sietske Fransen and Katherine M. Reinhart, "The Practice of Copying in Making Knowledge in Early Modern Europe: An Introduction," *Word & Image* 35 (2019): 211–22. On epistemic images, see Lorraine Daston, "Epistemic Images," in *Vision and Its Instruments: Art, Science, and Technology in Early Modern Europe*, ed. Alina Alexandra Payne (University Park: Pennsylvania State Univ. Press, 2015), 13–35; Alexander Marr, "Knowing Images," *Renaissance Quarterly* 69 (2016): 1000–13; Horst Bredekamp, *Theorie des Bildakts: Über das Lebensrecht des Bildes* (Frankfurt am Main: Suhrkamp, 2010); and Christoph Lüthy and Alexis Smets, "Words, Lines, Diagrams, Images: Towards a History of Scientific Imagery," *Early Sci. & Med* 14 (2009): 398–439.

translators of the seventeenth century looked for ways to translate European texts on anatomy faithfully into Japanese, a second wave of translators working in the early nineteenth century reinterpreted the images to fit Sino-Japanese medicine.[73] For example, the Japanese physician Kako Ranshū reused European anatomical images in his *Kaitai chin'yō* 解体鍼要 (Essentials of anatomy and acupuncture, 1819) to show important acupuncture points rather than blood vessels. To aid readers' understanding and to adapt the images to his text of traditional Sino-Japanese medicine, Kako used short labels to convey the new meanings afforded to the images. Trambaiolo's study offers new ways of parsing how images might have been "read" by Sino-Japanese practitioners, as well as new ways to look at how medicine was translated from context to context.

In the final article in this section, Projit Mukharji outlines how three translator-practitioners struggled with translating the concept of blood circulation to their communities in late nineteenth-century Bengal. He shows how this physiological concept was conveyed in three separate genres—a schoolbook, an Ayurvedic book in verse, and a "materio-spiritual" guide to the human body—and contends that social contexts determined how knowledge was translated to suit specific target audiences. Mukharji demonstrates that studying how different knowledge traditions were "braided" together can allow us to attend to the disarticulations of the various medical practices. For example, the concept of "death pulses"—a means of foretelling the date of an individual's death—is disarticulated from other types of pulse medicine in Ashutosh Mitra's *Nara Shareer Bidhan* (The system of the human body).[74] Mitra, who translated the notion of blood circulation for an explicitly Hindu Anglophone upper-caste audience, explained that the pulsation of blood is directly dependent on the beating of the heart. This was necessary to make clear that the concept he was elucidating was not related to the local tradition of pulse-diagnosis (*nadi-pariksha*). In this particular translation it is the Harveian notion of blood circulation that is articulated, at the cost of the traditional pulse theory. Mitra subsequently dismisses William Harvey as the discoverer of blood circulation by arguing that ancient Hindu physiologists already had this knowledge.

The focus on translation across media adds a tool to our metaphorical armamentarium that allows us to find the unwritten or silent voices of actors that were involved in the (daily) practice of medicine, while at the borders of these media we become aware of the ways in which traditions are braided into each other.

Section 3: Translator-Practitioners, Expertise, and Authority

In the final section the focus shifts from the texts, materials, media, and practices of medical translation to the experiences and identities of those who translated and those who healed. Exploring translation and medicine in tandem allows us to derive a number of insights related to the figure of the translator and the healer, pushing forward the existing literature relating both groups. First, these articles uncover the wide range of approaches used to translate expertise and to assert medical authority in ways that would be credible for new audiences. Translation and medicine could both be perilous

[73] Trambaiolo, "Translating the Inner Landscape" (cit. n. 67).
[74] Mukharji, "Casting Blood Circulations" (cit. n. 67).

pursuits. Healers attended those in the dangerous lands between sickness and health. Translators were often involved in the weird alchemy of converting concepts into another form, while maintaining as far as possible the substance and essence of the original. Second, these articles help us to reconstitute and analyze the key figure of the *translator-practitioner*, who, we argue, became a special type of mediator.

The case studies in these articles reflect a wide spectrum of medical expertise and practice and uncover multiple ways in which the identities and authority of healers were rearticulated and disarticulated through processes of translation. Montserrat Cabré's study traces how the figure of Trota of Salerno (fl. early 12th c.) and understandings about her general expertise in medicine were transformed through the translation of Latin treatises and the development of a corpus of late-medieval Catalan vernacular texts related specifically to women's health. Pablo Gómez "unmuffles" the voices of medical practitioner-translators in the early modern Caribbean, where healers of African and Amerindian descent developed new presentations of their medical expertise through translation. Tara Alberts's account examines the reinvention of a French surgeon in seventeenth-century Siam (Thailand), exploring how his identity and authority over medical matters were recast by the local context as he attempted to translate his expertise. Hansun Hsiung's article turns to eighteenth- and early nineteenth-century Japan. He uncovers the strategies of Japanese translator-practitioners and physicians to translate Dutch approaches to medical ethics into frameworks commensurate with Japanese moral systems, in an attempt to reconcile the invasive violence of European surgical techniques with the Neo-Confucian virtue of "humaneness" 仁 (C: *ren*; J: *jin*).

Through these case studies we see that there are a number of comparisons to be made between translators and healers, especially concerning articulations of claims to authority—textual or medical. For many healers and translators, authority was dependent on three key factors. First, in many cases, authority came from evidence of training or inculcation into a body of knowledge. During the period covered by these essays (thirteenth to eighteenth centuries), in many parts of the world, new ideas about the training, regulation, and licensing of healers created new structures of knowledge exchange and new categories of protected spheres of learning: from the attempted delineation of the boundaries of medicine and surgery in seventeenth-century France alluded to by Alberts, to the expansion of new schools for physicians in eighteenth-century Japan mentioned by Hsiung.[75] The practice of translation too was the focus of sustained reflection, theorizing, and debate as new cadres of specialists developed in a number of cultures.[76] In both, new norms and rules of practice emerged within

[75] See Tara Alberts, "Translating Alchemy and Surgery"; and Hansun Hsiung, "Use Me as Your Test!," both in this volume (*Osiris* 37). For a cross-cultural consideration of training and status in medicine, see also Pols, Thompson, and Warner, eds., *Translating the Body* (cit. n. 8).

[76] See especially Peter Burke, "Lost (and Found) in Translation: A Cultural History of Translators and Translating in Early Modern Europe," *European Review* 15 (February 2007): 83–94; Kapil Raj, "Go-Betweens, Travelers, and Cultural Translators," in *Companion to the History of Science*, ed. Bernard Lightman (Chichester, West Sussex, UK: Wiley-Blackwell, 2016), chap. 3; Dejanirah Couto, "The Role of Interpreters, or *Linguas*, in the Portuguese Empire during the 16th Century," *e-JPH* 1 (2003): https://digitalis.uc.pt/en/node/84960; Najaf Haider, "Translating Texts and Straddling Worlds: Intercultural Communication in Mughal India," in *The Varied Facets of History: Essays in Honour of Aniruddha Ray*, ed. Ishrat Alam and Syed Ejaz Hussain (Delhi: Primus Books, 2011), 115–24; and Federico M. Federici and Dario Tessicini, eds., *Translators, Interpreters and Cultural Negotiators: Mediating and Communicating Power from the Middle Ages to the Modern Era* (Basingstoke: Palgrave Macmillan, 2014).

specialized branches of practitioners; new forms of self-conscious identity developed.[77] In some cases authorities also sought to impose boundaries and rules on certain practices: from forbidding certain types of translation to policing healing practices. The African and Amerindian healers examined by Gómez, for instance, often fell afoul of the Inquisition as their ways of thinking about and treating the body were deemed incommensurate with European conceptions of licit healing. The renowned Amerindian healer Luis Andrea, for example, was banished and prohibited from curing.[78]

Second, and connected to the first point, translators and healers gained their authority because of their privileged access: they were both mediators between the individual seeking their aid and something otherwise unknown, incomprehensible, or unobtainable for their client. Translators claimed authority through their access to original texts or languages, and to essential cultural knowledge that their audiences did not have.[79] Healers similarly often had access to forms of practical, professional, and experimental knowledge, and to embodied or even cultural knowledge about techniques. The various groups of female experts—Jewish and Muslim women, Salernitan women—explored by Cabré, for example, were credited with especial expertise over matters of women's health.[80] Similarly, Inquisition testimonies uncovered by Gómez reveal the esteem in which some Black Caribbean ritual practitioners held Amerindian healers, due to the latter's greater perceived understanding and mastery of the local natural world and its spiritual entities.[81] Healers could also have special access to the body: as Cabré shows, women healers had access to female patients, which was not available to Salernitan male physicians.[82] In Siam, Alberts demonstrates, the king's physicians were almost uniquely able to approach and touch his sacred body, which was hidden even from the view of the rest of his subjects.[83] Paying attention to these forms of special access helps us to understand the position of both the healer and the translator.

Third, moreover, some translators and some healers had access to particularly esoteric, prestigious, and hidden knowledge. They had facility with sacred, ancient, or prestigious written languages, or had access to privileged knowledge residing in private manuscript collections and libraries. Alberts, for example, notes the importance of jealously guarded recipe collections and medical treatises, which contained scattered references and quotations in the ritual language Pāli. These were preserved in monastic and royal archives or handed down through medical lineages that claimed their origins lay with the Buddha's physician Jivaka. Esoteric knowledge could include any matters hidden from the uninitiated, from the inner workings of the human

[77] See, for example, David Gentilcore, *Healers and Healing in Early Modern Italy* (Manchester: Manchester Univ. Press, 1998); Sandra Carvallo, *Artisans of the Body in Early Modern Italy: Identities, Families and Masculinities* (Manchester: Manchester Univ. Press, 2007); Charles Burnett, ed., *Arabic into Latin in the Middle Ages: The Translators and Their Intellectual and Social Context* (London: Routledge, 2009); and Angela K. C. Leung, "A 'South' Imagined and Lived: The Entanglement of Medical Things, Experts and Identities in Premodern East Asia's South," in *Asia Inside Out: Itinerant People*, ed. Eric Tagliacozzo (Cambridge, MA: Harvard Univ. Press, 2019), 122–44.

[78] Gómez, "[Un]Muffled Histories" (cit. n. 69).

[79] See especially Lawrence Venuti, *Translator's Invisibility* (cit. n. 28); Jean Selisle and Judith Woodsworth, eds, *Translators through History* (Amsterdam: John Benjamins, 1995); and Liu, *Tokens of Exchange* (cit. n. 27).

[80] Cabré, "Female Authority in Translation" (cit. n. 15).

[81] Gómez, "[Un]Muffled Histories" (cit. n. 69).

[82] Cabré, "Female Authority in Translation" (cit. n. 80).

[83] Alberts, "Translating Alchemy and Surgery" (cit. n. 75).

body, to etiology, to the composition and use of certain cures. Hsiung, for example, alludes to the anesthetic developed by the physician Hanaoka Seishū (1760–1835), an "unwritten proprietary secret" revealed only to his disciples.[84] Together, these articles illuminate how a focus on translation allows us to interrogate and differentiate between these features of authority construction in histories of science and medicine.

Examining these processes can nuance our understanding of figures such as the "hybrid healer," the go-between, and the mediator, who have played a significant role in the historiography of global (scientific) exchange.[85] The essays in this section also show how a figure arose who was both healer and translator at the same time. These "translator-practitioners" emerge from our analysis as a special type of mediator, someone who drew on the traditions and norms of both translation and healing, but for whom moments and spaces of translation were also often opportunities to create new articulations of identity, authority, and expertise. Equally, moving into new arenas could prompt translator-practitioners to disarticulate their identities and expertise, deprioritizing certain medical skills, practices, or concepts. In short, we demonstrate how the translation of medicine provided special opportunities for certain individuals to completely reinvent themselves, in a process analogous to the other forms of translation already analyzed.

The two identities of healer and translator were co-constitutive and mutually dependent in the figure of the translator-practitioner. Translator-practitioners could claim privileged access in both spheres, linguistic and medical. In Gómez's article we see how Black Caribbean healers could, through access to a wide variety of African and Amerindian languages and to cross-culturally valent emergent diagnostic techniques concerning "bundles of disease," create new signifiers of expertise. In Hsiung's article, we see how translator-practitioners who were able to read texts of "Dutch Learning" (*rangaku*) integrated new ideas into existing moral and medical frameworks in order to promise new pathways to attempt cures of hitherto incurable diseases such as breast cancer. Drawing on the *Yōi shinso* (New book of surgical medicine), the physician Hanaoka Seishū (1760–1835) attempted the first surgical excision of a cancerous tumor from a patient in Japan. In the process he offered a new way of articulating the ethical relationship between patient and practitioner, and, subsequently, the forms of treatment that were morally permissible for a physician to "test" on his patient.

The pluralistic, polylingual worlds of the articles in this section underline how the human landscape of translator-practitioners remained complex. There was often a disjunction between contemporary normative and descriptive accounts of the worlds of translators and healers, and the messy complexity of reality. Those credited with the most authority by consumers could be outside systems of accreditation, and

[84] Hsiung, "Use Me as Your Test!" (cit. n. 75).

[85] See, for example, Kapil Raj, Simon Schaffer, Lissa Roberts, and James Delbourgo, eds., *The Brokered World: Go-Betweens and Global Intelligence, 1770–1820* (Sagamore Beach, MA: Science History Publications, 2009); Peter Burke, "The Renaissance Translator as Go-Between," in *Renaissance Go-Betweens*, ed. Andreas Höfele and Werner von Koppenfels (Berlin: De Gruyter, 2005), 17–31, see 57; Miles Ogborn, "'It's not what you know. . .': Encounters, Go-Betweens, and the Geography of Knowledge," *Mod. Int. Hist.* 10 (2013): 163–75; Anna Winterbottom, *Hybrid Knowledge in the Early East India Company World* (Cham, Switzerland: Springer, 2016); and Markku Hokkanen and Kalle Kananoja, "Healers and Empires in Global History: Healing as Hybrid and Contested Knowledge," in *Healers and Empires in Global History*, ed. Hokkanen and Kananoja (Cham, Switzerland: Palgrave Macmillan, 2019), 1–26.

"untrained" according to the standards of normative literature. Bans by the Inquisition did little to diminish the value of the Amerindian and African healers discussed by Gómez, for example. Prohibitions on practice were difficult to enforce, and such healers continued to enjoy high status and esteem among various communities.

Moreover, while discourses concerning both roles could increasingly emphasize the role of skilled, trained individuals practicing alone, in reality, collaboration, competition, and cooperation between multiple actors lay behind most undertakings in both medicine and translation.[86] These complex realities come into sharp focus in Gómez's reconstruction of the "cacophonous" early modern Caribbean, where "rich communal processes of translation" were developed by Amerindians and people of African descent concerning bodily matters. His article also underlines the communal aspect of healing, where disease could be seen as a matter of "the disequilibrium not of individuals' bodies, but that of communities," the resolution of which necessitated mediation between a vast array of persons, material and immaterial ancestors, and nonhuman elements. Alberts's article similarly uncovers the multilingual, cosmopolitan world of Ayutthaya, where healers from a wide range of traditions engaged in the exchange, translation, and reinvention of concepts, ideas, and materia medica.

Finally, the articles illustrate how this privileged access, and this role as a special kind of mediator, could also make translator-practitioners ambivalent figures. Possessors of esoteric knowledge, they could be feared and mistrusted at the same time they were being sought after and fêted. Knowing more than their client, they could be suspected of deceit. Ambivalence about healers finds expression in negative stereotypes in a wide range of cultures, from the trope of the atheistic, self-serving physician common in medieval and early modern European popular culture, to depictions of uncertainties over physicians' skills and expertise in Qing China.[87]

In colonial, semicolonial, or pericolonial spaces, understandings of these mediator figures can be particularly complex. In some reckonings, translation is the first, necessary stage of further destruction as territories are rendered comprehensible to the colonial gaze. The trope of the translator-as-traitor becomes particularly important in these narratives. We see this, for example, in the complex legacy of figures such as Malinche, Columbus's enslaved interpreter, who is at once celebrated, mourned, and vilified as empowered agent, victim, and facilitator of the colonial violence that destroyed her people.[88]

Translator-practitioners in this narrative can also serve as double agents, to be treated with suspicion by the powerful.[89] Indeed, their endeavors could be a means of rebellion and resistance, of empowerment for the disempowered. The complexity of these "power-creation dynamics" of medical translation processes are uncovered by Gómez, who

[86] See especially Cheung, "Chinese Discourse on Translation" (cit. n. 30), 56–72.

[87] See especially William Birken, "The Social Problem of the English Physician in the Early Seventeenth Century," *Medical History* 31 (1987): 201–16; Christi Sumich, *Divine Doctors and Dreadful Distempers: How Practicing Medicine Became a Respectable Profession* (Amsterdam: Rodolphi, 2013), 19–24; and Chu Pingyi, "Calendrical Learning and Medicine, 1600–1800," in *The Cambridge History of China, Vol. 9: The Ch'ing Dynasty to 1800, Part 2*, ed. Willard J. Peterson (Cambridge, UK: Cambridge Univ. Press, 2016), 372–411.

[88] María Laura Spoturno, "Revisiting Malinche: A Study of Her Role as an Interpreter," in Federici and Tessicini, *Translators, Interpreters and Cultural Negotiators* (cit. n. 76), 121–35.

[89] See especially Mona Baker, *Translation and Conflict* (cit. n. 38); Tymoczko and Gentzler, *Translation and Power* (cit. n. 27); and Rafael, *Contracting Colonialism* (cit. n. 38).

points out the longevity and "recalcitrant persistence" of the ideas of many African healers in the Caribbean, in the face of prosecution by ecclesiastical and colonial authorities.

Taken together, the essays in this volume demonstrate the rich gains to be made when we move away from a primary emphasis on the traditional "key movements" of textual translation into and between dominant languages (Arabic, Latin, Greek, Chinese, etc.), and onto the "key moments" of translation deemed important to the narrative of European or "Western" medical and scientific history. Our engagement with concepts and terms as they were used by our actors, and as they are used in contemporary scholarship, push us to consider anew the extent to which taxonomies of thought and lived experience translate over time and space. It is crucial to assess what was and what could be translated – and what was likely to be lost, or changed beyond all recognition, by these epistemic processes. Translation, we contend, was at once a process of creation and destruction which formulated new hybrids, even new languages, of cure and medical practice. By bringing into focus the importance of the diverse translation practices undertaken by a wide range of groups and individuals, and of languages and concepts hitherto marginalized in grand narratives, our volume offers new ways to think about the creation and blurring of boundaries of knowledge in moments of intercultural contact.

ARCHIVES AND THE AUTHORITY OF PRACTICE

Translation and the Making of a Medical Archive:

The Case of the Islamic Translation Movement

*by Ahmed Ragab**

ABSTRACT

Translation has occupied a central role in the historiography of Islamic science and medicine, and the "Translation Movement" from Greek to Arabic is often considered the birth moment of the "Golden Age." In this view, translation is understood as a transition in which knowledge moves across a linguistic divide. However, this translation-as-transition paradigm fails to capture the linguistic diversity that existed on both sides of this seeming divide, and the production and consumption of this translated knowledge and its diffusion beyond the spheres of learned scientific and medical practice. In this article, I look at translation in the history of Islamic medicine not as a transition but rather as a part of a larger and more comprehensive process of archive making. Through following the works of translators and historians, I investigate how translation contributed to the production of a particular form of learned medicine.

INTRODUCTION

It is hard to think of a concept or a historical moment more significant to the study of Islamic sciences and medicine than translation. Traditional accounts of Islamic sciences begin with the translation of classical Greek texts into Arabic, a process roughly dated to the first half of the ninth century.[1] The "Translation Movement" is seen as a moment of birth for learned Islamic cultures because it linked these cultures to a long history of Hellenism, which, the traditional narrative argues, had declined in the Byzantine context and survived only in the Islamic world, and in Arabic.[2] The conventional account perceives translation not as a process in which knowledge moves from one linguistic tradition into another but rather as a process of encapsulation. In this telling, Arabic was not a true home for Hellenistic knowledge but rather

* Department of the History of Medicine, Johns Hopkins School of Medicine, 1900 East Monument St., Baltimore, MD 21205, USA; aragab2@jhmi.edu.

[1] David Lindberg, *The Beginnings of Western Science: The European Scientific Tradition in Philosophical, Religious, and Institutional Context, Prehistory to AD 1450* (Chicago: Univ. of Chicago Press, 2007).

[2] Ibid.

Osiris, volume 37, 2022. © 2022 History of Science Society. All rights reserved. Published by The University of Chicago Press for the History of Science Society. https://doi.org/10.1086/719219.

a casing that protected this knowledge for centuries until it found a new home in Latin. The idea of translation as encapsulation legitimized and motivated a wealth of scholarly works on the survival of Greek texts in Arabic and on the spread of these texts. Although these traditional accounts have been critiqued and questioned by many scholars, they have continued to shape public conversation, teaching, and archival restoration projects with the result that the Translation Movement is still considered by most nonspecialized narratives as the true beginning of scientific and medical learned culture in the medieval Islamic world.[3] The effects of this focus become obvious when we observe the almost complete absence of pre-ninth-century, or pre-translation, history of Islamic sciences in most history of science textbooks, or when we compare the pages and scholarly energy dedicated to the periods before and after the translation.[4]

These traditional narratives have come under significant scholarly criticism. Abdelhamid Sabra has proposed *appropriation* as a different way of thinking about the relationship between the Greek and Islamic sciences.[5] Chase F. Robinson and Konrad Hirschler have provided compelling accounts of Islamic historiography that challenge Franz Rosenthal's thesis on the emergence of Islamic historiography from Greek origins.[6] Dimitri Gutas has convincingly demonstrated the long history of Islamic philosophy and its flourishing well into the period of "decline."[7] Peter Pormann and others have studied the making of long commentary traditions on Greek medical texts.[8] I have mentioned but a few of the many who criticize traditional narratives, yet *translation*, as a key moment and an analytic that organizes the history of Islamic sciences, endures in popular discourse, in the West and in the Islamic world alike, and also in scholarly writings, even by way of criticism. Translation, therefore, is not simply a historical moment that merits a degree of analysis, but it is in fact a *chronology*. As a chronology, the Translation Movement forces engagement, even by asking readers to rectify, rearrange, or reconsider. And in the process, the textual logic that governs the conception of the *translation* half of the phrase, and the transmission/transfer paradigm that underwrites the *movement* part, become key gravitational points in scholarship and public discourse alike.

[3] One of the earliest and most profound critiques of the place of the Translation Movement in the scholarship was penned by Abdelhamid Sabra, who introduced the notion of *appropriation* as an alternative for *transmission*; see Sabra, "The Appropriation and Subsequent Naturalization of Greek Science in Medieval Islam: A Preliminary Statement," *Hist. Sci.* 25 (1987): 223–43.

[4] For instance, see Justin Stearns's discussion of the need to direct scholarly energy to the study of the early modern period; Stearns, "Writing the History of the Natural Sciences in the Pre-modern Muslim World: Historiography, Religion, and the Importance of the Early Modern Period," *History Compass* 9 (2011): 923–51.

[5] Sabra, "Appropriation and Subsequent Naturalization" (cit. n. 3).

[6] Chase F. Robinson, "History and *Heilsgeschichte* in Early Islam: Some Observations on Prophetic History and Biography," in *History and Religion: Narrating a Religious Past*, ed. Bernd-Christian Otto, Susanne Rau, Jörg Rüpke, and Andrés Quero-Sánchez (Boston: De Gruyter, 2015), 119; Robinson, *Islamic Historiography* (Cambridge, UK: Cambridge Univ. Press, 2003); Konrad Hirschler, *Medieval Arabic Historiography: Authors as Actors* (New York: Routledge, 2006).

[7] Dimitri Gutas, "The Heritage of Avicenna: The Golden Age of Arabic Philosophy, 1000–ca. 1350," in *Avicenna and His Heritage: Acts of the International Colloquium Leuven—Louvain-La-Neuve, September 8–11, 1999*, ed. Jules Janssens and Daniel De Smet (Leuven, Belgium: Leuven Univ. Press, 2002), 81–98.

[8] N. Peter Joosse and Peter Pormann, "Decline and Decadence in Iraq and Syria after the Age of Avicenna? ʿAbd Al-Laṭīf Al-Baghdādī (1162–1231) between Myth and History," *Bull. Hist. Med.* 84 (2010): 1–29; Pormann and Kamran I. Karimullah, "The Arabic Commentaries on the Hippocratic Aphorisms: Introduction," *Oriens* 45 (2017): 1–52.

Certainly, translations occupied an important place in the medieval Islamic intellectual imaginary. Biographical dictionaries of scholars provided space for translators, and many historians transmitted accounts that narrated the first translations to Arabic (scholarly or otherwise). However, as I will argue, translation was more than a textual or learned process. It was an integral part of knowledge making that built on an already existing scholarly tradition in the Near East, of which Greek knowledge was a part. Moreover, and as scholars have shown, translations were not only from Greek to Arabic. They also involved many other languages and contributed to a larger process of linguistic landscape reorganization in the emerging Muslim polities.[9] From this perspective, written translations, which form the backbone of the Translation Movement and are the key marker of its chronology, are but a limited aspect of a more widespread process that extended over centuries in a multilingual region and intellectual tradition.

In this article, I investigate the translation of medicine in the ninth century as part of the extended development of Islamic Galenism. First, through looking at narratives on the origins of the Translation Movement, I discuss the meaning of translation as a socioepistemic process and its connections to the making of Islamic Galenism. Second, looking closely at Ḥunayn ibn Isḥāq (d. 873), one of the more prominent medical translators, I investigate his intentions and how his work contributed to the production of what I describe as the Islamic Galenic archive. Finally, I look at how translation and archive making extend beyond the Greco-Arabic dyad. In the conclusion, I consider the place of the Translation Movement in Euro-American and Global narratives about the history of medicine, how translations produce relations of priority and derivation in the postcolonial context, and how the study of medieval translations can be influential in these debates.

TRANSLATION AS ARCHIVE MAKING

The oldest example of a written translation of a medical text is attributed to a Jewish physician of Syriac-Persianate origin, by the name of Masārjawayh. He may have translated a *kunnāsh* (textbook) composed by Aaron of Alexandria from Syriac to Arabic under the reign of the Umayyad caliphs Marwān I (r. 684–5) or Umar II (717–20). In the account of the Andalusian physician Ibn Juljul (d. ca. 994), Umar II was responsible for disseminating the translated *kunnāsh*:

> Masārjawayh was Jewish Syriac. He is the one who undertook the explanation of Aaron of Alexandria's [Ahrūn ibn al-Qiss] book to Arabic during the Marwanid reign. Umar ibn ʿAbd al-ʿAzīz found [the book] in the book cabinets and ordered it to be taken out and put in his prayer hall. He then consulted God [through prayer] whether to bring it out to the Muslims to benefit from it. When this happened [i.e., he felt that God supported this effort], he released it to the people and disseminated it among them.[10]

[9] See, for example, M. Shefer-Mossensohn and K. Abou Hershkovitz, "Early Muslim Medicine and the Indian Context: A Reinterpretation," *Medieval Encounters* 19 (2013): 274–99.

[10] Ibn Juljul, *Ṭabaqāt Al-Aṭibbāʾ Wa-Al-Ḥukamāʾ* (Cairo: Institut Français d'archéologie Orientale, 1955), 60. My translation: unless otherwise noted, the translations in this article are mine. Al-Qifṭī maintained that Masārjawayh translated the text during the reign of Marwan I; see ʿAlī ibn Yūsuf al-Qifṭī, *Tārīkh al-Ḥukamāʾ* (Leipzig: 1912), 324–5. Ibn Juljul reported this account on the authority of Abū Bakr Muḥammad ibn ʿUmar ibn al-Qūṭiyyah (d. 978), who was a well-known scholar in al-Andalus but who most likely never encountered ʿUmar II. On Ibn Juljul's biographical dictionary, see Cristina Álvarez Millán, "Medical Anecdotes in Ibn Juljul's Biographical Dictionary," *Suhayl* 4 (2004): 141–58.

Ibn Juljul's account indicates that Masārjawayh completed his translation earlier than the reign of Umar II, perhaps during the longer and more stable reigns of ʿAbd al-Malik ibn Marwān (r. 685–705) and his son al-Walīd I (r. 705–15), during which the minting of a new Umayyad coin as well as the building of the Umayyad mosque, the Aqṣā mosque, and the Dome of the Rock took place.

One of the key sources that many historians, both medieval and contemporary, relied on in understanding and documenting the histories of translation was a book catalog authored by Abū al-Faraj Muḥammad ibn Isḥāq al-Nadīm (d. 995). Ibn al-Nadīm (or al-Nadīm)[11] was born in Baghdad around 932 and was trained by some of the more prominent scholars of religious and linguistic sciences there. While the list of his teachers varies a bit from one account to another, it is clear that he was in contact with key translators, logicians, linguists, and religious scholars in Baghdad and beyond. He likely traveled to Kufa and Basra, both important intellectual centers in tenth-century Iraq, as well as to Aleppo and Mosul, where he collected manuscripts and sought key scholars. The epithets "al-Nadīm" (lit. "boon companion"; normally used to refer to courtiers) and "al-warrāq" (bookseller), which was also attached to his name, indicate that he had a prominent career both as a courtier and as a bookseller in Baghdad. In 987 he started composing the book for which he would be known: *Al-Fihrist* (The catalog). Recording what he believed to be the most important books in every discipline, he listed hundreds of books with biographies of their authors, translators, and patrons. He also explained how disciplines related to one another, and how a given branch of knowledge first appeared in Islamdom and in Arabic. While it is hard to verify many of his accounts, they represent the state of knowledge about these disciplines, authors, and books in tenth-century Iraq, providing us with the chance to better understand the intellectual history and the history of translation at the time.

According to the *Fihrist*, the first translations of scholarly writings were patronized by Khālid ibn Yazīd ibn Muʿāwiyah (d. 709).[12] Khālid was the son of the second Umayyad caliph, Yazīd I (r. 680–83); he briefly sought the caliphate throne after the abdication of his brother Muʿāwiyah II but failed in his bid against Marwān ibn al-Ḥakam (r. 684–5). Khālid was known to be interested in the sciences (especially alchemy) and was given the title the Wiseman of the Umayyads.[13] Ibn al-Nadīm explained that Khālid patronized Greek-speaking scholars in Egypt, especially those knowledgeable in alchemy, and ordered them to translate books from Greek and Coptic to Arabic for his benefit. Whether Khālid's patronage of translations was part of a larger trend within the Arabic-speaking political elites is unclear from Ibn al-Nadīm's accounts. In his account, the passage following Khālid ibn Yazīd's translations was not directly related to sciences. Rather, it involved the translation of the *Dīwān*, or the chancery records, in Iraq (from Persian to Arabic) and in the Levant (from Greek to Arabic). The chancery records (*al-dīwān*) likely referred to accounting and tax documents, which included valuation of land productivity for the purposes of taxation, as well as other political and financial records. Because Arab Muslim rulers inherited

[11] George Saliba has explained that the author in question may indeed be called "al-Nadīm" as opposed to "Ibn al-Nadīm." See Saliba, *Islamic Science and the Making of the European Renaissance* (Cambridge, MA: MIT Press, 2014).

[12] Mohammed ibn Ishaq ibn al-Nadīm, *Al-Fihrist* (Cairo: Hayʾat Quṣūr al-Thaqāfah, 2006), 1:242.

[13] Ibn al-Nadīm calls him "the Wiseman of the Marwanids" (ibid., 1:242). However, Khālid did not belong to the Marwānid clan.

such records from the Byzantines, in Egypt and the Levant, and from the Sassanids, in Iraq and Iran, the documents were kept in Greek or Persian, which forced the new rulers to rely on bureaucratic dynasties also inherited from pre-Islamic rule. The translation of the chancery records, therefore, was an important moment in refashioning the bureaucracy, the state, and the system of governance, and aligning them more with the new rulers of the realm.

In his analysis of Ibn al-Nadīm's account, George Saliba argues convincingly that the translation of the chancery records may very well represent the first instance of systematic written translations.[14] Saliba accepts Ibn al-Nadīm's account that al-Ḥajjāj ibn Yūsuf al-Thaqafī (d. 714), the feared Umayyad general and governor of Iraq, ordered the translation of records there from Persian to Arabic.[15] The attribution of such a significant step to al-Ḥajjāj is reasonable given the importance of al-Ḥajjāj's reign in stabilizing Umayyad (and Arab-Muslim) control over Iraq and Fars.[16] In fact, Ibn al-Nadīm presented al-Ḥajjāj's effort in sponsoring translations as a move against local bureaucratic elites, who spoke Persian and controlled the workings of the chancery. It is reasonable to assume that al-Ḥajjāj, in his efforts to consolidate his control over Iraq, also attempted to extend his control over the chancery through Arabization, which would permit him to hire more trustworthy bureaucrats. Ibn al-Nadīm had a less clear idea about the equivalent chancery translations in the Levant. He attributed the effort to either Hishām ibn ʿAbd al-Malik (r. 724–43) or his father, ʿAbd al-Malik ibn Marwān (r. 685–705).[17] Saliba has demonstrated that the translation of the chanceries led to significant changes in the structure of the state bureaucracy. Indeed, Ibn al-Nadīm explained that the translations coincided with the introduction of new families into the bureaucracy, which had previously been controlled by old bureaucratic dynasties that had survived from pre-Islamic rule. Saliba has proposed that the chancery records were likely translated at the same time as were practical works in mathematics, astronomy, astrology, and other court- and chancery-related sciences. Hoping to regain their old positions, the old bureaucratic dynasties looked to translate key scientific writings to gain an edge over their competitors, sparking the Translation Movement in the process.[18]

The competition model Saliba has offered provides insights into the role professional considerations and priorities played in the making of translations. Just as competition in the court may have influenced the translation of books of court-related sciences, such as mathematics and astronomy, other forms of competition and the changing landscape of practice may have played a similar part in relation to medicine.[19] For learned elite physicians, being able to communicate with the new Arabic-speaking political and economic elites would have been crucial for maintaining their status and expanding their clientele. At another level, Saliba credits this competitive pressure with pushing the new Arabized elites to discover (or recover) scientific Greek texts that had been all but forgotten in the Byzantine context. Yet scholars have demonstrated that Byzantine

[14] Saliba, *Islamic Science* (cit. n. 11).
[15] Ibn al-Nadīm, *Al-Fihrist* (cit. n. 12), 1:242.
[16] See Z. I. Oseni, "The Military Engagements of Al-Hajjaj Ibn Yusuf as an Umayyad Governor of Iraq and the East," *Journal of Arabic & Religious Studies: JARS* 11 (1994): 60.
[17] Ibn al-Nadīm, *Al-Fihrist* (cit. n. 12), 1:242.
[18] Saliba, *Islamic Science* (cit. n. 11).
[19] For a more extended discussion of this question, see Ahmed Ragab, "'In a Clear Arabic Tongue': Arabic and the Making of a Science-Language Regime," *Isis* 108 (2017): 612–20.

scholarly circles interacted with and commented on classical works through the seventh century and beyond.[20] In other words, the elites described by Saliba and Ibn al-Nadīm, be they the old, established ones or the newly emerging ones, were already familiar with these texts. In the case of medicine, scholars and practitioners continued a long and consistent tradition of Galenic scholarship that had been based in the Near East and Eastern Mediterranean for centuries.[21]

It is therefore reasonable to look at the gradual Arabization of these elites themselves, both under the pressure of competition and as part of the overall Arabization of the high echelons of society. My use of the word Arabization here, as opposed to *translation*, is intended to highlight an incremental, cultural, and largely oral process whereby scientific and medical practitioners had to slowly but consistently adopt the language of their new patrons. The move to Arabic and the attendant translations were therefore a function of gradually increasing facility[22] with Arabic, which slowly climbed to the top of the linguistic regime over two or more generations. Physicians, astronomers, astrologers, and others needed to speak Arabic and to translate, albeit orally, their practice into Arabic to engage their new clients. In the same way, bureaucrats and state functionaries had to gain knowledge of Arabic to communicate with their new Arab lords. Translation was a daily exercise necessary for the proper functioning of market and state and had the effect of pushing these learned practitioners and functionaries toward more Arabization—understood as more facility in dealing with Arabic and thinking in Arabic as well.[23]

With this in view, I argue that translation should not be analyzed as a question of moving single texts, or even collections of texts, within a particular discipline or as part of the canon of a specific author. Instead, translation needs to be considered at the level of the archive. In his analysis of archives and memory, Derrida posits that "the technical structure of the archive also determines the structure of the *archivable* content even in its relationship to the future. The archivization produces as much as it records the event."[24] Here, an archive of Galenic medicine was one that endowed a particular set of texts, and a specific type of epistemic content, with value over others—in this case, humoralist Galenic medicine over other medical paradigms. Moreover, in its constantly dynamic life, this emerging archive invited completion, so as to render the totality of the Hippocratic and Galenic corpus knowable through acquisition and translation. It also invited conformity and uniformity in writing and orality, from language and style to categorization.

[20] See, among others, Stratis Papaioannou, *Michael Psellos: Rhetoric and Authorship in Byzantium* (Cambridge, UK: Cambridge Univ. Press, 2013); Matteo Martelli, "Greco-Egyptian and Byzantine Alchemy," in *A Companion to Science, Technology, and Medicine in Ancient Greece and Rome*, ed. Georgia Irby (Hoboken, NJ: Wiley, 2016), 217–30; Maria V. Mavroudi, *A Byzantine Book on Dream Interpretation: The Oneirocriticon of Achmet and Its Arabic Sources* (Leiden: Brill, 2002); and Paul Magdalino and Mavroudi, *The Occult Sciences in Byzantium* (Geneva: La Pomme d'or, 2006).

[21] Peter E. Pormann, *The Oriental Tradition of Paul of Aegina's Pragmateia*, Studies in Ancient Medicine 29 (Leiden: Brill, 2004).

[22] Here, I am using "facility" as opposed to "literacy" or "fluency" to refer to a more variable set of relationships with the language that encompasses the ability to talk, read, or write but without placing them in a particular order or hierarchy.

[23] Ragab, "'In a Clear Arabic Tongue'" (cit. n. 19).

[24] Jacques Derrida, *Archive Fever: A Freudian Impression*, trans. Eric Prenowitz (Chicago: Univ. of Chicago Press, 1996), 17 (emphasis mine).

For practitioners of medicine, knowledge of Arabic and the ability to communicate with clients was as important as it was for the bureaucrats and scribes of the treasury described by Ibn al-Nadīm. Narrative was key to the practice of humoralists on at least two different levels.[25] At one level, the complicated logical structure of humoralist practice and its reliance on treatment by opposites meant that the rationale for a practitioner's diagnoses and treatments might not be immediately apparent to the client. Practitioners thus needed to be able to effectively narrate their reasoning to convince patients that the recommendations were appropriate and accurate.[26] This was especially the case when dealing with elite patients, who routinely called on multiple physicians at a time.[27] At another level, narrative was key to humoralist diagnostic structures. Practitioners needed to be able to understand their patients and their complaints, the progress of their symptoms, and the terms that they used to describe their ailments, as well as their descriptions of their healthy habits and their bodies in health. Physicians also needed to communicate with other practitioners of the medical arts, from herbalists and pharmacists to cuppers and surgeons. In other words, the fact that medical practice was deeply connected to a market consisting of clients, merchants, and other actors meant that practitioners needed to maintain a flexible linguistic outlook in order to be able to navigate a complex and increasingly Arabized marketplace.

To achieve such goals, translating specific texts or the entire oeuvre of certain authors, even ones as important as Hippocrates or Galen, was less urgent than the need to translate concepts, symptoms, diagnostic categories, and elements of pharmacopeia, which existed across multiple texts. For instance, Hippocratic and Galenic practice paid close attention to temporality: diseases developed on a predictable timeline where key moments (such as days 4, 7, 10, and 14, based on fractions of a moon's cycle of 28 days) indicated disease progress. On these "crisis days," symptoms intensified or were resolved, thus denoting an auspicious or catastrophic disease course. In this context, it was more pressing to translate and communicate the term and meaning of *crisis*, which was translated to the Syriac term *buḥrān*, than to fully translate the theoretical corpus that conditioned the meaning of *crisis* in the humoralist context.[28] As such, oral translation provided for spotty and practice-oriented translations of ideas, concepts, and categories in ways that would later influence the work of important translators of texts, such as Ḥunayn ibn Isḥāq. In his analysis of Ḥunayn's translations of Galen's work on crises,

[25] On narrative and humoral medicine, see, among others, Lee T. Pearcy, "Diagnosis as Narrative in Ancient Literature," *Amer. J. Philol.* 113 (1992): 596–616; Gianna Pomata, "The Medical Case Narrative: Distant Reading of an Epistemic Genre," *Literature and Medicine* 32 (2014): 1–23; Or Hasson, "Between Clinical Writing and Storytelling: Alfonso De Santa Cruz and the Peculiar Case of the Man Who Thought He Was Made of Glass," *Hispanic Review* 85 (2017): 155–72; and Caroline Petit, *Galien de Pergame ou la rhétorique de la Providence: Médecine, littérature et pouvoir à Rome* (Leiden: Brill, 2018).

[26] Much has been written on logic and its connection to the practice of humoral and Galenic medicine, though not as much in relation to Islamic or Arabic Galenism. See, for example, Ian Maclean, *Logic, Signs and Nature in the Renaissance: The Case of Learned Medicine* (New York: Cambridge Univ. Press, 2007); and Nancy G. Siraisi, *Medieval and Early Renaissance Medicine: An Introduction to Knowledge and Practice* (Chicago: Univ. of Chicago Press, 2009).

[27] See al-Rāzī's instructions to his student in the famed *Epistle to a Student: Abū Bakr Muḥammad ibn Zakarīyā al-Rāzī, Akhlāq Al-Ṭabīb: Risālah Li-Abī Bakr Muḥammad ibn Zakarīyā Al-Rāzī ilā Ba'ḍ Talāmīdhih* (Cairo, 1977).

[28] See Glen Cooper, ed., *Galen, "De Diebus Decretoriis," from Greek into Arabic: A Critical Edition, with Translation and Commentary, of Ḥunayn ibn Isḥāq, "Kitāb ayyām al-buḥrān"* (Farnham, Surrey, UK: Ashgate, 2011).

Glen M. Cooper describes Ḥunayn's method of translation as reader-oriented because Ḥunayn's style paid little, if any, attention to preserving the integrity of the original text and instead focused primarily on conveying concepts and thoughts to his readers—medical practitioners.[29] Cooper correctly notes that many concepts are translated using various words, from Arabic, Syriac, Persian, or even transliterated Greek, without a clear or consistent logic behind these choices, which Ḥunayn and other translators hardly explained.[30] In other words, it appears that Ḥunayn and other translators relied on a wealth of translated concepts and terms that existed in practice well before the translated texts came to catch up to them.

Cooper discusses a passage on translation by Khalīl ibn Aybak al-Ṣafadī (1296–1363) that was commonly cited by medieval scholars. Al-Ṣafadī, who was a well-known scholar and historian from Safad (modern day Israel/Palestine), explained that translations developed from focusing on translating word for word to translating thoughts. Cooper shows that this evolution described by al-Ṣafadī is difficult to trace historically in the lives and careers of early translators.[31] Yet al-Ṣafadī's characterization describes an evolution not only in the history of written translations but one in translation tout court. Word-for-word translations were necessary to allow physicians and other practitioners to make sense of the symptoms described by their patients and to explain their approach to them. While this word-for-word translation may have influenced the early works of written translations, it was less needed in later works by Ḥunayn and others—translations that ended up dominating Arabic humoralism. Yet these later translations depended on the infrastructure produced orally.

In Ibn al-Nadīm's account, disgruntled Persian bureaucrats fearful of being replaced by Arabic speakers tried to prove that the translations were not accurate, or that the chosen terms fell short of conveying the complex technical meanings of the original Persian. The key test that they proposed was to ask about specific words and terms, and see how effectively they were translated.[32] In translating these concepts and terms (or, at times, transliterating or adapting them into Arabic), the chancery records became legible to the Arabic-speaking patrons, and the translation of the various mathematical and astronomical texts would naturally follow as the bureaucratic elites became more and more Arabized and started to seek materials that facilitated and consolidated their bilingual practice. In this case, and in medicine, the point of translation was not to move materials or concepts to Arabic but to make them legible and comprehensible to the new clientele. Whether this was achieved through adopting Arabic words, creating new ones, or Arabizing Persian or Greek words was of less significance.

Thinking of translation in terms of archive making admits the rationale expressed by Ibn al-Nadīm's own narrative. Putting aside the exact chronology of translation, Ibn al-Nadīm placed the accounts of translations in a subchapter titled "Mentioning the reason why the books of philosophy are numerous in these lands."[33] In his narrative, the movement of texts into Islamic domains included both their translation into Arabic

[29] Glen M. Cooper, "Ḥunayn ibn Isḥāq's Galen Translations and Greco-Arabic Philology: Some Observations from the *Crises* (*De crisibus*) and the *Critical Days* (*De diebus decretoriis*)," *Oriens* 44 (2016): 1–43; see 5–6.
[30] Ibid., 11–2.
[31] Ibid., 7.
[32] Ibn al-Nadīm, *Al-Fihrist* (cit. n. 12).
[33] Ibid., 1:243.

and their physical movement and acquisition. He described how the Abbasid caliph al-Maʾmūn (r. 813–33) saw Aristotle in a dream and was thereafter inspired to acquire as many classical Greek texts as possible. Ibn al-Nadīm then explained the efforts of many other patrons of translation, such as the three brothers Muḥammad, Aḥmad, and al-Ḥasan Banū Shākir al-Munajjim—astronomers, mathematicians, courtiers, and patrons of sciences:[34]

> Muḥammad, Aḥmad and al-Ḥasan banū Shākir al-Munajjim were some of those who were interested in obtaining books from the land of the Romans [the Byzantine empire]. They spent what is dear and precious [to buy and acquire these texts]. They sent Ḥunayn ibn Isḥāq and others to the Romans to bring them remarkable books in philosophy, engineering, music, arithmetic, and medicine. Qusṭā ibn Lūqā also brought books with him and translated them or had them translated.[35]

Ḥunayn ibn Isḥāq and Qusṭā ibn Lūqā were two of the more prominent and celebrated translators of scientific and medical texts. In Ibn al-Nadīm's account, they were also tasked with acquiring texts, not simply translating them.

Ibn al-Nadīm's focus on and interest in writing and written texts are naturally connected to his career as a bookmaker and seller. For Ibn al-Nadīm, acquiring books, forming libraries, and collecting materials were necessary for knowledge making, and also a sure sign of the greatness of different rulers and the magnificence of their reigns. At the same time, Ibn al-Nadīm's recounting of the history of knowledge in the lands of Iraq, Persia, and the Near East was deeply connected to the movement of books: that is, the ability of different sovereigns and patrons to collect, acquire, and protect them. This history is inflected by his and his contemporaries' view on the rise of learning in Iraq under the Abbasids. In other words, the Abbasid efforts to collect books and to patronize authors, translators, and practitioners were seen as part of a pattern of great rulers sponsoring and patronizing knowledge making, and another stage in the waxing and waning history of knowledge in the Near East.

In this narrative, Alexander the Great loomed large as a key figure in pre-Islamic history and in the history of knowledge. In Ibn al-Nadīm's view, Alexander's invasion led to a significant decline in learning in Iraq and Iran because he seized troves of books and materials there and sent them back to his Greek domains, where learning and knowledge grew as a result.[36] Similarly, the rise of learning in Alexandria, a school regarded with reverence among Islamicate Hellenists, was connected to a similar process of book acquisition. Ibn al-Nadīm recounted that King Ptolemy II Philadelphus of Egypt (r. 283–46 BCE) ordered some of his associates to collect as many books as possible. They "collected fifty-four thousand and a hundred and twenty books and said that even more existed in Iran, India and China."[37]

Throughout, oral translation was implied in the process of collecting and classifying. For instance, in each of the several incidents that Ibn al-Nadīm narrated where troves of books were found, he mentioned whether there were people able to read the languages in which these books were written and explain their contents.[38] In some

[34] On Banū Mūsā ibn Shākir al-Munajjim, see Donald P. Hill, *The Book of Ingenious Devices/Kitāb Al-Ḥiyal: Kitāb Al-Hiyal by the Banū (Sons of) Mūsā ibn Shākir* (Springer, 2012).
[35] Ibn al-Nadīm, *Al-Fihrist* (cit. n. 12), 1:243.
[36] Ibid., 1:239.
[37] Ibid., 1:239–40.
[38] Ibid., 1:240

cases, the discovered books could not be fully understood because they were written in old or unknown languages but were kept until someone who could read them could be found.[39] Written translation, which has consistently dominated modern scholarly discussions, was but a stage in the larger process described by Ibn al-Nadīm. Thinking this way, in terms of archive making, recognizes written translation as a gradual process, secondary to the acquisition and circulation of texts, and part of the development of the book market and the scientific elite's increasing facility with Arabic.

HUNAYN'S ARCHIVE

Few names are more connected to the history of Greco-Arabic translations, and particularly medical translations, than that of Ḥunayn ibn Isḥāq al-Ibādī (d. 873). An Arab Nestorian Christian from al-Ḥīra in southern Iraq, Ḥunayn was a native Arabic speaker, fluent in Syriac (the liturgical language of the Eastern Church) as well as Greek. After migrating to Baghdad, he studied under the famous physician Yuḥannā ibn Masāwayh (d. 857), a Syriac physician hailing from the city of Gundishapur in southwest Iran. Ibn Masāwayh seems to have grown tired of Ḥunayn, whose ethnic and tribal origin was seen as inferior, and ultimately chased him out of Baghdad. After some absence, during which Ḥunayn presumably traveled in the Levant and in Asia Minor honing his Greek skills, he returned to Baghdad and rose to the highest echelons of the medical community there. His reputation as one of the more important translators continued to grow, and his medical translations came to dominate the medical field for centuries to come. Moreover, Ḥunayn worked with and trained a number of students (including his son), who established a stellar reputation as translators in their own right.[40] In a treatise that he wrote to enumerate Galen's available books and to specify which of them were translated, Ḥunayn perhaps came closest to explaining his methods and to reflecting on the translation process from within.[41] Originally commissioned to be a catalog of all Greek medical writings, the treatise offers an important view of how Ḥunayn understood the medical archive, which he had been instrumental in producing. In this section, I will look closely at two stages in the life of his treatise, its original composition in Syriac and then its translation to Arabic, to understand the production of the medical archive from the standpoint of Ḥunayn, his patrons, and readers.

Ḥunayn's treatise was not limited to listing his own translations or those of his students and associates but also included all of Galen's works that were known and available at the time, including ones translated by others and ones that were not yet translated. Ibn al-Nadīm cited and relied on the treatise in his *Fihrist*, and the famous physician Abū Bakr al-Rāzī commented on and completed the treatise, adding works that Ḥunayn did not include. Following Gotthelf Bergsträsser's edition of the treatise in 1925, scholars looked to the text as an explanation of Ḥunayn's method in translation

[39] Ibid., 1:238–40.

[40] Much has been written about Ḥunayn ibn Isḥāq's translations. For some of the more recent works, see Cooper, "Ḥunayn Ibn Isḥāq's Galen Translations" (cit. n. 29); and Oliver Overwien, "The Paradigmatic Translator and His Method: Ḥunayn Ibn Isḥāq's Translation of the Hippocratic Aphorisms from Greek via Syriac into Arabic," *Intellectual History of the Islamicate World* 3 (2015): 158–87.

[41] Ḥunayn ibn Isḥāq al-ʿIbādī, Gotthelf Bergsträsser, and Ayasofya Kütüphanesi, *Hunain ibn Ishaq über die syrischen und arabischen Galen-Übersetzungen: Zum ersten Mal herausgegeben und übersetzt von G. Bergsträsser* (Leipzig: Brockhaus, 1925).

and a map of medical translators during this period.[42] Yet at another, less-investigated level, the treatise provides us with important information about Ḥunayn's work in constructing an archive of medical knowledge and gives us a glimpse into the intellectual and patronage project of which this treatise was part. It is important to note that the treatise we have today is but a poor substitute for a much loftier project that Ḥunayn failed to accomplish. Standing as evidence of failure, the treatise allows us to better understand the making of scientific knowledge and the potential, logic, and intent that the failed project embodied.

In his introduction, Ḥunayn explained the purpose of the treatise and the difficulties that surrounded its composition. Ḥunayn wrote this Arabic treatise at the request of some of his main patrons, the previously mentioned Shākir ibn al-Munajjim brothers. The Shākir brothers (Banū Shākir)[43] were themselves prolific scholars and powerful courtiers who grew up in the Abbasid court after their father entrusted them to the care of his patron and friend the Abbasid caliph al-Maʾmūn. Banū Shākir rose to the forefront of the Islamic Hellenistic culture that flourished in Baghdad in the middle of the ninth century. They were also known as eloquent authors and talented illustrators and were major patrons of arts and sciences in Abbasid Baghdad, collecting books, attracting and recruiting scholars, and contributing to scholarship themselves.

The treatise commissioned by the Shākirs was not meant to be restricted to Galen's works. Instead, it was supposed to include "what was proven to be useful of the ancients' books on medicine."[44] Ḥunayn was to list the main purpose for each book and explain why a student or a reader might need it. The entry was also supposed to include information about the chapters or treatises included in the book and the different questions discussed in each, "so that it becomes easier for one studying a particular question to locate it, and to know in which book [to find it], in which treatise in that book and in which part of the said treatise."[45] The goal was therefore to create a finding aid of sorts that would enable scholars and students to locate answers to their questions more easily. As such, the treatise would transform the ancients' disparate collection of medical texts into a veritable archive that was internally coherent and thus could be studied not only by author or text but also by topic.

The extent of the treatise's consolidating mission is better grasped when viewed in light of some of Ḥunayn's remarks about his own translations. In the treatise and elsewhere, Ḥunayn explained that he often filled in the blanks in his source texts and included, in his translated texts, comments and explanations based on his other readings, and his own medical knowledge and practice.[46] Conversely, Ḥunayn also omitted repetitions and sometimes discarded entire treatises or works when he found them unsatisfactory or offering little value because the content was better presented elsewhere.[47] While some incidences of Ḥunayn's editorial interventions were visible

[42] For a listing of these works and an analysis of their contributions, see Oliver Overwien, "The Art of the Translator, or: How Did Ḥunayn ibn ʾIsḥāq and His School Translate?," in *Epidemics in Context: Greek Commentaries on Hippocrates in the Arabic Tradition*, ed. Peter E. Pormann (Boston: De Gruyter, 2012), 151–70. For a discussion of some of the more important works on this topic, see Uwe Vagelpohl, "In the Translator's Workshop," *Arab. Sci. Phil.* 21 (2011): 249–88.

[43] *Banū* is plural of *Ibn*. *Banū Shākir* translates as "the sons of Shākir"; see Hill, *Book of Ingenious Devices* (cit. n. 34).

[44] al-ʿIbādī, Bergsträsser, and Kütüphanesi, *Hunain Ibn Ishaq* (cit. n. 41), 1.

[45] Ibid.

[46] Ibid., 1. See also Overwien, "Art of the Translator" (cit. n. 42), 157–9.

[47] Overwien, "Art of the Translator" (cit. n. 42), 158.

and easily detectable, Oliver Overwein, Uwe Vagelpohl, and others have convincingly argued that Ḥunayn's practices likely extended well beyond these clearly marked or observable instances.[48] Moreover, there is no reason to assume that this practice was restricted to Ḥunayn. It is likely that other translators, about whose methods we know less, took similar approaches and edited the works that they translated. In its totality, translation was thus also a practice of editing and explaining. The goal was not simply to move texts as faithfully as possible from one language to another but rather to contribute to editing, distilling, and consolidating, to create a body of knowledge in which repetitions (especially inferior ones) were omitted, contradictions resolved, and texts properly cross-referenced. In this context, a treatise like Ḥunayn's would serve as a guide to the consolidated body of knowledge produced by different translators. Moreover, the critical evaluation of the works of different translators/editors/movers was also important because it demonstrated the progress of this process.

Ḥunayn offered his apologies to the Shākirs because he was not able to compose the desired treatise, despite his belief in its importance. He had lost his library, and he could not recall all these works from memory.[49] Instead of the desired all-encompassing finding aid, Ḥunayn offered a more modest one that focused only on Galen, and that relied on an earlier treatise he had prepared in Syriac for another patron shortly after losing his library. In the introduction to the old treatise, which Ḥunayn translated or summarized, he further explained why this project was important. Although he would later commit to the Shākir project, he had been rather skeptical when his former patron had first approached him. He told his Syriac patron that Galen had already written a treatise in which he enumerated his works, and which had been translated under the title *Fihrist.* Galen also wrote another shorter treatise that included instructions on how and in what order one should read his works. As such, Ḥunayn reasoned, "Learning about Galen's book from Galen is certainly superior to learning about them from me."[50] The patron responded, as quoted by Ḥunayn:

> "While what you [argue is reasonable], the people interested in this art [medicine] and who read books in Syriac or Arabic are in need of knowing which books have been translated to Arabic or Syriac and which have not [. . .], and which Greek texts were located and which were not, so that effort can be extended to translate those that were found, and to seek those that were not."[51]

Ḥunayn was won over by this argument. His enthusiasm for the Shākir project was evidently rooted in this previous conversation.

In this explanation, two important tasks were at stake: (1) collecting Greek materials; and (2) translating them to Syriac or Arabic. While the scope of the Syriac treatise was smaller, both patrons wanted to take stock of the expanding archive of medical texts with the explicit intent of further expanding the archive and facilitating access to it. In this context, Galen's own catalog was useful but hardly sufficient. It was a finding aid for a different archive. The new Islamicate archive, produced

[48] Ibid.; Vagelpohl, "In the Translator's Workshop" (cit. n. 42).
[49] al-ʿIbādī, Bergsträsser, and Kütüphanesi, *Hunain Ibn Ishaq* (cit. n. 41), 1.
[50] Ibid., 2.
[51] Ibid., 2–3.

in Syriac, Arabic, or both, required a different organization, one that explained its contents and adjudicated its worth.

The archive, in this sense, is not a concrete collection of materials housed or preserved in a space. While Derrida's archive was a site (a place) of commencement and commandment, this archive of classical medical texts was rather a moment (a temporality) of such commencement and commandment. Here, "archive" describes not simply a collection of documents that record events but rather a collection of materials that invite a commencement: an action into the future. The archive of scientific classical texts enabled this moment of (re)commencement of commentaries, critique, and practice. This impulse to (re)commence was evident in the need for the two treatises that Ḥunayn described. The Syriac project looked to (re)commence the evaluation of the archive's materials and the completion of the translations, a process necessary for the full inclusion of ancient materials. The Arabic treatise looked to (re)commence the function of the archive by facilitating the use, study, and analysis of different ancient medical texts perceived in their totality as a coherent whole. This moment of, or invitation to, commencement is unmistakably linked to an actual commandment: Ḥunayn was commanded to perform this task of collection, evaluation, and arrangement.

In the same vein, the archive provides for a measure of uniformity and "commands" a degree of coherence that is constantly sought but almost never achieved. Following Foucault's argument that the archive is "the system which governs the appearance of statements as unique events,"[52] I argue that the archive, as a dynamic collective, produces gravitational power that pulls for completion as well as uniformity. Along with their interests in acquiring missing texts and adding to the body of translated materials, both patrons were also interested in an evaluative project—one that aimed to purify and rectify the expanding archive. The Syriac patron asked Ḥunayn to help evaluate different translations by making their attributions explicit and comparing them with one another. The Shākir patrons pushed for full consolidation of the ancient corpus, which could only be achieved by pushing forward Ḥunayn's and his colleagues' efforts in editing Greek materials, explaining them, and cutting redundancies. In both cases, the composition of the treatises was symptomatic of the organizational, gravitational power at the heart of the archive.

Moreover, the project intended to rearrange the corpus of Greek texts in a manner that would go beyond authors and their works to address diseases, conditions, and other related practical concerns. In that sense, the project would fundamentally intervene in the coherence of the archive and its constituent texts. Ḥunayn was to break down texts, disrupt their coherence, and establish a new level of organization whereby discrete textual boundaries melted, giving rise to a larger whole. This logic relied on the assumed commensurability of the constitutive texts and materials. It could work only if these authors and their texts addressed a similar conception of illness, healing, the human body, and medical practice, among other considerations. The project, therefore, indexes the view of medicine, and in particular Galenic medicine, as a coherent whole, where breaks may exist but can be remedied, and where texts and authors agreed on the fundamental foundations of the medical art. More fundamentally, the project offers evidence that

[52] Michel Foucault, *The Archaeology of Knowledge*, trans. Alan Sheridan (New York: Pantheon Books, 1972), 129.

Ḥunayn and his patrons and readers believed that such foundations exist. In other words, if the discrete nature and internal logic of each text was to be disarticulated, it was for the sake of a higher order of coherence—that of the medical art. Not only was this coherence assumed, it was desired.

At another level, entertaining such a project demonstrated a specific view shared by Ḥunayn and his patrons on the sheer size of available knowledge. Ḥunayn was aware of the first Hippocratic aphorism, which he had translated, that stated "Life is short and the art is long."[53] If the art is indeed long, too long for a lifetime, how could such a project be feasible or even reasonable to entertain? The length of the art was connected to the variations of human bodies and to the need for long experience. However, as this endeavor demonstrates, it was not related to the size of the Greek corpus or the fundamental knowledge that physicians needed to know. As such, the project operated with the view of two types of knowledge: one that is limited and fully digestible in a single finding aid, and another that is more variable, changing, and too long for a single or many human lives. The first is precisely the archive that translations intended to move, edit, and consolidate—an inherited textual corpus that outlined the fundamentals of the art and upon which practical and experiential knowledge could be built. Not only is this corpus by definition limited, it is limited enough for one person, possibly with students and aides, to undertake cataloging and indexing.

Finally, the failed project operated with a skeptical view of the corpus and the authority of its authors. On one hand, and as mentioned before, the coherence of specific texts, and therefore the authorial/authoritative voice of their authors, was less important. On the other, the project intended to overcome the perceived unwieldy nature of the corpus. Ḥunayn was to organize it, trim its excesses, and offer a clear and concise way to approach it. In this attempt to create the catalog to end all catalogs, the corpus is seen as suffering from redundancies, problems, breaks, and useless information. The new archive is to be trim, useful, and more easily navigable. This brings us back to the more modest project that Ḥunayn achieved. There, too, and as mentioned before, Galen's own catalog was not sufficient, as it indexed a different archive. In this view, the translations that Ḥunayn and others engaged in were indeed a process of making a new medical archive that was built on the practice that physicians engaged in, and was meant to facilitate such practice. As such, I argue that understanding these efforts needs to start not from the Greek text or its Arabic counterpart but from the practical categories that these texts underlined and indexed.

In offering his apologies for not completing the project, Ḥunayn expressed his hopes that he might retrieve his confiscated library with the aid of his influential patron. The physical library occupied a key position in the production of this catalog. The catalog project was simply unfeasible in its totality without the presence of the physical library that Ḥunayn had assembled in his various travels. Ḥunayn's library, we come to find out, contained a collection of Greek manuscripts by many authors and from different sources. It also included several copies of many texts, since he came to

[53] On the first aphorism, see Franz Rosenthal, "'Life Is Short, the Art Is Long': Arabic Commentaries on the First Hippocratic Aphorism," *Bull. Hist. Med.* 40 (1966): 226–45. On the aphorisms and the commentary traditions in Arabic, see, among others, Pormann and Karimullah, "Arabic Commentaries on the Hippocratic Aphorisms" (cit. n. 8); and Rosalind M. Batten, "The Arabic Commentaries on the Hippocratic Aphorisms: Arabic Learned Medical Discourse on Women's Bodies (9th–15th Cent.)" (PhD diss., Univ. of Manchester, Manchester, UK, 2018).

use these copies for verifying and correcting the Greek text before translating it.[54] The centrality of the physical library demonstrated Ḥunayn's view of his role not simply as a translator for hire but as a collector and mover of knowledge who spent much of his energy traveling and seeking manuscripts on behalf of his patrons and for his own benefit. The physical library has a predictable dialectical relationship to memory: the absence of one highlights the importance of the other. This highlights how the project itself was one intended not to replace memorization but precisely to complement practices of memorization with those of referencing and looking up.

The content of Ḥunayn's treatise offers a more nuanced picture of the translation archive. For Ḥunayn, and perhaps others too, translation often came after, and was intertwined with, acquiring, comparing, and correcting Greek texts. Additional copies and better manuscripts were sought and, as Overwien has suggested, earlier translations may have been consulted as well.[55] Translation was key for the full habilitation of a text within the emerging archive, but no single translation was perceived as a definite last step. For one, translations occurred not just from Greek into Arabic but also into Syriac, from Syriac into Arabic, and, in some rare instances, from Arabic into Syriac.[56] This constant movement of materials not only flowed outward from the Greek but also flowed back to the Greek as translated texts helped translators correct, verify, and complete Greek texts. In other words, the archive was a dynamic environment where texts and knowledge passed in multiple directions, with the aim of consolidating knowledge, filling in gaps, and resolving contradictions.

The iterative nature of translation was key to such a dynamic archive and contributed to its constant internal mobility. Ḥunayn explained how translators edited each other's work, whether motivated by cooperation or competition; how they relied on one another's works; and how they collaborated in translating some texts by creating a multistep translation (from Greek to Syriac and then to Arabic, for example) to complement their varying linguistic skills. This iterative nature was fostered by the multiplicity of patrons, some of whom may have had favorite translators or favorite topics and disciplines. Moreover, medical practitioners such as Ḥunayn and some of his patrons, as well as the readers and consumers of these texts, contributed to the spread of particular texts and therefore to the reputation of certain translators. In addition to the expanding written translations, the oral translations embedded in the practice allowed for the gradual, and often uneven, production of specific terms in Arabic that were used to translate Greek technical terms. In the case of medical translation, translators created these terms and made their choices with an eye toward the habits, traditions, and preferences of medical practitioners, and these terms came to spread across the written archive through the combined influence of practice and this iterative process of translation. This contributed to the uniformity of the archive and to its consistency.

As explained before, the archive was connected to the narrative nature of the humoralist practice. In her article "The Sciences of the Archive," Lorraine Daston identifies how particular scientific disciplines in the early modern period developed as an enterprise dependent on archival and archived knowledge.[57] In astronomy, for instance, historical observations were necessary to make sense of contemporary observations and

[54] al-ʿIbādī, Bergsträsser, and Kütüphanesi, *Hunain Ibn Ishaq* (cit. n. 41).
[55] Overwien, "Art of the Translator" (cit. n. 42), 152.
[56] al-ʿIbādī, Bergsträsser, and Kütüphanesi, *Hunain Ibn Ishaq* (cit. n. 41).
[57] Lorraine Daston, "The Sciences of the Archive," *Osiris* 27 (2012): 156–87.

to create astronomical knowledge. In these disciplines, Daston argues, the library and the archive were physically centered within the spaces of the laboratory or observatory or other similar scientific institutions. Daston's observations allow for deeper considerations of the role played by archives, physical and metaphorical, in the construction of specific scientific practices. If this is the case, what role did archives play in the construction of the version of Galenic medicine under consideration here? And was this Galenic medicine also "a science of the archive"?

Indeed, this version of Galenic medicine was deeply invested with a particular form of historical consciousness, and the importance of historical knowledge. Ḥunayn's failed project relied on an understanding of medical knowledge as commensurable, which rendered the inherited Greek knowledge important and useful. Treating a largely unchanged human body, physicians and medical practitioners engaged actively with historical observations. Perhaps a key example comes from another text, one that was translated by Ḥunayn and remained exceedingly popular, namely, the *Epidemics*. Built on a collection of cases attributed to Hippocrates, the text continued to serve as a clear example for proper practice and to inspire a significant body of commentaries for centuries to come.[58] The staying power of *Epidemics* is perhaps one of the clearest examples of belief by Galenics in the validity of such "historical" knowledge even at the practical level.

Yet this archive is significantly different from contemporary astronomical archives, for instance, which offer a picture close to what Daston describes. As explained, the medical archive was not all written; it was also an oral archive constructed around the experiences of patients, physicians, and other medical practitioners. As such, it was not an archive that could be stored in a single place or kept in a library. Ḥunayn's lost library represented the written part of an archive that was to be organized, arranged, and indexed to follow the logic of the largely oral practice. Moreover, medical practice in the Near East was consistently multilingual, as physicians dealt with patients, traders, practitioners, and others who spoke in various languages and vernaculars. This multiplicity provided for even more complexity of the archive and resulted in the mix of different languages that appear in medical writings. Yet, as Ḥunayn's project demonstrated, the archive pulled for uniformity and coherence. This pull is motivated by its users, who seem to have consistently valued more accessible and easily digestible content.

SINGULAR AND MULTIPLE ARCHIVES

Making an archive also reflects a mode of identity formation. Producing a medical archive based on Greek humoralist materials engendered and emphasized the identities of specific groups of practitioners and favored their brand of medicine. Admittedly, this was a project championed by Galenic practitioners, some of whom hailed from former Byzantine territories in Egypt and the Levant, and others from the Hellenized Syriac communities in former Sassanid territories. But while both groups of physicians traced their intellectual genealogy to the same set of ancient texts and to generally similar groups of successive commentators, they were not the same. Moreover, in the beginning

[58] Peter E. Pormann, ed., *Epidemics in Context: Greek Commentaries on Hippocrates in the Arabic Tradition* (Boston: Walter de Gruyter, 2012), especially Uwe Vagelpohl, "Galen, *Epidemics*, Book One: Text, Transmission, Translation," 125–50.

of the ninth century, the medical scene was even more complex than a simple struggle between two factions of competing Galenics.

Ibn al-Nadīm's accounts from the middle of the tenth century included three different groups of practitioners: Greek humoral practitioners, who were likely the more numerous and influential in his time; Indian practitioners; and Persian practitioners. Among the Persians, Ibn al-Nadīm mentioned only two physicians. The first was Tiādurūs, who presumably served the Sassanid emperor, Sabūr II (r. 309–79), and for whom the famous emperor built a monastery. Tyādurūs also wrote a textbook on medicine (*kunnāsh*), which was translated into Arabic and seems to have circulated in the ninth and tenth centuries. The second was called Tyādūq and was reported to have served al-Ḥajjāj ibn Yūsuf.[59] Ibn al-Nadīm's account of Indian physicians was more elaborate, although still significantly shorter than his report on the humoralists. Instead of focusing on names of physicians, Ibn al-Nadīm opted to enumerate "the names of Indian books on medicine that are found in the language of the Arabs." He listed twelve books, which included one *kunnāsh*; a book that included a summary of one hundred diseases and treatment; a drug formulary; a book on women's diseases; a number of specific books on snake poisons and on medicinal preparations useful for pregnant women; and a book on "imagination [causing] ailments."[60] Three of these twelve books were connected to the Bīmāristān (hospital) built by the Barmakids, the vizir dynasty of the ninth century.

Similar to, but more prominent than, the Banū Shākir, the Barmakids were major patrons of scientific and literary activity. The Barmakids claimed to have descended from a prominent dynasty of Zoroastrian or Buddhist priests. Under the Umayyads, some members of the family served in the bureaucracy, but they rose to prominence under the Abbasids after becoming supporters of the Abbasid revolt in 750. Eventually, they came to be the vizier dynasty of the early Abbasid empire, with their members becoming the close friends, mentors, and confidants of the princes of the Abbasid dynasty. They reached the apogee of their influence under Hārūn al-Rashīd (r. 786–809). It was also al-Rashīd who put an abrupt end to their influence in the events known as the Fall of the Barmakids (*nakbat al-barāmikah*) in 803. Although al-Rāshīd confiscated much of their property and arrested the more significant members of the dynasty, their influence endured through their clients and protégés, such as al-Faḍl ibn Ḥasan, who was the strong vizir of al-Ma'mūn (r. 810–33).[61]

According to Ibn al-Nadīm, the Barmakids were particularly interested in patronizing Persianate and Indian scholars and translators who could translate from Pahlavi (middle Persian) and Sanskrit (often through Pahlavi) to Arabic.[62] Participating in the emerging tradition of hospital building in the Abbasid metropole, the Barmakids built

[59] Ibn al-Nadīm, *Al-Fihrist* (cit. n. 12), 1:303.
[60] Ibid.
[61] See Kevin van Bladel, "The Bactrian Background of the Barmakids," in *Islam and Tibet—Interactions along the Musk Routes*, ed. Anna Akasoy, Charles Burnett, and Ronit Yoeli-Tlalim (Burlington, VT: Ashgate, 2016), 43–88; C. Edmund Bosworth, "Abū Ḥafṣ ʿumar Al-Kirmānī and the Rise of the Barmakids," *Bull. Sch. Orient. Afr. Stud.* 57 (1994): 268–82; Tayeb El-Hibri, *Reinterpreting Islamic Historiography: Harun al-Rashid and the Narrative of the Abbasid Caliphate* (Cambridge, UK: Cambridge Univ. Press, 1999); and Mohammed Didaoui, "Translation and Textual Incongruity: The Background for Al-Jahiz's Rhetorical Work," in *Proceedings of the International Conference on Similarity and Translation: Bible House, New York City, May 31–June 1, 2001*, 2nd ed., ed. Stefano Arduini and Robert Hodgson Jr. (Rome: Edizioni di storia e letteratura, 2007), 427–48.
[62] Ibn al-Nadīm, *Al-Fihrist* (cit. n. 12).

a hospital where medical authors and translators worked and produced several texts of Indian medicine. In the same vein, Ibn al-Nadīm mentioned that a physician called Mankah translated an Indian *kunnāsh* for Yaḥya ibn Khālid al-Barmakī (d. 806), who was a governor under the Abbasid caliph al-Manṣur (r. 754–75) and a vizir under al-Rashīd.[63] Members of the Abbasid ruling dynasty seem to have patronized some translations as well. These include ʿAbd Allāh ibn ʿAlī (d. 764), who was the uncle of Abū ʿAbbās al-Saffāḥ, the first Abbasid caliph. ʿAbd Allāh ibn ʿAlī was a central figure in the Abbasid push to defeat the Umayyads. He was also known to have been a patron of authors and scholars.[64] Another translation was patronized by Isḥāq ibn Sulaymān ibn ʿAbd Allāh, who was a cousin of the first caliph, Abū al-ʿAbbās al-Saffāḥ, and of ʿAbd Allāh ibn ʿAlī.[65] Common among these patrons of Persian scholars and translations is their connection (through lineage or through their career) to the Persianate component of the empire and to the early Abbasid revolt, which was supported by Persianate subjects.

For a number of Indian texts, it appears that the translations occurred through Persian. This pattern fits well with a Sanskrit-Pahlavi (middle Persian) literary connection that had existed since the Arab conquest of Iraq and Fars. For instance, Ibn al-Muqaffaʿ (724–59), a famous bureaucrat and author, made a career for himself as a secretary under the Umayyads and the Abbasids. Ibn al-Muqaffaʿ, himself a descendant of noble Persian families, was likely among many others who joined the Arab court to support state administration in former Sassanid territories in Iraq and Iran. His claim to fame is partly based on his translation of the animal fables *Kalilah wa Dimnah*, originally a Sanskrit collection which was translated to Pahlavi (Middle Persian) and that he translated into Arabic.

The famous fables were not the only book or collection that Ibn al-Muqaffaʿ may have translated. *The Khudaynāmah* (Book of kings), *The Āʿīn-nāmah* (The book of manners), *Kitāb al-Tāj* (The book of the crown), *Kitāb Mazdak* (The book of Mazdak) and *The Letter of Tansar*, all of Sanskrit/Indian origin, were also attributed to him.[66] Ibn al-Muqaffaʿ's work at such an early period of Abbasid rule was probably a prelude to more translations and to further integration of Persian and Indian writings into the new Islamicate archives. Such efforts may have been supported by the Barmakids, themselves Persians of Indian stock.[67]

In the field of medicine, Indian and Persian physicians competed for recognition and for clients with the humoralists during this period—such competition would have included attempts to translate and popularize medical theories and medical recommendations, in order to influence the habits of the potential clients who belonged to the

[63] Shefer-Mossensohn and Hershkovitz, "Early Muslim Medicine" (cit. n. 9).

[64] Ibn al-Nadīm, *Al-Fihrist* (cit. n. 12), 1:303. Ibn al-Nadīm mentioned that ʿAlī ibn ʿAbd Allāh translated a book on Indian medicine from Persian to Arabic. Although ʿAlī ibn ʿAbd Allāh probably knew Persian, as he led Persian armies during the Abbasid revolt, it is more likely that he sponsored the translation rather than performed it himself. See Muḥammad ibn Aḥmad ibn ʿUthmān al-Dhahabī, *Siyar Aʿlām Al-Nubalāʾ* (Beirut: Muʾassasat al-Risālah, 2001).

[65] Ibn al-Nadīm, *Al-Fihrist* (cit. n. 12).

[66] J. D. Latham, "Ibn Al-Muqaffaʿ and Early ʿAbbasid Prose," in *ʿAbbasid Belles-Lettres*, ed. Julia Ashtiany, Latham, R. B. Serjeant, and G. Rex Smith (Cambridge, UK: Cambridge Univ. Press, 1990), 48–77.

[67] Dominique Sourdel, *Le vizirat ʿabbāside de 749 à 936 (132 à 324 de l'Hégire)* (Damas: Institut français de Damas, 1959).

Abbasid elites.[68] However, most of these Indian and Persian physicians disappeared from surviving sources in the early decades of the ninth century. Perhaps the fall of the Barmakids as key patrons, the rising tensions between Arabs and Persians—as part of the *Shuʿubiyyah* (nationalism) controversies, which culminated in the civil war between al-Maʾmūn, whose mother was Persian, and his older brother al-Amīn, whose mother was an Abbasid Arab—and al-Maʾmūn's own interest in Greek materials contributed to the temporary demise of this intellectual tradition.

In all these cases, the process of building a medical archive was connected to, and undertaken by, groups of learned practitioners that hailed from particular intellectual and ethnic origins and whose work and fortunes were connected to the rise and fall of their intellectual and ethnic groups. In other words, the process of making a humoralist archive in Arabic was connected not only to the organic process of Arabization of the practice and practitioners but also to the consolidation of an identity in Arabic—in this case, the Syriac humoralists. The consolidation of the archive meant the consolidation of such an identity—not as a foreign imported one, dependent on visitors and emigres, but rather as a local Arabized identity. At stake was the emergence of the Arabic humoralist, which could only materialize in the shadow of an Arabic humoralist archive.

The medical archive is therefore not only a collection of medical texts but also the stories, mythologies, narratives, and genealogies that described these texts and legitimized the said archive. In Derrida's theoretical formulation, the archive conditioned the central events that legitimized its production. The production of a humoralist Arabic archive emphasized the centrality of Hellenistic learning and of humoral medicine at the same time that it relied on such centrality to legitimize itself. As such, as Ḥunayn and others were laboring to consolidate the burgeoning archive, they were also building a historical narrative and a genealogical structure that emphasized the centrality of Greek medicine, its connection to Syriac Hellenism, and its legitimacy as the key medical practice of the emerging learned body politic of the Islamicate landscape. The eventual success of this process—that is, the making of an Arabic humoralist archive—meant that the genealogical mythology and the identity narrative adopted by the humoralist physicians, the archive makers, was also disseminated, adopted in the larger learned circles and considered to be the central narrative describing the history of medicine in the Islamicate landscape. Here, I argue that the production and consolidation of a medical archive was necessarily accompanied by the production of a historical narrative and an identity that mirrored such an archive and legitimized its existence. Once the process was completed, a humoralist from Damascus, Cairo, or Granada became the descendent of Hippocrates, Galen, and Ḥunayn, and the guardian of this legacy as materialized in the archive.

CONCLUSION

The modern scholarly focus on the life of Greek texts in the Islamic context engendered the notion of "Greek heritage" as a marker of a unique intellectual historical

[68] See Shefer-Mossensohn and Hershkovitz, "Early Muslim Medicine" (cit. n. 9); and S. A. Husain and P. K. Subhaktha, "Ayurveda during Abbasid's Period," *Bulletin of the Indian Institute of History of Medicine (Hyderabad)* 30 (1999): 27–34.

trajectory.[69] "Greek heritage" often referred not only to Greek writings but also to ideas and views that were seen as emerging from writings and discussions in classical antiquity. The use of "Greek heritage" as an analytical concept relied on a particular chronological arrangement that placed temporal and intellectual distance between commentators and authors of Hellenistic texts, who are makers of such heritage, and those who contributed to its life in the Islamic context.[70] In this context, the Translation Movement became a marker of chronology, creating not a continuity between Greek materials and their Arabic counterparts but rather a clear line of demarcation that seals off the Greek heritage as "Greek" and inaugurates Islamic Hellenism as "Arabic" or "Greco-Arabic." As such, "Greek heritage" becomes a repository of meaning, and a locus of intellectual worth. To be sure, Abdelhamid Sabra's foundational work on the insular nature of Greek heritage has been subject to significant rebuttals and deep questioning by many scholars, who have argued that Muslim and Arabophone authors appropriated and fully integrated Greek works. This argument has lent legitimacy to the nomenclature "Greco-Arabic" as descriptive of science and medicine during the medieval period.

My concern here, however, is not with the traditional narratives where Islamic sciences are seen as a prelude to European knowledge, or as Greek knowledge put on ice. Rather, I am concerned with the place of the Translation Movement as a chronological marker that separates "Greek" from "Islamic" or even from "Greco-Arabic." Here, translation is a key moment in Islamic history. It was the reason Greek knowledge seemed to encounter "Islam." The outcome of the translation—namely, the encounter between a coherent and foreign Greek heritage and an equally coherent indigenous Islamic core— becomes the entire intellectual history of Islam: a series of exchanges that extends from ninth-century debates on science and religion to twentieth-century debates on secularism and modernity. This chronological marker affects not only debates, public and scholarly, but also allocations of funds, archival restorations, and hiring and teaching in higher education. In other words, the chronology recreates itself.

In his *On the Postcolony*, Achille Mbembe rehabilitates the notions of *âges* and *dureés* as markers of chronology in the postcolony:

> By focusing the discussion on what I have called the "postcolony," the aim was not to denounce power as such, but rather to rehabilitate the two notions of *âge* and *durée*. By age is meant not a simple category of time but a number of relationships and a configuration of events—often visible and perceptible, sometimes diffuse, "hydra-headed," but to which contemporaries could testify since very aware of them. As an age, the postcolony encloses multiple *durées* made up of discontinuities, reversals, inertias, and swings that overlay one another, interpenetrate one another, and envelope one another: an entanglement.[71]

[69] See, most importantly, Franz Rosenthal, *The Classical Heritage in Islam* (Berkeley: Univ. of California Press, 1975). The notion of Greek heritage was and continues to be deployed in relation to local discourses around modernization. See, for example, Fauzi M. Najjar, "Ibn Rushd (Averroes) and the Egyptian Enlightenment Movement," *Brit. J. Mid. East Stud.* 31 (2004): 195–213.

[70] Take for instance the example of Aaron of Alexandria or Paul of Aegina. While both figures lived in the seventh century and may have lived and worked under Muslim rulers in Egypt and the Levant, they are studied as part of a Greek heritage that comes to Islam only a century later when the translation movement takes place. See also Sarah Stroumsa, *Freethinkers of Medieval Islam: Ibn al-Rawāndī, Abū Bakr al-Rāzī, and Their Impact on Islamic Thought* (Leiden: Brill, 1999).

[71] Achille Mbembe, *On the Postcolony* (Berkeley: Univ. of California Press, 2001), 14.

Mbembe's reformulations are useful in understanding the production of Islamic intellectual historical narratives. Here, "Islam" is posited as a set of relationships and a configuration of events that delimit a period in the "Medieval," characterizing it as particularly Islamic. Yet privileging these relations by way of constructing the category of Islam requires severing other preexisting relations, echoing the oft-criticized F. Renan, who explained that "being a Muslim is different from being anything else."[72] Indeed, the construction of the *âge* of Islam is a process through which Islam is rarefied as unlike anything else.

In contrast, a *durée* is ultimately a thematic construction, a period defined not by continuity but by affinity. A *durée* is not structured around chronology. Instead, it serves to reproduce chronology around a particular concept or locus of worth, the coherence of which is preserved by the *durée* itself. In this view, the study of Islamic sciences is organized around a *durée* of translation. Here, translation does not have a particular end, and is not plotted on a chronological scale. Rather, it is constructed as an entanglement that preserves a series of chronological interruptions and reversals and maintains the centrality of the Translation Movement as a maker and repository of intellectual value. In writing the history of science in the *durée* of translation, the historian is forced to contend with the unending demands of such entanglement. There is simply no escape from the centrality of "Greek heritage," as problematic a category as it may be.

While historians continue to critique the notion of "Islamic Sciences" as Greek knowledge kept on ice, they are forced to contend with a chronology that encapsulates their object between two translations, and that anchors their narratives to these signposts. Even more profoundly, the archive of Islamic sciences, be it metaphorical or physical, is also organized around translations. On the physical level, the centrality of the "Golden Age," as the product of the Translation Movement, conditions maintenance and preservation efforts in archives across the Middle East and Islamic world. Similarly, this centrality influences practices of acquisition by collectors and libraries—practices that are problematic on their own but that also profoundly influence the direction of scholarship in the West and beyond. More theoretically, the centrality of translation continues to push scholars into additional investigations of the preservation and transmission of Greek texts at the expense of other endeavors. Dismantling such chronology is necessary to fully appreciate the history of Islamic sciences in the ninth century and beyond, and to comprehend the meaning of translation and knowledge transmission in the medieval Islamic context.

It would be a mistake to think that the entanglement of the *durée* of (Greek-Arabic) translation is effectual only within the corners of the field dedicated to Islamic science and medicine. The Translation Movement plays a key part in the master narrative of the history of (Western) science and medicine. In fact, Sabra's work, which remains one of his most cited works among nonspecialists, and solidified the concept of appropriation along with other work proving the "contributions" of Islamic sciences, has further highlighted the central role of this "movement" as a beginning of the "Islam" chapter. In the background, this view maintains a certain level of stability and coherence to scientific and medical practice, which is capable of moving from one place to another through a series of translations. In other words, the centrality of translation as a

[72] Ernest Renan, *L'islamisme et la science* (Paris: Calmann-Lévy, 1883).

chronology-defining event obscures the ongoing processes of translations that occur consistently within scientific and medical practice. It also obscures the incoherent, dynamic, and iterative process of knowledge production, which cannot and should not be summarized or contained in a "movement" of translations. As the articles in this volume demonstrate more lucidly than I could, translation has always been a dynamic and integral part of knowledge production in different parts around the world and in many periods.

Translations do not stop. In the postcolony, translation is fraught with trouble. It is a process whereby a new linguistic order is established and consolidated. This linguistic order is not necessarily one in which the destination language resides at the top. In fact, in many cases and especially in scientific translations, the consistent and diligent efforts to translate scientific writings from the language of the former metropole(s) to the local language of the colony becomes a ritualized act of remembrance of how the metropolitan language remains, and will always be, on top of the global linguistic regime. Translation here is also an act of subscription, whereby the colony subscribes to the global economy of scientific knowledge, paying its dues in the form of journal subscriptions, international rights for books, and translators' salaries. In some cases, such as in medical education in Egypt, the unending cost of such subscription, and the unrelenting pace required for effective translations, motivated a shift in the direction of translation. Instead of medical books being translated into Arabic, students and patients are translated into English. The dream of cheaper and fuller subscriptions continues to crash into the reality of a new, hybrid Egyptian Anglicized Arabic in which Egyptian medical education persists. It may be acknowledged that this process is also one of archive making (or, at least, it aims to be). But this archive is often forced into a derivative position. It relies on the language on top and looks to replicate such archives in the colony's language.

The archive of medieval Islamic medicine was anything but derivative. It did not look to catch up to or to replicate a Greek archive. Instead, it regarded the classical materials as resources or raw materials, which, precious as they may have been, required purification, disciplining, and organization, and which invited intellectual activity. As such, the task of language crossing (or translation) was only part of the larger process of archive construction. In fact, the legacy of prominent translators such as Ḥunayn ibn Isḥāq is one of rejecting word-by-word translations and embracing editing, completing, revising, and translating based on meaning. It was a process through which a body of knowledge was reorganized and by which a new linguistic regime was developing, with Arabic ever so briefly at the top. As such, I argue that the study of translations in the Islamic context and beyond needs to be part of a larger investigation of archive making, in which the efforts of scholars, collectors, patrons, and translators are studied as parts of a larger whole.

Unveiling Nature:
Liu Zhi's Translation of Arabo-Persian Physiology in Early Modern China

*by Dror Weil**

ABSTRACT

This article examines the multifaceted approach to the translation of medicine as it appears in the works of Liu Zhi, a seventeenth-century Chinese-Muslim translator from Arabic and Persian into Chinese. Through empire-wide journeys to recover manuscripts, the building of an archive of Arabo-Persian knowledge on the natural world, and the application of various methods to produce coherence, authority, and compatibility with local epistemes, Liu assembled translations that presented early modern Chinese readers with new insights into the structure and operation of the human body. Liu Zhi's translations provide a rare glimpse into a cross-Asian circulation of knowledge on the human body and add a philological dimension to the premodern knowing of the body.

In 1704, Liu Zhi 劉智 (1660–1730), a Chinese-Muslim scholar, published a book with the title *Human Nature and Cosmic Principles in Islam* (*Tianfang xingli* 天方性理).[1] In this book, and in his subsequent two books, Liu endeavored to translate what he considered fundamental Islamic theories on the structure and operation of the natural world, including those related to the human body, for the benefit of his fellow Chinese readers. To that end, Liu undertook empire-wide journeys to recover Arabic and Persian texts forgotten in private libraries or newly introduced by foreign visitors. He scrupulously copied, collated, and compared manuscripts, and later painstakingly studied their contents, marking portions that were relevant to his study. His philological scrutiny, however meticulous, was subsumed under his more general interest in knowledge making, and by his genuine attempt to reconcile Arabo-Persian[2] and Chinese theories and concepts. Liu's utmost ambition, as well as his greatest predicament,

* Faculty of History, University of Cambridge, West Road, Cambridge, CB3 9EF, UK; dw612@cam.ac.uk.

[1] Liu Zhi 劉智, *Tianfang xingli* 天方性理 [Human nature and cosmic principles in Islam], in *Qingzhen Dadian* 清真大典 [The complete corpus of Chinese Islamic literature], ed. Huang Xiefan 周燮藩 et al. (Hefei shi: Huangshan shushe, 2005), 17:1–136. Unless otherwise noted, translations in this article are mine. Throughout, I have used quotation marks to indicate literal translations. I have capitalized the Chinese terms for bodily systems to distinguish them from Western and biomedical anatomical terminology.

[2] Labels and cultural categories pose a challenge for historians writing on cross-cultural subject matters. Throughout this article I will use "Arabo-Persian" to label the corpus of texts in Arabic or Persian

Osiris, volume 37, 2022. © 2022 History of Science Society. All rights reserved. Published by The University of Chicago Press for the History of Science Society. https://doi.org/10.1086/719220.

was how to achieve what many of his predecessors had failed to do: carry the nuanced philosophical insights on the natural world he found in these Arabic and Persian texts into Chinese.

Liu Zhi was not a practicing physician, nor did he seem to aspire to compile books exclusively on medical issues.[3] He was a Chinese literatus, mainly interested in philosophical questions and matters related to the theology and practice of Islam. Yet, his works abound with insights on the structure of the human body, its generation and physiological operation, the human psyche, and some pathological conditions that risk human well-being, which he drew from reading Arabic, Persian, and Chinese texts.

Liu Zhi's works spotlight the richness of information, theories, and concepts on the human body in texts outside the purview of practicing physicians.[4] His investigation of the human body sheds light on a philological dimension in the premodern knowing of the body. Descriptive medical knowledge, as seen in his works, emerged from textual analysis and intertextual coherence, rather than from bedside observation or medical treatment. A series of cognitive practices and textual manipulations substituted for clinical experience and professional training as the main routes to understanding the body and its operation.

The circulation of Liu Zhi's works among Chinese literati, some of whom cited them for presenting unique theories on the human body,[5] brings to light the blurriness between the professional and philosophical makers of medical knowledge. This blurriness is likewise echoed in the respectful place that Liu Zhi's theoretical explanations and vocabulary still receive today in curricula of *Huiyi* 回醫 (lit. "Arabo-Persian medicine" or, alternatively, "Islamic medicine")—a newly established ethno-medical expertise offered in some universities and medical institutions in contemporary China.[6] One of the objectives of this essay is to flag the potential in expanding the study of medicine and healthcare beyond the archives of medical professionals and to highlight the role of philosophers in producing, digesting, and spreading medical knowledge even during the early modern period.[7]

---

circulated in premodern China in order to mirror the integrated Chinese view of scholarship in both languages and the corpus's inclusivity of scientific and religious texts.

[3] By literature for medical practitioners, I especially refer to the genre of *yishu* 醫書 (medical treatises) and *fangshu* 方書 (formularies). The former are compilations that narrate physicians' theories and experiences in treatment, and the latter include medicinal recipes.

[4] On the intertwining between medical practice and Confucian philosophy, see Peter K. Bol, *Neo-Confucianism in History* (Cambridge, MA: Harvard Univ. Asia Center, 2008), 174–5; Paul U. Unschuld, *Medicine in China: A History of Ideas* (Berkeley: Univ. of California Press, 2010), 154–88; Charlotte Furth, "The Physician as Philosopher of the Way: Zhu Zhenheng (1282–1358)," *Harvard J. Asia. Stud.* 66 (2006): 423–59; and Hong Yu and Deyuan Huang, "'All Things Are Already Complete in My Body': An Explanation of the Views of the Taizhou School on the Human Body," *Frontiers of Philosophy in China* 5 (2010): 396–413.

[5] Readership of Liu Zhi's works went beyond Chinese Muslim communities and also included some "mainstream" literati. For example, the early nineteenth-century philologist Yu Zhengxie 俞正燮 (1775–1840) cites Liu Zhi's works in discussing embryogenesis; Yu Zhengxie, *Guisi cungao* 癸巳存稿 (Shanghai: Shang wu yin shu guan, 1957), 4:6.

[6] See Shan Yude 單于德, *Huiyi yaoxue mianmian guan* 回醫藥學面面觀 (Yinchuan: Ningxia renmin chubanshe, 2016); He Xiaohui 賀曉慧, Jia Xusheng 賈戌生, and Jia Menghui 賈孟輝, "Huiyi zhuanye jichu lilun jiaoxue zongjie yu sikao" 回醫專業基礎理論教學總結與思考, *Zhongguo minzu yiyao zazhi* 中國民族醫藥雜誌 25 (2019): 70–2.

[7] Buddhist texts served as vehicles for transmitting medical concepts and theories into China during the first millennium CE and as such constituted an earlier parallel of the Islamic case; C. Pierce

Liu Zhi's works display a complex relationship between the explanatory and the explained—a relationship that is inherent to acts of translation. The translational acts display a Hegelian oscillation between a broadening of the episteme to accommodate new imported insights on one end, and the use of the existing local episteme as an explanatory frame on the other. Accordingly, a translation act could be viewed at once as a reconfiguration of form and content orchestrated by individual idiosyncrasy, or a product of epistemic and linguistic regimes set by the particular intellectual landscape in which the translation takes place.[8] This relationship suggests a multifaceted approach to translation that accounts for conventions and originality that reconfigure the translated message at the semantic/lexical and epistemic levels.

The structure of this article seeks to reflect the multifaceted approach to translation as it is manifested in the case of Liu Zhi's translation of Arabo-Persian physiology in early modern China. It begins by situating Liu Zhi within the intellectual landscape of Islamic scholarship in early modern China, introducing the presence of Arabic and Persian texts in China and some of the reading practices that Liu Zhi's predecessors employed in their reading of these foreign texts, and describing aspects in Liu Zhi's translation as a historical continuity. It then proceeds to investigate three unique features of Liu Zhi's translation. First, the article will explore the ways in which Liu built an epistemic foundation through searching for, collecting, and selecting Arabic and Persian manuscripts, and using hermeneutical and interpretive methods to domesticate foreign ideas and produce coherence and discursive familiarity for the benefit of the Chinese reader. Second, it will examine Liu Zhi's tactics of lexical and semantic rendition and the ways he articulated western medical theory–based concepts in Chinese. Finally, the article will shed light on some of Liu's insights on the structure and the operation of the human body, and the contributions to the contemporary Chinese medical discourse that Liu Zhi's translations were able to make by juxtaposing Arabo-Persian and Chinese medical theories and negotiating ways to bring meaning across cultures.

Salguero, *Translating Buddhist Medicine in Medieval China* (Philadelphia: Univ. of Pennsylvania Press, 2014). On the role of religion as a vehicle for introducing Galenic medicine to various parts of Asia, see Fabrizio Speziale, "The Relation between Galenic Medicine and Sufism in India during the Delhi and Deccan Sultanates," *East and West* 53 (2003): 149–78; Jennifer W. Nourse, "The Meaning of Dukun and Allure of Sufi Healers: How Persian Cosmopolitans Transformed Malay-Indonesian History," *J. Southeast Asian Stud.* 44 (2013): 400–22; and Ronit Yoeli-Tlalim, "Galen in Asia," in *Brill's Companion to the Reception of Galen*, ed. Petros Bouras-Vallianatos and Barbara Zipser (Leiden: Brill, 2019), 594–608. The role of Jesuits and other European missionaries in introducing medical theories and concepts has received somewhat more scholarly attention; Henri Bernard, "Notes on the Introduction of the Natural Sciences into the Chinese Empire: Cultural Contacts between China and the West," *Yenching Journal of Social Studies* 3 (1941): 944–65; Daniel Asen, "'Manchu Anatomy': Anatomical Knowledge and the Jesuits in Seventeenth- and Eighteenth-Century China," *Soc. Hist. Med.* 22 (2009): 23–44; Chu Ping-yi 祝平一, "Tianxue yu lishi yishi de qianbian—Wang Honghan de 'Gujin yishi'" 天學與歷史意識的變遷-王宏翰的'古今醫史,' *Lishi yuyan yanjiuso jikan* (*Bulletin of the Institute of History and Philology Academia Sinica*) 77 (2006): 591–626; Noël Golvers, "The Jesuits in China and the Circulation of Western Books in the Sciences (17th–18th Centuries): The Medical and Pharmaceutical Sections in the SJ Libraries of Peking," *EASTM* 34 (2012): 15–85; Nicholas Standaert, "Medicine," in *Handbook of Christianity in China, Volume One: 635–1800*, ed. Standaert (Leiden: Brill, 2001), 786–802.

[8] On idiosyncrasies and norms in translation, see Gideon Touri, *Descriptive Translation Studies and Beyond* (Amsterdam: Benjamins, 1995); and Lawrence Venuti, *The Translator's Invisibility: A History of Translation* (Hoboken, NJ: Taylor and Francis, 2012).

ARABIC AND PERSIAN TEXTS IN EARLY MODERN CHINA

Liu Zhi's discourse on the natural world can best be described as a nexus between the scholarship produced by early modern Chinese Muslims, which promoted immersion in the study of Arabic and Persian languages and texts—a theme that has only recently begun to receive scholarly attention[9]—and the vibrant intellectual landscape of the Jiangnan region in South China, which witnessed the refashioning of Confucian, Buddhist, and Daoist philosophies and the introduction of European thought.[10] The Arabic and Persian texts Liu Zhi read and some of the methods he applied to analyze their content were predicated on a philological-pedagogical enterprise that emerged in China during the last decades of the sixteenth century among Chinese Muslims, and sought to promote the study of Arabic and Persian texts.[11] At the same time, Liu read these texts through the analytical lens and learned practices of his contemporary Confucian classical learning and saw his fellow literati as the target audience of his publications.

Growing up amid the turbulent times of the mid- and late seventeenth century, Liu Zhi witnessed the thriving cultural scene that characterized his city of Nanjing, a longstanding cultural and intellectual metropolis. The city changed its face as the Ming empire fell to the hands of the Manchu conquerors, yet resiliently kept its cultural and intellectual vibrancy. The booming book culture brought to his attention a plethora of literatures and writings.[12] These include the Jesuits' translations of Western works as well as philosophical treatises by Confucian, Daoist, and Buddhist authors.[13] He learned as well of the attempts by Chinese Muslims of his father's generation to produce Islamic literature in Chinese with the aim of introducing Islamic ideas and theories

[9] On that theme, see Zvi Ben-Dor Benite, *The Dao of Muhammad* (Cambridge, MA: Harvard Univ. Press, 2005); Dror Weil, "Islamicated China—China's Participation in the Islamicate Book Culture during the Seventeenth and Eighteenth Centuries," *Intel. Hist. Islam. World* 4 (2016): 36–60; and Kristian Petersen, *Interpreting Islam in China: Pilgrimage, Scripture, and Language in the Han Kitab* (Oxford: Oxford Univ. Press, 2018).

[10] On the intellectual landscape in late imperial China, see Benjamin A. Elman, *From Philosophy to Philology: Intellectual and Social Aspects of Change in Late Imperial China* (Cambridge, MA: Council on East Asian Studies, Harvard Univ., 1984); Yü Chün-fang, "Ming Buddhism," in *The Cambridge History of China, Volume 8: The Ming Dynasty, 1368–1644, Part 2*, ed. Denis C. Twitchett and John K. Fairbank (Cambridge, UK: Cambridge Univ. Press, 1998), 893–952; Dewei Zhang, *Thriving in Crisis: Buddhism and Political Disruption in China, 1522–1620* (New York: Columbia Univ. Press, 2020); Golvers, "Jesuits in China" (cit. n. 7); and various chapters in *The Cambridge History of China, Volume 9, Part 2: The Ch'ing Dynasty to 1800*, ed. Willard J. Peterson (Cambridge, UK: Cambridge Univ. Press, 2016).

[11] On the ways Arabic and Persian texts were read in late imperial China, see Weil, "Islamicated China" (cit. n. 9), 36–60; and Weil, "Literacy, in Arabic and Persian, in Late Imperial China," in *Encyclopaedia of Islam: Three*, ed. Kate Fleet et al. (Leiden: Brill, 2020), 92–5.

[12] On the surge of printed books in the late Ming and early Qing, see Cynthia J. Brokaw and Kai-wing Chow, eds., *Printing and Book Culture in Late Imperial China* (Berkeley, CA: Univ. of California Press, 2005); Brokaw, "Commercial Publishing in Late Imperial China: The Zou and Ma Family Businesses of Sibao, Fujian," *Late Imperial China* 17 (1996): 49–92; Kai-wing Chow, "Writing for Success: Printing, Examinations, and Intellectual Change in Late Ming China," *Late Imperial China* 17 (1996): 120–57; Tobie Meyer-Fong, "The Printed World: Books, Publishing Culture, and Society in Late Imperial China," *The Journal of Asian Studies* 66 (2007): 787–817; Wu Kuang-Ch'ing, "Ming Printing and Printers," *Harvard J. Asia. Stud.* 7 (1943): 203–60; and Ōki Yasushi 大木康, "Minmatsu kōnan ni okeru shuppan bunka no kenkyū" 明末江南における出版文化の研究, *Hiroshima daigaku bungakubu kiyō* 50 (1991): 1–176.

[13] On the interaction between Jesuits and Chinese Muslims, see Zvi Ben-Dor Benite, "Western Gods Meet in the East: Shapes and Contexts of the Muslim-Jesuit Dialogue in Early Modern China," *Journal of the Economic and Social History of the Orient* 55 (2012): 517–46.

to Chinese readers. Liu Zhi's father, a Chinese-Muslim educator, shared the perplexity of many other Chinese Muslims of the period, torn between a commitment to perpetuating traditional Muslim education and the wish to engage with the non-Muslim intelligentsia, or, alternatively framed, between a scholarship of transregional, cross-Asian dimensions and the local Chinese intellectual discourses.

Liu Zhi's father was a member of a long-lasting empire-wide network of Chinese Muslims invested in the linguistic study of Arabic and Persian texts. This network emerged in the northwestern province of Shaanxi in the mid-sixteenth century and soon spread to other parts of China. It brought together people who practiced Islam, descendants of Muslim migrants, and people interested in learning Islamic texts.[14] Hu Dengzhou 胡登州 (1522–97), a Shaanxi Chinese Muslim, is widely accredited as the founder of this network. Hu Dengzhou's fascination with the nuanced philosophy of nature he found in Islamic texts led him to develop a pedagogical program for reading Arabic and Persian texts and subsequently inspired him to establish a school dedicated to that cause.[15] His disciples, and following generations of disciples, created a network of schools and scholars equally devoted to training in Arabic and Persian yet differing from one another in their selection of texts and thematic foci. Members of Hu Dengzhou's network selectively collected and copied Arabic and Persian manuscripts on themes such as theology, Islamic law, grammar, and logic for their teaching curricula, and produced by the late sixteenth century an impressive cross-country archive of manuscripts that provided unprecedented access to texts produced across the Islamicate world.

While oral interpretation of Arabic and Persian texts must have taken place, members of Hu Dengzhou's network placed most of their scholarly emphasis on the analysis of grammatical structures and literal glossing of the Arabic and Persian texts, followed by an investigation of their rhetorical and logical patterns and a thorough scrutiny of their contents. The scholars copied and collated Arabic and Persian manuscripts, compared versions, marked textual divergences, parsed sentences, identified their various syntactical components, and supplemented them with interlinear glossing and marginal commentaries. Difficult Arabic and Persian words were explained by other, apparently easier, Arabic or Persian words, or vernacular Chinese.[16] Textual variance and intertextuality were crammed into dense marginalia around the main text.[17]

By the early seventeenth century, the economic prosperity and attendant sociocultural consequences that swept China's metropolises produced a vibrant book culture and invigorated intellectual activity in the wealthy Jiangnan area. This was also the period when Jesuit missionaries moved to the region and ushered in the proliferation

[14] On that movement, its geographical spread in China, and its pedagogies, see Ben-Dor Benite, *Dao of Muhammad* (cit. n. 9); and Weil, "Islamicated China" (cit. n. 9).

[15] The earliest historical record on Hu Dengzhou and his network is Zhao Can's 趙燦 *Jingxue xichuanpu* 經學系傳譜 [The genealogy of classical learning], which was composed in the 1660s, almost half a century after Hu Dengzhou's death; Zhao Can, "Jingxue xichuanpu," in Huang Xiefan et al., *Qingzhen Dadian* (cit. n. 1), 20:23–166.

[16] *Vernacular* here refers to the spoken register of Chinese, which differed between locales. This register was rarely put into writing. Writing was mainly the domain of Classical Chinese, which differs in grammar and vocabulary from the local spoken idioms. One example of written vernacular Chinese in Arabic script that annotated manuscripts from Northern Chinese is called *xiao'er jing* 小兒經; Weil, "Islamicated China" (cit. n. 9), 55–6.

[17] For more on the ways Arabo-Persian texts were read in late imperial China, see ibid., 49–57.

of Christian-Western ideas and texts through lectures and translations. Chinese-Muslim scholars, inspired by the intellectual possibilities that the local print culture offered, and aspiring to popularize the insights they encountered in their study of Arabic and Persian texts, began to publish translations and summaries of Arabo-Persian texts in Chinese. Many of these scholars found this new form of scholarship a useful platform from which to socialize and engage intellectually with the non-Muslim local intelligentsia. By the 1640s and 1650s, they had produced a plethora of printed Chinese treatises on various themes, including Islamic philosophy of nature, Islamic law and religious practice, and translations of Arabic and Persian literature and history. Their shared aspiration to align their scholarship with the contemporary Chinese discourse and style resulted in a variety of methods for iterating Arabo-Persian ideas in Chinese. Some published literal translations of complete Arabo-Persian texts, others merged translated excerpts and paraphrases into monographs; some put together lecture notes in a vernacular register, others published eloquent Chinese iterations of Arabic and Persian ideas.[18]

Descriptions of the natural world and explanations of natural phenomena, including phenomena related to the operation of the human body, were integral aspects in the study of Islamic theologies and cosmologies, in the expounding of Islamic law, and even in the narration of prophetic histories. Consequently, they constituted central themes in the published works of Chinese Muslims from the mid-seventeenth century onward. The writers of these texts faced, however, a critical methodological challenge in reconciling the epistemic, conceptual, and terminological disparities between their sources and their contemporary Chinese discourses. The grammatical focus that had been a central method in approaching and deciphering Arabic and Persian texts among members of Hu Dengzhou's network was no longer useful when writing in Chinese, and it was replaced with a "philosophical" investigation of Arabo-Persian texts that could spotlight distinct ideas and compared them to other philosophical traditions.[19] This shift had broader implications in terms of the analytical lenses through which Arabo-Persian ideas were read and the ways coherence and authority were conveyed to Chinese readers. This shift in focus also prompted Chinese-Muslim scholars to experiment with various strategies to reconfigure epistemic boundaries, their styles of presentation, and methods of conveying their ideas to their target audiences.[20] Furthermore, moving away from scrutiny of linguistical patterns to exposition

[18] An example of such a case is Wu Zunqi's 伍遵契 (1598–1698) translation of Najm al-dīn Rāzī's (d. 1256) mystical treatise, *Mirṣād al-ʿibād* [The path of God's bondsmen] into Chinese, completed in 1678 and titled *Guizhen yaodao* 歸真要道 [The essential way to submit to the Truth].

[19] Elman points to an opposite trend among mainstream Chinese scholars during the seventeenth and eighteenth centuries, in which philosophical discussions on the principles of nature were replaced by new forms of "evidential investigations" (*kaozheng* 考證, sometimes referred to as "philology"); Benjamin A. Elman, "Philosophy (I-Li) versus Philology (Kao-Cheng): The Jen-Hsin Tao-Hsin Debate," *T'oung Pao* 69 (1983): 175–222; Elman, *From Philosophy to Philology: Intellectual and Social Aspects of Change in Late Imperial China* (Cambridge, MA: Council on East Asian Studies, Harvard Univ., 1984).

[20] For English translations of some of these Chinese-Islamic works, see Sachiko Murata, *Chinese Gleams of Sufi Light: Wang Tai-Yü's "Great Learning of the Pure and Real" and Liu Chih's "Displaying the Concealment of the Real Realm"* (Albany: State Univ. of New York Press, 2000); Murata, William C. Chittick, and Tu Weiming, *The Sage Learning of Liu Zhi: Islamic Thought in Confucian Terms*, Harvard Yenching Institute Monograph Series (Cambridge, MA: Harvard Univ. Press, 2009); Murata, *The First Islamic Classic in Chinese: Wang Daiyu's "Real Commentary on the True Teaching"* (Albany: State Univ. of New York Press, 2017).

of ideas allowed Chinese-Muslim scholars who did not master the Arabic and Persian languages to participate in this scholarship and contribute their share in producing coherence and authority.

A common tactic applied by Chinese-Muslim scholars to lend coherence and authority to their works was to adopt concepts, theories, terms, and even epistemological theories from the contemporary intellectual Chinese discourse. Being part of the literati class in China, these Chinese-Muslim scholars received Chinese classical education and were acquainted with discourses on the natural world. In particular, they had knowledge of the Confucian cosmological framework, expounded in the works of Zhu Xi 朱熹 (1130–1200) and the Cheng 程 brothers, which constituted the orthodox view during the Ming and early Qing periods (known as the Cheng-Zhu 程朱 school). Many of them were also aware of the heated debates among Chinese philosophers on the metaphysical and epistemological grounds of that framework.[21] The cosmological framework left its mark in the works of Chinese-Muslims in their adoption of vocabulary, such as the foundational concepts of *xing* 性 ("[Human] Nature"), *li* 理 ("Principle" or "Coherence"), *xin* 心 ("Heart" or "Mind"), *qi* 氣 ("Energy Matter" or "Matter Influence"), the analytical frameworks of yin 陰 and yang 陽 and the Five Phases (*wuxing* 五行),[22] or the selective adaptation of aspects from Confucian epistemological theories such as *gezhi* 格致 ("The Extension of Knowledge thorough the investigation of things"), *daxue* 大學 ("the Great Learning") and *liangxin* 良心 ("Innate Intuition"). These terms and frameworks were important tools for iterating, naturalizing, and explaining some of the ideas Chinese-Muslim scholars encountered in their reading of Arabic and Persian texts.

Moreover, the borrowing of terms and frameworks from the Confucian philosophical discourse, as well as alluding to major Confucian works in their titles, lent an intellectual identity and scholarly authority to these Chinese-Islamic works and situated them within the purview of Chinese orthodoxy, in particular that of the Cheng-Zhu school.[23] For example, the seventeenth-century Chinese-Muslim scholar Wang Daiyu 王岱輿 (ca. 1570–ca. 1660) named one of his major works on Islamic theology *The Pure and True Great Learning* (*Qingzhen daxue* 清真大學), explicitly linking the study of Islam (or, as it was commonly known in Chinese, "the pure and true teaching") and Confucianism by using the term *daxue* 大學 "The Great Learning"—the title of one of the Confucian Four Books that was greatly promoted by Zhu Xi and his school as the epitome of Confucian self-cultivation and its epistemological program.[24] Similarly, Liu Zhi clearly sought to situate his scholarship as an extension

[21] Chinese-Islamic works show some engagement with the debate between the Cheng-Zhu orthodoxy and the school of Wang Yangming王陽明.

[22] The complex and changing contents of these concepts made their definition and translations matters of great debate among historical actors and contemporary scholars alike. While I propose some common translations here, they should be taken with a grain of salt. For representative discussions on the definitions of these core principles in Chinese natural philosophy and medicine, see Nathan Sivin, *Health Care in Eleventh-Century China* (Cham: Springer, 2016); Bol, *Neo-Confucianism in History* (cit. n. 4); and Unschuld, *Medicine in China* (cit. n. 4).

[23] See, for example, Qin Huibin, "On Cosmology and Tawhid in the Works of Wang Daiyu," in *Islam*, ed. Jin Yijiu and Ho Wai Yip (Leiden: Brill, 2017), 245–72; and Murata, *First Islamic Classic in Chinese* (cit. n. 20).

[24] On the Four Books and Zhu Xi's rearrangement of *The Great Learning*, see Daniel K. Gardner, *Chu Hsi and the Ta-hsueh: Neo-Confucian Reflection on the Confucian Canon* (Cambridge. MA: Harvard Univ. Press, 1986); and Bol, *Neo-Confucianism in History* (cit. n. 4). By the seventeenth century,

of the Confucian orthodoxy by naming his treatise, *Human Nature and Cosmic Principles in Islam*, after the fifteenth-century imperially-endorsed compendium of Confucian cosmology, *The Great Compendium on Human Nature and Cosmic Principles* (*Xingli daquanshu* 性理大全書). By using such titles for their works and structuring their texts after renowned orthodox Confucian classics, Chinese-Islamic authors sought to naturalize their scholarship and present their works as legitimate components of the contemporary Chinese discourse on the natural world and human society.

At the turn of the eighteenth century, about half a century after the first work on Islam appeared in Chinese, Liu Zhi published *Human Nature and Cosmic Principles in Islam*. Building on past scholarship and his own new translations, the book sought to introduce the foundation of Islamic cosmology to a Chinese reader. Together with his subsequent two publications, *Annotated Selection on Islamic Norms and Rites* (*Tianfang dianli zeyao jie* 天方典禮擇要解), a work dedicated to the study of Islamic practice, and *Veritable Records of Islam's Most Venerable* (*Tianfang zhisheng shilu* 天方至聖實錄),[25] a chronological account of Prophet Muhammad's life, these works constituted a trilogy dedicated to what Liu perceived to be the three axes of the natural world: the Cosmos, Human Society, and History.

References to physiological and pathological processes and descriptions of human and animal anatomy are scattered throughout Liu Zhi's works, drawing on and referring to Arabic, Persian, and Chinese sources. The references in *Human Nature and Cosmic Principles in Islam* and *Annotated Selection on Islamic Norms and Rites* are very different in character, both in terms of the sources they draw on and their roles in Liu's larger intellectual projects. Whereas Liu Zhi widely quotes passages from identified Arabic and Persian texts to construct his expositions, in the latter work, quotes from identified Chinese sources are provided as cross-references to and support for his exposition on Islamic practices. In *Human Nature and Cosmic Principles in Islam*—a work that focuses on the various correspondences between the universe as a macrocosm and the human body as a microcosm—a full chapter is dedicated to discussing embryogenesis, whereby synopses, expositions, and illustrations of the various phases of embryonic development and the functions of bodily organs, based on references from Arabic and Persian sources, are provided. Conversely, in *Annotated Selection on Islamic Norms and Rites*, physiological and anatomical explanations in the form of excerpts from Chinese medical and pharmaceutical literature are provided as justification and underlying rationales for Islamic ritual behavior, dietary restrictions, hygienic requirements, and methods of slaughter.

Similar to his Chinese-Muslim predecessors, Liu Zhi sought to present a discourse on the natural world that could at once convey coherence, authority, and compatibility with Confucian philosophy. Challenged by the nature of his core sources, Liu Zhi sought to develop a translational program that would bring across the meanings of

---

Zhu Xi's reorganization and interpretation of the text of *The Great Learning* was heavily contested by rival Confucian schools; Wang Huaiyu, "On Ge Wu: Recovering the Way of the 'Great Learning,'" *Philosophy East and West* 57 (2007): 204–26; Bruce Rusk, "Not Written in Stone: Ming Readers of the 'Great Learning' and the Impact of Forgery," *Harvard J. Asia. Stud.* 66 (2006): 189–231.

[25] Liu Zhi 劉智, *Tianfang dianli zeyao jie*天方典禮擇要解 [Annotated selection on Islamic norms and rites], in *Qingzhen Dadian* [The complete corpus of Chinese Islamic literature], ed. Huang Xiefan 周燮藩 et al. (Hefei shi: Huangshan shushe, 2005), 15:46–190; Liu Zhi 劉智, *Tianfang zhisheng shilu* 天方至聖實錄 [Veritable records of Islam's most venerable], also in *Qingzhen Dadian*, 14:1–365.

his foreign sources without communicating a sense of foreignness, idiosyncrasy, or iconoclasm. For that purpose, Liu adopted a guiding strategy that worked on two levels: the epistemic and semantic/lexical. The former included the establishment of an archive of Arabic, Persian, and Chinese texts and the application of a series of commentarial tools to explain the contents of the Arabic and Persian texts and to lend authority to their readings. The latter included various translational tactics to bridge the linguistic gap, to produce a sense of compatibility with other Chinese discourses, and to bring across meanings. These two strategies will be further examined in the following pages of this article.

BUILDING A TEXTUAL ARCHIVE

Translation of discourses on the natural world, such as the one carried out in Liu Zhi's works, entailed negotiation between two sets of textual corpuses representing two distant epistemic systems: the one that is mirrored in the sources and the one held by the translator and readers. To produce a sense of commensurability and affinity between these two systems, Liu Zhi created an archive of Arabic, Persian, and Chinese texts on the natural world and employed a series of translation methods and reading practices to construct a coherent discourse in Chinese on what he saw as the Islamic view of the natural world. The archive, which consisted of two distinguished, yet intertwined, sections—Arabic-Persian and Chinese texts—allowed Liu Zhi to push the boundaries of the Chinese canon, yet at the same time to display the coherence and authority of his new synthetic discourse. The Arabic and Persian texts served as repositories for new insights that could be assimilated into, and verified against, the Chinese understanding of the natural world. The Chinese texts constituted a verification mechanism that attested to the authenticity of the assimilated knowledge.

For the purpose of establishing his archive, Liu Zhi set out on cross-country journeys in search of relevant texts. He described in his works the laborious task of obtaining such Arabic and Persian texts in China during the late seventeenth and early eighteenth centuries. He also recounted his visits to multiple private libraries throughout China, where he happened on Arabic or Persian manuscripts of interest; copied them himself or asked a local literatus with the required linguistic skills to assist; and thereafter carried out a critical investigation of their contents, comparing different editions and marking central points or odd parts. Using methods of excerption, paraphrasing, and summarization, he then incorporated parts of the texts he encountered into his works on the natural world.

Vestiges of Liu Zhi's archive, which might have been once a physical library at his residence in Nanjing, are manifested in the form of identified and unidentified references scattered in his works. Two bibliographical lists—paratextual elements that are rarely seen in contemporary Chinese works—comprising forty and forty-six titles were included in *Human Nature and Cosmic Principles in Islam* and *Annotated Selection on Islamic Norms and Rites* correspondingly. Each entry in these lists begins with a transliteration of the original Arabic or Persian title in Chinese characters, followed by a short explanatory comment in Chinese. These bibliographies list only Arabic and Persian texts, and entirely omit any indication to the Chinese texts that Liu Zhi used, references to which appear only in the body of the works. In some sections of the works, such as in the first part of *Human Nature and Cosmic Principles in Islam*, known as "The Root Classic" (*benjing* 本經), references to the sources are given at the end of each passage in a manner similar to modern footnotes. Likewise, references to the Chinese sources appear at the beginning or end of the quotes that Liu Zhi incorporated in his works.

Liu Zhi's archive includes multiple Arabic and Persian texts on theology, natural philosophy, Islamic jurisprudence (*fiqh*), astral and earth sciences, and language. While there is no indication to any work that is exclusively medical in Arabic or Persian, or to works that focus on the study of the human body, Liu quotes from Chinese medical and pharmaceutical works and includes passages from Neo-Confucian philosophies and popular encyclopedias. Liu Zhi employs cross-textual readings in which he extracts information on physiology, anatomy, and pathology from Arabic and Persian sources and reads them against discussions of medicinal properties of substances and physiological processes in Chinese materia medica literature; foundational theories of the structures and operations of the natural world and the human body expounded in Neo-Confucian philosophical works; and related references in popular encyclopedias. Such cross-textual reading performed two functions for Liu Zhi's translation: it provided an explanatory framework and a method for authenticating and authorizing the ideas presented in Arabic and Persian texts. Liu Zhi employed Chinese framing of subject matter, theories, and terms to make sense of, explain, and render in Chinese what he encountered in Arabic and Persian texts. At the same time, he used cross-textual reading to provide evidence for the veracity of the accounts by pointing to existing parallels in the related Chinese literature.

Among the Arabic and Persian[26] works on theology that Liu Zhi obtained for his archive are ʿAbd al-Raḥmān Jāmī's (d. 1492) *Ashiʿāt al-lamaʿāt* (Rays of the flashes) and *Lawāʾiḥ* (Gleams), Najm al-Dīn Rāzī's (d. 1256) *Mirṣad al-ʿibād* (The path of God's bondsmen), ʿAzīz al-Dīn Nasafī's (fl. 13th c.) *Maqṣad-i aqṣá* (Furthest goal), and an unidentified work titled *Mawāqif* (Stations) and a commentary on it.[27] These texts presented adaptations of Aristotelian and Neo-Platonic cosmological theories, including views on the relationship between human soul and human body and descriptions of human cognition. They were also instrumental in providing glimpses into Hippocratic and Galenic views on the structure and operation of the human body, such as the four humors, the phases of embryonic generation, and the functions of bodily parts, as well as the dichotomy of human soul and body that were deeply embedded in Islamic theological works. Legal manuals, such as Burhān al-Dīn Marghīnānī's *Al-Hidāya fī sharḥ al-bidāya* (Guidance in the commentary of al-Bidāya, comp. 1178 in Samarqand), Tāj al-Sharīʿa al-Maḥbūbī's (d. 1344) *Sharḥ wiqāyat al-riwāya fī masāʾil al-hidāya* (Commentary on The Protection of the narration in matters of the *Guidance*, comp. 1342, a commentary on Marghīnānī's work) and *Kāfī dar fiqh* (The sufficient work on jurisprudence, comp. fourteenth c., originally in Persian), were important

[26] Liu Zhi does not differentiate Arabic from Persian texts. In some cases, it is unclear if he used an Arabic work or its Persian translation. In order to maintain actors' categories, my discussion will treat Arabic and Persian texts as a single literary and linguistic category; Weil, "Literacy, in Arabic and Persian" (cit. n. 11), 92–5.

[27] A work by the title of *Mawāqif* and its commentary are listed in the bibliographies and are referenced in *Human Nature* (cit. n. 1). A possible identification might be al-Ījī's (d. 1355) important work on logic, titled *Mawāqif*, and al-Jurjānī's (d. 1423) commentary on it. I was, however, unable to locate the quoted passages in these works. As Murata suggests, it is likely that the titles refer to other works; Murata, Chittick, and Tu, *Sage Learning of Liu Zhi* (cit. n. 20), 14.

sources on the practicality and applicability of Islamic views on bodily purity and impurity, including the states of menstruation, sperm and blood discharges, hygienic practices, nourishment and diets, animal anatomy, and the physiological aspects of animal slaughter.

Using a series of translation methods that worked on both the epistemic and semantic/lexical levels, Liu Zhi sought to explore the ideas he found in his collected texts and present them as compatible with, and supplementary to, the contemporary Chinese discourses on the natural world and the human body. These methods will be the foci of the following sections.

EPISTEMIC CONFIGURATIONS

In their introduction to *Canonical Texts and Scholarly Practices: A Global Comparative Approach*, Anthony Grafton and Glenn W. Most tell us that producing a canonical text in the premodern period was often a project that involved choice and authorization, followed by domestication techniques with the aim of making the texts seem familiar, relevant, and up to the readers' standards and tastes.[28] Similarly, Alisha Rankin and Elaine Leong elsewhere in this volume show how translation tactics and learned practices were specifically chosen by the translator to suit the target audience.[29] These perceptions could equally apply to Liu Zhi's translation. As suggested in the prefaces to his translated works, the methods to domesticate Arabo-Persian ideas and produce coherence, authority, and discursive familiarity for the benefit of his Chinese readers were central concerns for Liu Zhi. Accordingly, he employed strategies in structuring his translations to overcome the epistemic gap between his foreign sources and the intended Chinese audience, to infuse his translations with authority and legitimacy, and to demonstrate the compatibility of Islamic and Chinese theories.

The core objective behind Liu Zhi's translations seems to have been his wish to demonstrate the compatibility of Islamic thought with the Neo-Confucian philosophy of the Cheng-Zhu school. Liu Zhi wrote that soon after he began to study Arabic and Persian works, he realized that they shared many of their essential ideas with "the teachings of Confucius and Mencius" and that anyone who delved into Islamic texts would surely find that "the texts are Islamic, but the [underlying] principles are universal."[30] This realization informed Liu Zhi's main strategy in presenting Islamic ideas: emphasizing the compatibility of Islamic theories with the Confucian philosophical framework. On that ground, Liu Zhi defined his investigation of Islamic natural philosophy and Islamic praxis in Confucian terms, labeling the former "the study of [Human] Nature and Cosmic Principles" (*xingli* 性理) and the latter "Norms and Rites" (*dianli* 典禮): both phrases carry loaded Neo-Confucian connotations and would be linked by any Chinese reader to Confucian cosmology and ethics. Moreover, this strategy lent itself to the application of particular translational methods that Liu Zhi described as "excerpting [passages] from multiple [Arabic and Persian] classics where the principles

[28] Anthony Grafton and Glenn W. Most, "How to Do Things with Texts: An Introduction," in *Canonical Texts and Scholarly Practices: A Global Comparative Approach*, ed. Grafton and Most (Cambridge, UK: Cambridge Univ. Press. 2016), 1–13.
[29] See Alisha Rankin, "New World Drugs and the Archive of Practice"; and Elaine Leong, "Translating, Printing, Reading," both in *Osiris* 37.
[30] *Human Nature* (cit. n. 1), 17:13.

are the same [as in Confucian philosophy] and the meanings match, and then compiling them into a single text."[31]

Producing compatibility between Islamic and Confucian theories could not be achieved without making adjustments to the epistemic grounds and lexicon. To that end, Liu Zhi's strategy was to use "thick translation,"[32] that is, to accompany the literal rendering of excerpts from the source texts with rich explanatory notes and glosses. In his preface to *Human Nature and Cosmic Principles in Islam*, Liu Zhi provided the following reason for this strategy: "The statements and meanings in this book are all extracted from Islamic classics. Among them are texts that are difficult to render literally into Chinese and require interpretations based on other texts. The texts might differ, but their meanings are identical."[33]

Liu Zhi's model for compiling *Human Nature and Cosmic Principles in Islam* sought to reconfigure the epistemic grounds of his readers by splitting his translation into three distinct, yet intertwined, parts: the first, titled "The Root Classic" (*Benjing* 本經), includes five concise chapters that are composed of translated excerpts from identified Arabic and Persian texts. The second part, a series of illustrations (*tu* 圖), aims at presenting a graphical representation of the main principles; and a third part, the Commentary (*zhuan* 傳), provides a thick and elaborate translation of the five chapters in the Root Classic. These three parts enabled Liu Zhi to expand his discussion beyond the original text and provide graphical representations of theories and concepts. The illustrations lent further clarity and coherence to Liu Zhi's translations of Arabo-Persian cosmological theories, including the theories on the structure and operation of the human body.

In his *Annotated Selection on Islamic Norms and Rites* Liu Zhi sought to translate and explain the underlying rationale of Islamic religious practices and daily routines through the prism of contemporary Chinese scientific discourse. As a means to control the readers' economy of attention and aid the reception of ideas,[34] Liu Zhi and the subsequent editors of his works interpolated different hermeneutical and commentarial devices to amplify certain ideas, clarify potential difficulties, further domesticate the discourse, and produce coherence and discursive familiarity for the benefit of the Chinese reader.

In his preface to *Annotated Selection on Islamic Norms and Rites*, Liu Zhi provided a rare insight into the range of hermeneutical devices and thick translation that he used in compiling the work. He listed the following devices: "interpretations" (*jie* 解), "major commentary" (*dazhu* 大註), "minor commentary" (*xiaozhu* 小註), "substantive meaning" (*shiyi* 實義), "expanded meaning" (*guangyi* 廣義), "evidential verification" (*kaozheng* 考證), "additional references" (*jilan* 集覽), "frequently asked questions" (*wenda* 問答), and "further discussion" (*fulun* 附論). According to Liu's explanation, the "expanded meaning" device offers examples when the underlying principle is clear but not concrete, and "substantive meaning" is used to generalize the underlying

[31] Ibid.

[32] I borrow Appiah's definition of *thick translation* to denote "translation that seeks with its annotations and its accompanying glosses to locate the text in a rich cultural and linguistic context"; Kwame Anthony Appiah, "Thick Translation," *Callaloo* 16 (1993): 808–19.

[33] *Human Nature* (cit. n. 1), 17:2.

[34] I borrow the phrase *economy of attention* from Daston's discussion of the cognitive practices entailed by note taking; Lorraine Daston, "Taking Note(s)," *Isis* 95 (2004): 443–8.

principle. "Additional references" and "evidential verification" are provided for the sake of readers who doubt the veracity of the presented theses and include supporting citations from Chinese sources (*ruzhe zhi yu* 儒者之語).[35] These devices sought to gain the readers' confidence regarding the epistemic grounds of the theories presented and to assert the accuracy of Liu Zhi's translation of the sources.

It was in these commentaries that Liu Zhi made references to Chinese texts on medicine and materia medica. These references include quotes from major Chinese materia medica works, such as Shizhen's 李時珍 (1518–93) *Bencao gangmu* 本草綱目 (Detailed outline of materia medica); Liao Xiyong's 繆希雍 (1546–1627) *Shennong bencao jingshu* 神農本草經疏 (Commentary on "The divine husbandman's materia medica classic," pub. 1625); and Wang Ang's 汪昂 (1615–94) *Zengding bencao beiyao* 增訂本草備要 (Expanded edition of Materia medica essentials, pub. 1694);[36] medical compendia such as *Yijing bielu* 醫經別錄 (Miscellaneous records from medical classics); the works of the famous physician Sun Simiao 孫思邈 (d. 682); and a certain unidentified professional work on women's menstruation and gestation (*funü jingchan* 婦女經產).

Adding illustrations and various forms of commentary to his translated text and providing cross-textual references were Liu Zhi's main devices for making sense of and conveying the meaning of the foreign ideas he translated. At the same time, these devices served to demonstrate how compatible these Islamic ideas were with Chinese discourses on the natural world and the human body. Whereas these methods played out at the epistemic level, Liu Zhi employed a series of tactics at the semantic/lexical levels to render specific Arabo-Persian concepts and terms in Chinese to further produce discursive familiarity and coherence. These tactics will be the focus of the following section.

TACTICS OF LEXICAL AND SEMANTIC RENDITION

At the lexical and semantic levels, Liu Zhi applied a series of tactics to render Islamic physiological, anatomical, and pathological concepts and terms in Chinese, including loan translation (or calque), equating Arabo-Persian with existing Chinese terms, and coining new Chinese terms to translate Arabo-Persian concepts.[37] These tactics further helped to domesticate Islamic theories and construct compatibility with Chinese philosophical and medical discourses.

The tactic of loan translation renders the literal meaning of an Arabo-Persian term in Chinese while ignoring the semantic and theoretical differences between the two discursive systems in the use of the term. Liu Zhi used loan translation to render, for example, the Aristotelian Four Elements. He translated the Arabo-Persian terms of the substances Soil, Water, Air, and Fire with the Chinese terms *tu* 土, *shui* 水, *feng* 風

[35] *Annotated Selection* (cit. n. 25), 15:57.

[36] On Li Shizhen's *Bencao gangmu* and the other two materia medica works, see Unschuld, *Medicine in China* (cit. n. 4); Carla Nappi, *The Monkey and the Inkpot: Natural History and Its Transformation in Early Modern China* (Cambridge, MA: Harvard Univ. Press, 2009); and Georges Métailié, "The Bencao gangmu of Li Shizhen: An Innovation in Natural History?," in *Innovation in Chinese Medicine*, ed. Elisabeth Hsu (Cambridge: Cambridge Univ. Press, 2001), 221–61.

[37] There is a great resemblance between Liu Zhi's translation tactics and those used to translate Buddhist texts more than a millennium earlier. See Salguero, *Translating Buddhist Medicine* (cit. n. 7), 55–60.

(or *qi* 氣), and *huo* 火.[38] Of these four, the translation of *Air* is the most problematic as the two terms *feng* and *qi,* which Liu Zhi alternately used to translate the Arabo-Persian concept of Air, carry very different meanings in Chinese philosophy and medicine.[39] It is hard to see how *feng*, which plays an important role in Chinese pathology, and *qi*, which is a core aspect of Chinese physiology, could convey the meaning of Air as a substance.[40] It is worth mentioning here that Liu Zhi had difficulty in rendering the general concept "Element" for the Aristotelian Four Elements, and that he employed a number of translation tactics for doing so. In some cases, he equated the Aristotelian concept with the existing Chinese term "Phase" (*xing* 行)—a central concept in the Chinese analytical framework that investigates the natural world through a series of five types of transformations. Equating these concepts, Liu passed over the significant difference between the Chinese Phase and the Aristotelian Element—they differ in their conceived materiality and analytical functions. Elsewhere, he used a literal translation of the Arabo-Persian term for Element (*'unṣur*) and rendered the Four Elements as *siyuan* 四元 ("The Four Original Components")—a term that might better convey to a Chinese reader the physical dimension of the Aristotelian concept.[41]

An interesting case of a loan translation are Liu Zhi's renderings of the concepts "Natural Disposition" (*ṭabī'a*) or "Temperament" (*mizāj*). These are two central concepts in the Four Element theory and have important implications in Greco-Arabo-Persian medical and pharmaceutical theory and practice. The underlying theory suggests that every substance consists of a certain balance, called Temperament, of the Four Elements that determine its natural qualities and inclinations. In translating these concepts, Liu Zhi used the Chinese term *xing* 性. This term is originally a central concept in Chinese moral philosophy, and in that context can be translated into English as "Human Nature." It was later adopted into Chinese cosmology, where it played a major role in defining the correspondence between the Cosmos and Man, and it was even incorporated into medical discourses on human generation, vitality, and self-cultivation.[42] Establishing the actual relations between this concept and the

[38] A similar set of four elements was introduced to China via Buddhist translations during the first millennium CE and then again in the translation of the Jesuits. These terms would have been quite widely known to literati in the seventeenth and eighteenth centuries. Liu Zhi testifies in one of his prefaces that he studied Buddhist and European texts. It is therefore likely that he encountered this terminology. On the Buddhist translation of these terms, see Salguero, *Translating Buddhist Medicine* (cit. n. 7), 58–9. On the Jesuit translation of the four elements, see Hsu Kuang-Tai, "Four Elements as *Ti* and Five Phases as *Yong*: The Historical Development from Shao Yong's *Huangji jingshi* to Matteo Ricci's *Qiankun tiyi*," *EASTM* 27 (2007): 13–62.

[39] On the terms *qi* and *feng* and their usages in Chinese philosophy of nature and medicine, see Unschuld, *Medicine in China* (cit. n. 4), 67–73; Shigehisa Kuriyama "The Imagination of Winds and the Development of the Chinese Conception of the Body," in *Body, Subject, and Power in China*, ed. Angela Zito and Tani E Barlow (Chicago: Univ. of Chicago Press, 1994), 23–41; and Bol, *Neo-Confucianism in History* (cit. n. 4), 64–9 and 170–2.

[40] Matteo Ricci faced a similar dilemma in using the term *qi* to render the Aristotelian element of Air; Hsu, "Four Elements" (cit. n. 38), 51.

[41] The Jesuits, facing a similar challenge, translated the four elements as *Si yuanxing* 四元行 (Four fundamental elements). It is plausible that Liu Zhi borrowed this term from Jesuit works in Chinese. On the Jesuit translation of the four elements, see ibid. The Jesuit translation also bears some resemblance to the Buddhist rendering of the Four Elements; Salguero, *Translating Buddhist Medicine* (cit. n. 7), 72–3.

[42] See Michael Stanley-Baker, "Health and Philosophy in Pre- and Early Imperial China," in *Health: A History*, ed. Peter Adamson (Oxford: Oxford Univ. Press, 2019), 36–7.

other core cosmological concepts of *li* 理 "Cosmic Principle", *xin* 心 "Heart/Mind," and *qi* 氣 was a matter of harsh debate between different philosophical schools during Liu Zhi's time.[43] By borrowing the term *xing* to translate "Temperament," Liu Zhi embraced the view of *xing* held by the orthodox Cheng-Zhu school, which defined it as the grasp of the Cosmic Principle within the human Heart/Mind and "the innate and unchanging norms according to which that thing operates,"[44] instilling moral philosophy and philosophy of the mind into the theory of the Aristotelian Four Elements.

By equating Arabo-Persian concepts with existing Chinese terms, Liu Zhi was able to pass over significant theoretical differences and produce constructed commensurability between the two systems. Liu used this method to translate human physiology and anatomy and ignore the differences between the Islamic and Chinese conceptualizations of the body. Blood and blood vessels, such as arteries and veins, for example, were an integral part of the Islamic constitution of the human body yet lacked exact parallels in Chinese. To translate these two terms, Liu equated the Islamic concepts with the Chinese physiological terms *qixue* 氣血 and *jingluo* 經絡 respectively. The former, which is sometimes translated into English as "*qi* and Blood," refers to a pair of primary vitalities in Chinese physiology, and the latter designates the conduits and network vessels that transport *qi* and Blood between different bodily systems and loci.[45] This way of equating the Chinese and Arabo-Persian concepts appears in Liu's description of the movement of blood in the fetus's body, where he explains that "*qi* and Blood flow through the various conduits and network vessels and prevent the decaying [of the body]."[46] Equating the Chinese pair of primary vitalities that circulate in the body and enable the operation of the various physiological systems with the Greco-Arabo-Persian notion of blood as a bodily flood may overlook the theoretical difference in quality and function between the two concepts, but it provides a useful way to explain the operation of the human body for a Chinese reader.

Similarly, Liu employed this method of equating concepts in his translation of the following Persian passage from Nasafī's *Maqṣad-i aqsá* (*Furthest goal*): "When the fetus begins to require nutrition, it extracts blood [*khūni*] that was accumulated in the mother's womb by way of the navel. When it enters the fetus's stomach, it goes through processes of digestion."[47] Liu rendered this description into Chinese, writing, "When the child absorbs *qi* and Blood, which enter the stomach by way of the navel, firmness and consolidation [of the fetus's body] begin; this is the 'mineral nature'; it supplies the one Hundred Bodily Members."[48]

[43] See Qian Mu 錢穆, *Xueshu sixiang yigao* 學術思想遺稿 (Taipei: Lantai chubanshe, 2000), 212–3; Willard J. Peterson, "Arguments over Learning Based on Intuitive Knowing in Early Ch'ing," in *The Cambridge History of China, Volume 9, Part 2: The Ch'ing Dynasty to 1800*, ed. Willard J. Peterson (Cambridge: Cambridge Univ. Press, 2016), 458–512.

[44] Bol, *Neo-Confucianism in History* (cit. n. 4), 165. On the concept of *xing* in Chinese philosophy, see Bol again (69–71).

[45] On the concepts of *qi* and Blood, see Charlotte Furth, *A Flourishing Yin: Gender in China's Medical History* (Berkeley: Univ. of California Press, 1999), 46–8; Unschuld, *Medicine in China* (cit. n. 4), 75–9.

[46] *Human Nature* (cit. n. 1), 17:75

[47] 'Azīz al-Dīn Nasafī, *Maqṣad-i aqsá* [Furthest goal] (1351; Tehran: Kitābkhānih-yi 'ilmīyah-yi Ḥāmidī, 1972), 58.

[48] I use here Murata's translation with some modification; Murata, Chittick, and Weiming, *Sage Learning of Liu Zhi* (cit. n. 20), 127.

When Liu Zhi failed to find suitable Chinese terms that could capture the meaning of an Arabo-Persian concept, he coined new terms in Chinese. A Chinese reader might find such terms unfamiliar but would be able to make sense of them. Liu used this method to translate fundamental concepts such as Hippocratic humorism or the Arabo-Persian concept of *rūḥ* ("soul" or "*pneuma*").

The theory of Hippocratic-Galenic humorism stands at the core of Islamic medicine and constitutes a central analytical framework for the discussion of human physiology and pathology. Liu Zhi rendered the general theory of humorism with a Chinese term he coined—*Siben* 四本 (lit. "Four Sources"). In his terminology for the four individual humors, Liu stressed their fluid nature and their associated colors. He translated black bile, yellow bile, blood, and phlegm as *heiye* 黑液 ("black fluid"), *huangye* 黃液 ("yellow fluid"), *hongye* 紅液 ("red fluid"), and *baiye* 白液 ("white fluid") respectively.[49] In contrast to the use of the humors in Greco-Arabo-Persian physiology, Liu seems not to identify "red fluid" with blood, or "white fluid" with phlegm (*tan* 痰). His terms, however, are surprisingly similar to those that the Jesuits used to translate the four humors into Chinese. Giulio Aleni's (1582–1649) *Xingxue cushu* 性學觕述, a Chinese adaptation of the Coimba commentary on Aristotle's *De Anima*, used the exact four terms in its exposition of the four humors.[50]

The translation of the concept of the Soul (*rūḥ* or *Nafs*) is another example of the invention of new Chinese vocabulary to translate Islamic concepts. The concept of the Soul constituted a major divide between the Islamic and Chinese views of the human body and its operation. The Islamic concept of the Soul juxtaposes religious and biological functions, whereby the Soul is seen as an emanation of the Divine and a link between the cosmos and the human body. Its biological function corresponds to the Galenic concept of *pneuma* (lit. "breath")—that which gives the human body its vitality. Greco-Arabo-Persian medical theories refer to three manifestations of the Soul in human physiology: the "Vegetative Soul" (Greek: *pneuma physicon*, Persian: *rūḥ-i nabātī*)—the source of the vegetative processes, such as growth and digestion, whose seat in the human body is the liver; the "Vital Soul" (Greek: *pneuma zoticon*, Persian: *rūḥ-i ḥaywānī*)—the source of living processes that regulate the innate heat and vital conditions of the body, whose seat is the heart; and the "Psychic Soul" (Greek: *pneuma psychicon*, Persian: *rūḥ-i nafsānī*)—the source of emotions and movement, whose seat is the brain. Translating this concept was a great challenge for Liu Zhi, who eventually decided to use the loaded term *xing* 性 ([Human] Nature)—a term he used also to translate the concept of Temperament. By doing so, he presented the concept of the Soul as a form of endowed character or personal inclination, infusing the religious Islamic definition of the concept with elements from Confucian moral philosophy.

[49] Interestingly, Hei Mingfeng 黑鳴鳳, who edited and published an edition of Liu Zhi's *Human Nature and Cosmic Principles in Islam* with short commentaries on the text, commented that this theory of the four liquids resembles the four humors theory (*siye zhi shuo* 四液之說) that he had encountered in a European book; Liu Zhu, *Human Nature* (cit. n. 1), 17:71.

[50] Thierry Meynard and Dawei Pan, *A Brief Introduction to the Study of Human Nature: Giulio Aleni* (Leiden: Brill, 2020), 163–4. On the Chinese reception of the Jesuit medical translations, see Chu Ping-yi 祝平一, "Shenti, linghun yu tianzhu: Mingmo Qingchu xixue zhongde renti shengli zhishi" 身體靈魂與天主: 明末清初西學中的人體生理知識, *Xin shixue* 7 (1996): 47–98; and Benjamin A. Elman, *On Their Own Terms: Science in China 1550–1900* (Cambridge, MA: Harvard Univ. Press, 2005), 63–221.

For translating the physiological manifestations of the soul, Liu coined new terms that use the Chinese term *xing* with particular qualifications according to what he viewed as the defining physiological features of each manifestation. Accordingly, he coined *muxing* 木性 (lit. "tree nature") for the "Vegetative Soul," *shengxing* 生性 (lit. "living nature") for the "Vital Soul," and *juexing* 覺性 (lit. "conscious nature") for the "Psychic Soul." To these three, he added a fourth type that does not appear in Greco-Arabo-Persian tradition: *jinxing* 金性 (lit. "mineral nature"), which he defined as the source of *qi* and Blood to the various bodily organs.[51] While these translations convey the multiplicity and the particular functions of the faculties of the human soul, a Chinese reader would not necessarily be able to link these terms with their assigned physiological processes.

SYNTHESES OF ISLAMIC AND CHINESE MEDICAL THEORIES IN LIU ZHI'S TRANSLATION

Synthesizing Islamic and Chinese discourses, Liu Zhi's translation was able to make some important claims about the structure and operation of the human body, supplementing and offering alternative theories to both Islamic and Chinese traditions. His charting of human generation provides an insightful framework for discussing human anatomy and the functions of bodily organs, one that juxtaposes the Hippocratic theory of the four humors with the Chinese theories of yin and yang and the Five Phases. Liu Zhi also links the physiological functioning of the Brain with its relation to the Heart, and even descriptions of the physiological processes of nourishment, digestion and growth. These joinings produced intriguing theories that transcend both Islamic and Chinese discourses.[52]

Liu Zhi's account of the nine months of embryonic development in *Human Nature and Cosmic Principles in Islam* includes an interesting description of the four humors and their role in developing human anatomy. The four humors, Liu explains, are the root of the "human Body, Blood, Flesh and Vital *qi*"; are divided into Clear (*qing* 清) and Turbid (*zhuo* 濁); and are distinguished by their colors.[53] The particular set of properties of an individual humor, Liu suggests, emerges from the burning effect of the womb's yin fire (*yinhuo* 陰火[54]) and the humor's position in the womb. These sets of properties associate each humor with one of the Four Elements (*sixing* 四行): Wind/Air, Fire, Water, and Soil. During the second month of gestation the humors are

[51] This type of nature does not appear in the passage in *Maqṣad-i aqṣá* from which Liu Zhi translated this part; Murata, Chittick, and Weiming, *Sage Learning of Liu Zhi* (cit. n. 20), 373n2.

[52] For a comparison of Liu Zhi's embryogenesis scheme with other Chinese theories, see Stephen R. Bokenkamp, "Simple Twists of Fate: The Daoist Body and Its Ming," in *The Magnitude of Ming: Command, Allotment, and Fate in Chinese Culture*, ed. Christopher Lupke (Honolulu: Univ. of Hawaii Press, 2005), 151–68; Furth, *A Flourishing Yin* (cit. n. 45); and Yi-Li Wu, *Reproducing Women: Medicine, Metaphor, and Childbirth in Late Imperial China* (Berkeley: Univ. of California Press, 2010). On the Islamic theories of gestation, see Nahyan Fancy, "Generation in Medieval Islamic Medicine," in *Reproduction: Antiquity to the Present Day*, ed. Nick Hopwood, Rebecca Fleming, and Lauren Kassell (Cambridge: Cambridge Univ. Press, 2018), 129–40; and B. F. Musallam, "The Human Embryo in Arabic Scientific and Religious Thought," in *Islamic Medical and Scientific Tradition*, ed. Peter Pormann (London: Routledge, 2010), 2:317–31.

[53] *Human Nature* (cit. n. 1), 17:70.

[54] Yin fire (*yinhuo* 陰火) is a Chinese concept that draws on a bifurcation of the Five Phases into yin and yang manifestations. In contrast to the common usage in Chinese cosmology, Liu uses the term here in what seems to be a reference to an actual substance.

distinguished according to the tendencies of their associated elements, and form the Outward (*biao* 表) and Inward (*li* 裏) frames of the body.[55] The humor that is associated with Soil makes up the Flesh; that which is associated with Water, the Vessels (*mailuo zhi lu* 脉絡之路); that which is associated with Air, the Heart; and the humor that is associated with Fire becomes the Apertures of Awareness (*lingming zhi kong* 靈明之孔) that stand on the left and right sides of the Heart.

The anatomical structure of the body, according to Liu Zhi's accounts, involves a division into four *zang* 藏 and six *fu* 府 elements. It is unclear from Liu's description whether these are systems or specific organs.[56] An abstract anatomical sketch suggests that the four *zang* elements include the Lung, Liver, Spleen, and Kidney, and the six *fu* include the ears, eyes, mouth, nose, four limbs and the "Hundred Members" (*baiti* 百體). The *zang* and *fu* categorization draws on a view commonly used in Chinese medical texts which divides bodily systems into five *zang* (including Liver, Heart, Spleen, Lung, and Kidney) that are in charge of generating and storing vital *qi*, and six *fu* (Stomach, Small Intestine, Large Intestine, Urinary Bladder, Gallbladder, and Triple Burner) that are in charge of transmitting vital *qi* and its digestion.[57] Liu's use of the terms *zang* and *fu*, however, greatly departs from these meanings, and seems to mark a certain binary of internal and external body parts. The four *zang* systems, Liu explains, are linked to one another in the space between the Heart and outer frame of the body, and they constitute the particular lodges of each of the Four Elements. According to Liu, the functionality of the body as a whole lies in "the linkage and coordination" (*guanhe* 關合) of body members and apertures (*tiqiao* 體竅) carried out by the *zang* and *fu* systems. Each of these systems is in charge of specific body members and apertures, while the Brain (*nao* 腦) is that which oversees the linkage and coordination of the entire body.

A major feature of Liu Zhi's synthesis is its assertion of the centrality of the Brain over the Heart. Liu explains that the Brain links together "the spiritual *qi*" (*lingqi* 靈氣) of the Heart and the "vital *qi*" 精氣 of the body and transforms both of them, suggesting a psychological-physiological interaction that takes part in the brain.[58] He suggests that the Brain is the main source of the various blood vessels, the enabler of cognition, sensation, and motion. According to Liu's description, the Brain collects and stores what the eye sees, the ear hears, and the heart cognizes. At the same time, the Brain acts as a sensory central command (*zongjue* 總覺), and through sinew networks (*jinluo* 筋絡) that link the Brain to the specific organs, it empowers the eye, ear, mouth, and nose to see, hear, taste, and smell accordingly, and produces sensation. Embedding this theory in Chinese medical theory, Liu further explains that the Liver has its aperture in

[55] In Chinese medicine, the pair *li* 裏 (internal) and *biao* 表 (external) represent a binary division of bodily organs. There are, however, different definitions of this binary. In some cases, they are used to mark the difference between external and internal organs. The former includes skin, flesh, hair, and blood vessels, and the latter internal organs. Alternatively, they are used to distinguish fu 腑 organs (defined as *biao*) and zang 臟 organs (defined as *li*); Manfred Porkert, *The Theoretical Foundations of Chinese Medicine: Systems of Correspondence* (Cambridge, MA: MIT Press, 1985), 162.

[56] In this and the following paragraphs, I follow the convention of capitalizing the names of systems to distinguish them from simple organs.

[57] Some Chinese texts speak of the six *zang* organs with the addition of the Pericardium; Unschuld, *Medicine in China* (cit. n. 4), 77–83.

[58] See also Zhu Yongxin, "Historical Contributions of Chinese Scholars to the Study of the Human Brain," *Brain and Cognition* 11 (1989): 133–8.

the eye, yet it is the Brain that empowers the eye to see. Similarly, it is the Brain that empowers the ear, mouth, and nose, the apertures of the Kidney, Spleen, and Lung accordingly, to hear, taste, and smell. This sensory network connecting the Brain with the various organs allows the hand to hold things, the feet to walk. and the "Hundred Members" to feel pain and itch.

The Heart/Mind (*xin* 心) was a subject of major debate among Chinese philosophers and medical theoreticians. In his works, Liu Zhi discussed various features of the Heart/Mind at great length, pointing to the role of Heart/Mind as the seat of emotions and thoughts, a complement of *xing* ([Human] Nature) and the microcosm parallel to macrocosm Heavens—the abode of the cosmic principles. In describing the role of the Heart/Mind in the operation of the human body, Liu Zhi placed it secondary to the Brain. Liu explains in *Human Nature and Cosmic Principles in Islam* that although the Heart is the seat of Awareness (*lingming* 靈明), it relies on the Brain. When the Brain is well nourished, Liu suggests, the Heart/Mind's Awareness is solid, but when the Brain lacks proper nourishment, the Heart's vigor wanes. The relationship between the Brain and the Heart/Mind, according to Liu, is one of a planner and executer, where the Brain commands the Hundred Bodily Members and apertures to carry out what the Heart/Mind wishes to do.

Liu Zhi's translation provides various descriptions of physiological processes such as nourishment, digestion, and growth. These descriptions synthesize Islamic and Chinese theories. Nourishment, for example, is described as taking place in two ways: through the navel and in the Gallbladder. The former represents an Islamic description that is found in the Arabo-Persian sources of Liu Zhi, and the latter is borrowed from Chinese medical theory. According to Liu Zhi, the navel conducts *qi* and Blood from the mother's womb into the fetus's stomach where nourished substances are extracted; the gallbladder separates the transported *qi* and Blood into benevolent and malicious substances, further transporting the benevolent and storing the malicious.

These various theories are the outcome of Liu Zhi's attempts to read Greco-Arabo-Persian physiology through the prism of Chinese medical terminology and theories. Juxtaposing the two medical traditions, Liu Zhi was able to point to differences and similarities between the ways each tradition explains the structure and operations of the human body. By extracting parts of these explanations, matching concepts, and synthesizing theories in his translation, Liu Zhi was able to come up with fresh and innovative views of the human body and its operation.

* * *

Liu Zhi's translations and their descriptions of the human body and physiological processes were further reconfigured by a list of editors, readers, and publishers that took an active role in shaping the works' contents and formats long after Liu's death. At least two editions of *Human Nature and Cosmic Principles in Islam* include editorial interventions and the commentaries of Hei Mingfeng 黑鳴鳳 (b. 1673), a high-ranking official and a military *jinshi*-degree holder. Hei seems not to have had the linguistic skills to read original Arabic and Persian texts but was immersed in Chinese classical learning and familiar with some of the published works of the Jesuits. His commentaries expanded Liu Zhi's discussions on the human body and added further coherence and authority to the works. Hei's high-ranking status also lent the works further visibility among Chinese literati. In 1868, Ma Dexin 馬德新 (1794–1874), a

Chinese Muslim scholar who spent a long time in the Middle East and even studied Islamic theology in Al-azhar University in Cairo, published a commentary in Arabic on Liu Zhi's *Human Nature and Cosmic Principles in Islam*, seeking to explain how Liu Zhi's thought conforms to the Islamic theological system, including Liu's interpretations of physiology.[59] His student Ma Lianyuan 馬聯元 (1841–1903) published a subcommentary on *Human Nature and Cosmic Principles in Islam* under the title *Sharḥ al-laṭā'if* (Explanation of the Subtleties).[60] These editions and extensions not only attest to the impact of Liu Zhi's works on the accommodation of Islamic philosophies of nature and views of the human body in China but also display the open-endedness of the translation of such branches of knowledge and the continuous interaction between authors, commentators, and readers in their search for coherence and meaning.

As the various essays in this volume show, as early modern texts moved across geographies and languages, they carried with them local experiences and implanted them in new environments. Translation in that regard can be described as the collective textual acts and forms of articulation that are required for such imported experiences to become coherent, authoritative, and relevant for a target audience. Liu Zhi's translation shows such movement of ideas and theories on the human body and its operation from the Islamicate world, broadly defined to incorporate the sum of the various Greco-Arabo-Persian cultures and traditions as viewed from the perspective of an early modern Chinese, eastward into the Yangtze Delta, the heartland of China's Ming and Qing empires.

Isolated from the centers of Islamic scholarship in other parts of Asia, and with only limited access to Arabo-Persian texts, Liu undertook empire-wide journeys to recover manuscripts and build an archive of Arabo-Persian knowledge on the natural world, scrupulously excerpting, interpreting, and translating their contents. His translation provides us a rare glimpse into a global circulation of knowledge on the human body that is neither enabled by colonial violence, commercial activities, or imperial impositions nor embedded in texts produced by or for medical practitioners. It is the meticulous work of a philosopher whose investigation of the natural world sheds light on a philological dimension in the premodern knowing of the body. Rethinking the economies and mechanism of medical translation, this essay brings to light the contribution of religious texts, hermeneutical methods, and translation tactics in the global dissemination of views on the body and physiological theories.

[59] On Ma Dexin's translation, see Wang Xi 王希, "Ma Fuchu 'Benjing wuzhang yijie' chutan" 馬復初《本經五章譯解》初探, *Journal of Hui Muslim Minority Studies* 3 (2012): 44–9; and Petersen, *Interpreting Islam in China* (cit. n. 9).

[60] On that work, see Matsumoto Akiro 松本耿郎, "Ma Lianyuan cho 'Tianfang xingli awen zhujie' no kenkyū" 馬聯元著『天方性理阿文注解』の研究, *Tōyōji kenkyū* 58 (1999): 176–211; and Hu Long 虎隆, "Ma Lianyuan de zhushu yanjiu - shang" 馬聯元的著述研究上, *Zhongguo musilin* 5 (2018): 27–34.

New World Drugs and the Archive of Practice:
Translating Nicolás Monardes in Early Modern Europe

*by Alisha Rankin**

ABSTRACT

This essay examines European vernacular translations of the Spanish physician Nicolás Monardes's treatises on drugs from the New World. I use the concept of an "archive of practice," a set of original sources based in practices rather than texts, to examine how translators articulated or disarticulated indigenous Amerindian knowledge in Italian, English, French, Dutch, and German sources. Examining practice as a kind of archive allows us to consider sources captured only peripherally in written texts. This concept is especially important for Amerindian medical practices, which had no written source material. All Monardes sources, including his original publications, functioned as highly mediated translations of indigenous practices, and vernacular translations were even more distant from the primary sources. Nevertheless, I argue, the foundation of New World drugs on an archive of indigenous practice formed an important backdrop to vernacular translations. The empirical, non-European origins of the new knowledge gave translators significant freedom to adapt Monardes's texts to their specific intellectual, political, professional, and commercial interests.

In the summer of 1589, a provincial Bavarian physician named Johann Wittich published two separate German books on remedies from the New World. Both works were partial translations of the Spanish physician Nicolás Monardes's popular treatises on New World drugs, but they bore little resemblance to one another. A slim pamphlet, *On Guaiac Wood*, described four medicinal trees from the Americas thought to be good against the "unchaste pox." Addressed specifically to surgeons and barbers, it promised to explain the way "the Indians" prepared and used these drugs, "which accomplishes more than we Germans." Wittich claimed that his book would provide a comparison "between the two methods, the Indian and the German" to clear up the confusion among professionals about how best to administer the drugs. In contrast, Wittich's other book, *On the Wondrous Bezoar Stone*, enticed a broader readership with the "unheard-of and unbelievable healing powers" of various gems, herbs, minerals, and animal materials

* Department of History, Tufts University, East Hall, 6 The Green, Medford, MA 02155, USA; alisha.rankin@tufts.edu.

Osiris, volume 37, 2022. © 2022 History of Science Society. All rights reserved. Published by The University of Chicago Press for the History of Science Society. https://doi.org/10.1086/719221.

from both the New World and lands to the east. These drugs, he boasted, were "unknown to the ancient and new scribes" and had "only been brought in the last thirty years from the Oriental and Occidental Indies . . . and have never before been put into German."[1] While *On Guaiac Wood* gave practical advice, *On the Wondrous Bezoar Stone* presented exotic novelties—yet both books brought Monardes's optimism about New World drugs to a German audience.

Wittich came late to the party. Fifteen years earlier, the Dutch physician Carolus Clusius had published a Latin translation of Monardes's first two books on New World drugs, which quickly became the dominant version of Monardes among European scholars.[2] By 1580, versions of Monardes had appeared in French, Italian, English, German, and Dutch, in addition to the many editions in Latin and in Monardes's original Spanish (see table 1). This intense translation activity underscored the avid interest in New World drugs around Europe, and it turned Monardes into a household name in medical circles. Historians have already devoted significant attention to Clusius's Latin translation (1574) and John Frampton's English translation (1577), which took vastly different approaches to Monardes's original Spanish text.[3] While Frampton translated Monardes's work more or less faithfully, Clusius altered it significantly, creating more of an interpretation of Monardes than a direct translation. Wittich's approach was different still: he translated specific portions from Clusius for his two very divergent intended readerships. In fact, all Monardes translations stand out in their extreme variability. Translators made very different decisions about which parts of Monardes to transmit, ranging from direct translations from the original Spanish, to translations from Latin, to the piecemeal translation of specific sections, often with further modifications. This use of different tactics for different audiences was not unusual, as a large body of literature on the cultural history of translation has made clear.[4]

Yet Monardes himself was a translator more than an author. As Wittich pointed out, he offered knowledge "unknown to ancient and new scribes," with "unheard-of and unbelievable healing powers." His knowledge was gleaned not from books but from

[1] Johann Wittich, *Von dem ligno guayaco, Wunderbawn, res noua genandt* [On guaiac wood] (Leipzig, 1592), a3r-v; Wittich, *Bericht von den wunderbaren bezoardischen Steinen* [On the wondrous bezoar stone] (Leipzig, 1589),)a(1r-)a(3v. The 1589 edition of the *Ligno guayaco* is very scarce, and I will cite the more common 1592 edition here. Unless otherwise noted, the translations in this article are mine.

[2] Nicolás Monardes, *De simplicibus medicamentis ex occidentali India delatis quorum in medicina usus est*, trans. Carolus Clusius (Antwerp, 1574). For a list of Monardes's works by year, language, and translator, see table 1.

[3] On this point, see especially José Pardo-Tomás, "Two Glimpses of America from a Distance: Carolus Clusius and Nicolás Monardes," in *Carolus Clusius: Towards a Cultural History of a Renaissance Naturalist*, ed. Florike Egmond, Paul Hoftijzer, and Robert Visser (Amsterdam: Koninklijke Nederlandse Akademie van Wteneschappen, 2007), 173–93; and Antonio Barrera-Osorio, "Translating Facts: From Stories to Observations in the Work of Seventeenth-Century Dutch Translators of Spanish Books," in *Translating Knowledge in the Early Modern Low Countries*, ed. Harold John Cook and Sven Dupré (Zürich: LIT Verlag Münster, 2012), 321–5.

[4] See especially Peter Burke and R. Po-chia Hsia, eds., *Cultural Translation in Early Modern Europe* (Cambridge, UK: Cambridge Univ. Press, 2007); Burke, "Cultures of Translation in Early Modern Europe," in *Cultural Translation in Early Modern Europe*, ed. Burke and Po-chia Hsia (Cambridge, UK: Cambridge Univ. Press, 2007), 7–38; Cook and Dupré, *Translating Knowledge* (cit. n. 3); and Andreas Höfele and Werner von Koppenfels, *Renaissance Go-Betweens: Cultural Exchange in Early Modern Europe*, Spectrum Literaturwissenschaft 2 (Berlin: Walter de Gruyter, 2005). On the specific relevance of translation practices to early modern science and medicine, see especially Sietske Fransen, Niall Hodson, and K. A. E. Enenkel, eds., *Translating Early Modern Science*, Intersections 51 (Leiden: Brill, 2017); and Dupré, "Introduction: Science and Practices of Translation," *Isis* 109 (2018): 302–7, https://doi.org/10.1086/698234.

Table 1. *Translations of Monardes's works*

Date	Language	Translator	Text(s) Translated	Full or partial translation (Material included if partial)
1572	French	Jacques Gohory	*Dos libros*	Partial (Mechoachan)
1574	Latin	Carolus Clusius	*Dos libros, Segunda parte*	Full but altered
1575	Italian	Giodarno Ziletti	*Dos Libros, Segunda parte, Libro de Nieve*	Full
1576	Italian	Annibale Briganti	*Dos libros*	Full
1577	English	John Frampton	*Dos libros*	Full
1580	English	John Frampton	*Historia medicinal* (1574)	Full
1580	German	Anonymous	*Historia medicinal* (1574)	Partial (Sassafras)
1580	Dutch	Nicolaes vander Woudt	*Historia medicinal* (1574)	Partial (Tobacco)
1580	German ms.	Johannes Strupp	Clusius (likely 1574)	Partial (Sassafras)
1580	German ms.	Anonymous	*Dos libros* (1565)	Partial (Bezoar treatise; sassafras)
1582	Latin	Carolus Clusius	*Historia medicinal* (1574)	Partial (Third part on New World drugs)
1589	German	Johann Wittich	Clusius (1579)	Partial (Sassafras, guaiac, sarsaparilla, China)
1589	German	Johann Wittich	Clusius (1579)	Partial (Numerous substances)
1602	French	Antoine Colin	Clusius (1579)	Full, combined with da Orta and Acosta
1605	Latin	Carolus Clusius	*Historia medicinal* and da Orta	Partial, mishmash
17th c.	Italian ms.	Annibale Briganti	*Dos libros* (1565)	Full

informants who brought him reports from the New World, and from his own experience trying out the "unbelievable" new drugs on his patients. The ur-sources of Monardes's information, as Wittich recognized, were the indigenous practitioners who had long used the substances in their own medicine. His archive was an archive of practice. As other historians have noted, Monardes adapted and Europeanized these nuggets of Amerindian practical knowledge, while Clusius downplayed the role of indigenous practitioners and expunged many of these anecdotes.[5] Less widely

[5] On Monardes, see José Pardo-Tomás, "Obras españolas sobre historia natural y materia médica americanas en la Italia del siglo XVI," *Asclepio, archivo iberoamericano de historia de la medicina y antropología médica* 43 (1991): 51–94; Pardo-Tomás, "East Indies, West Indies: Garcia da Orta and the Spanish Treatises on Exotic Materia Medica," in *Medicine, Trade and Empire: Garcia de Orta's Colloquies on the Simples and Drugs of India (1563) in Context*, ed. Palmira Fontes da Costa (Farnham, Surrey, UK: Ashgate, 2015), 195–212; Pardo-Tomás, "Two Glimpses" (cit. n. 3); Barrera-Osorio, "Translating Facts" (cit. n. 3); Daniela Bleichmar, "Books, Bodies, and Fields: Sixteenth-Century Transatlantic Encounters with New World Materia Medica," in *Colonial Botany: Science, Commerce, and Politics in the Early Modern World*, ed. Londa L. Schiebinger and Claudia Swan (Philadelphia: Univ. of Pennsylvania Press, 2005), 83–99; and Donald Beecher, "Nicolás Monardes, John Frampton and the Medical Wonders of the New World," in *Humanismo e ciência: Antiguidade e*

studied, however, have been the wildly differing ways that Monardes's vernacular translators conveyed his works and the indigenous practices they described. Their translations focused not only on the writings of one Spanish physician but also (and often primarily) on the materials and practices signified by that text.

This article uses the concept of an "archive of practice" to examine how European vernacular translators articulated or disarticulated indigenous Amerindian knowledge. By an "archive of practice," I mean a set of original sources based in practices rather than texts. As Ahmed Ragab argues in this volume, translations were only one step in a broader process of archive making. Examining practice as a kind of archive allows us to consider sources captured only peripherally in written texts. This concept is useful for various arenas of premodern medicine that stand outside of documents. Alison Walker has described pharmacopoeia annotations as archives of practice, and recipe collections similarly represent written artifacts from practice-based knowledge.[6] It is especially important for indigenous Amerindian practices, which had no written source material. Pablo Gómez's essay in this volume calls for an "unmuffling" of Amerindian and African voices in the history of translation. Monardes's texts and their translations demonstrate exactly why such an unmuffling is necessary. As José Pardo-Tómas has noted, Monardes translated indigenous practice from the stories of European informants, adding at least one additional level of translation from the original archive, and his interventions tended to downplay indigenous authority.[7] At the very least, vernacular translators translated Monardes's translation of his sources' translations of indigenous practice, and some translators added one more layer by using Clusius's Latin rather than the original Spanish. We can only assume that this elaborate game of telephone rendered the original indigenous practices—the primary sources—nearly unrecognizable.

Nevertheless, I argue in this essay, the foundation of New World drugs on an archive of indigenous medical practice formed an important backdrop to vernacular translations. The empirical, non-European origins of the new knowledge gave translators significant freedom to adapt Monardes's texts to their specific intellectual, political, professional, and commercial interests. This impulse was aided by the complex publishing history of Monardes, who developed his three books on the New World over time, between 1565 and 1574, and divided them into chapters highlighting each medical substance. This "modular" approach meant that within each book rested dozens of self-standing treatises, which could easily be separated from the whole. There were no established standards for scientific translation in early modern Europe, as Sietske Fransen has pointed out, and the malleable Monardes text gave translators significant leeway to pick and choose the portions they deemed most relevant and useful.[8]

---

renascimento, ed. António Manuel Lopes Andrade, Carlos de Miguel Mora, and João Manuel Nunes Torrão (Coimbra: Imprensa da Universidade de Coimbra, 2015), 141–60, https://doi.org/10.14195/978-989-26-0941-6_6.

[6] Ahmed Ragab, "Translation and the Making of a Medical Archive," in this volume (*Osiris* 37); Alison Walker, "Collecting Knowledge: Annotated Material in the Library of Sir Hans Sloane," in *Archival Afterlives: Life, Death, and Knowledge-Making in Early Modern British Scientific and Medical Archives*, ed. Vera Keller, Anna Marie Roos, and Elizabeth Yale (Leiden: Brill, 2018), 222–40, esp. 233–7.

[7] Pablo F. Gómez, "[Un]Muffled Histories," also in this volume (*Osiris* 37); Pardo-Tomás, "Obras españolas" (cit. n. 5); Barrera-Osorio, "Translating Facts" (cit. n. 3), 317–32.

[8] Sietske Fransen, "Exchange of Knowledge through Translation: Jan Baptista Van Helmont and His Editors and Translators in the Seventeenth Century" (PhD diss., The Warburg Institute, Univ. of London, 2014), 4.

While Monardes's Italian translators framed his text as a purely intellectual contribution to European humanist knowledge, all other translations engaged with the archive of practice head on, albeit in very different ways. Some, like Frampton, highlighted the threat of Spanish commercial dominance. Others, like the Frenchman Jacques Gohory, used indigenous practice to diminish the role of Spain. Several translators, including Wittich, used their translations to assert authority over a different archive of practice: that of local empirics and patients who had begun to use New World drugs. A side-by-side analysis of Monardes translations across multiple languages demonstrates the full complexities of translating from practices rather than texts—especially indigenous practices, which by their very nature provoked engagement with thorny professional, intellectual, and geopolitical conflicts. In nearly all cases, vernacular Monardes translations used the archive of practice not to recognize indigenous practitioners but to shore up the authority of European communities against threats both local and foreign.

THE ORIGINAL SOURCES

Monardes's true original sources—his archive—were the anecdotes of practice he received from travelers arriving in Seville and, as his fame grew, from informants' letters. This archive of practice drew from diverse areas of the Spanish colonies, and Monardes often tied the drugs he described to specific places: the provinces of Galisco, Mechoachan, Honduras, Nicaragua, and Peru; the port cities of Havana, San Juan de Puerto Rico, Quito, and Cartagena; and the appealing new territory of Florida. Sitting in Seville, he gathered anecdotes of indigenous practice from this wide swath of territory and wove them together into a portrayal of "New Spain" and its wealth of efficacious drugs. As Pardo-Tomás has noted, Monardes privileged the experience of European colonizers and his own tests of the drugs on his patients in Seville over the indigenous archive of practice, and he valued New World drugs especially as substitutions (*succedaneum*) for exotic cures from Asia.[9] Yet his inclusion of indigenous practices, however flawed, gave later translators space to articulate (or further disarticulate) this practice.

What translators meant by "Monardes" varied from the moment they picked a source from which to translate. Monardes completed his original Spanish works in stages. His interactions with New World drugs began as a business venture, in which enslaved Africans were traded for Amerindian dyes and medicaments. After that business declined, he fell into financial trouble.[10] In 1565, he published a slim volume titled *Dos libros* (Two books), which included a treatise on the marvelous virtues of New World drugs and a treatise on poison antidotes. Bankrupt and trying to avoid debtors' prison, Monardes achieved enough success with this work that he published a second treatise (*Segunda parte*) on New World materia medica in 1569, which included his effusive account of tobacco; and a medical treatise on snow, in 1571. He then composed a third book on New World medicines and in 1574 published his *Historia medicinal de las cosas que se traen de nuestras Indias Occidentales que seruen en medicina* (Medicinal history of the things found in our West Indies that

[9] Pardo-Tomás, "Two Glimpses" (cit. n. 3), 181.
[10] Ibid., 178.

are useful in medicine). This compilation included all three books on New World drugs, the books on poison antidotes and snow, and a treatise on the medicinal uses of iron.[11] Monardes's publishing strategy was both complex and constantly in flux until the full *Historia medicinal* appeared.

That instability was compounded by Carolus Clusius's Latin translation, which became the dominant edition of Monardes across Europe and the source for several vernacular translations. In 1574, Clusius translated Monardes's first two books on New World drugs, from the *Dos libros* (1565) and the *Segunda parte* (1569), in a work he called *De simplicibus medicamentis ex occidentali India delatis* (On simple medicaments brought from West India).[12] The title was meant literally: Clusius focused entirely on drugs from the New World. He did not translate the other half of the *Dos libros*, on poison antidotes, or the treatise on snow published in 1571. In 1582 he added a stand-alone translation of Monardes's third book of New World drugs. In contrast to Monardes's more wide-ranging introduction of multiple medical topics, Clusius focused solely on drugs from the Spanish colonies.

As already mentioned, Clusius took great liberties in his rendition of Monardes. He combined the two parts, truncated many entries, split other entries into more than one section, and added new headings. He also added his own annotations and illustrations while leaving out some of Monardes's original images.[13] Clusius took these steps deliberately and, in fact, explained to the reader the reasoning behind his approach. He noted that similar medicines were dispersed across Monardes's two parts and that there was significant repetition. So that he could discuss the medicines in a logical progression, he explained, "I have made one [book] out of two." This approach reduced the two lengthy Monardes treatises to a slim volume of eighty-eight pages, even though Clusius added annotations and new woodcuts. His overall goal, he stated, was to explain the valuable new plants to the reader who did not know Spanish.[14] Because he published in Latin, he specifically spoke to Europe's intellectual elite.

The emphasis on plants was telling. As Pardo-Tómas has noted, the erudite Clusius exhibited a "manifest lack of interest in all that Monardes conveys concerning the knowledge and practices of the Amerindians."[15] In his translations, the archive of practice all but disappeared, with no real sense of the indigenous practitioners, or even of Monardes's many medical experiments on his own patients. Clusius instead assimilated the exotic drugs into the Galenic system of European materia medica and highlighted his own web of connections among scholars across Europe. He took a similar approach in his translations of Garcia da Orta's 1563 Portuguese treatise on South Asian drugs and Cristóbal Acosta's Spanish treatise on New World plants, both frequently published or

[11] Nicolás Monardes, *Dos libros. El uno trata de todas las cosas que traen de nuestras Indias Occidentales, que sirven al uso de medicina . . . El otro libro, trata de dos medicinas maravillosas que son contra todo veneno, la piedra Bezaar, y la yerva Escuerçonera* (Seville, 1565); Monardes, *Segunda parte del libro, de las cosas que se traen de nuestras Indias Occidentales, que siruen al vso de medicina* (Seville, 1571); Monardes, *Libro qve trata dela nieve, y de sus propiedades* (Seville, 1571); Monardes, *Primera y segunda y tercera partes de la historia medicinal de las cosas que se traen de nuestras Indias Occidentales que siruen en medicina* (Seville, 1574).

[12] Monardes, *De simplicibus medicamentis* (cit. n. 2), title page.

[13] Pardo-Tomás, "Two Glimpses" (cit. n. 3), 185–7.

[14] Monardes, *De simplicibus medicamentis* (cit. n. 2), 4.

[15] Pardo-Tomás, "Two Glimpses" (cit. n. 3), 192. Antonio Barrera-Osorio has argued that Clusius did highlight experience in his headings, but not indigenous experience; Barrera-Osorio, "Translating Facts" (cit. n. 3), 321–5.

bound with the Monardes.[16] In all these works, the substances took precedence over any sort of practice.

European authors translating Monardes into vernacular tongues thus had several paths they could follow, depending on their own linguistic abilities and access to the source texts. They could translate from any variety of the original Spanish texts, or they could use Clusius's Latin. If they used Clusius, they could translate Monardes alone, or they could add the translation of da Orta. One might expect translations drawing on Clusius to contain far fewer references to the archive of indigenous practice than translations using Monardes's Spanish. In practice, however, the source text did not determine the centrality of the archive of practice to vernacular translators. Some translators giving extensive space to Amerindian practice used Clusius, while the archive of practice appeared in variable levels in translators drawing on Monardes's original Spanish. This variation suggests that the *concept* of an archive of indigenous practice existed independently of the texts. In fact, the translators who devoted the least amount of discussion to the archive of practice were the three complete translations (two in Italian, one in English) from the original Spanish.

"LIGHT FROM THESE LEARNED DOCTORS"

The Venice-based printer Giodarno Ziletti made his readers acutely aware of the difficulties in translating Monardes.[17] In 1575, Ziletti issued an edition of everything Monardes had published up to 1571, translated directly from Spanish to Italian. He claimed to have received the *Dos libros* from a Spanish friend and "decided to communicate it to you, in our language," suggesting that he had translated the work himself for the benefit of readers. Just as he was about to print it, he received the *Segunda parte* and the treatise on snow. Not wanting "my readers to lack for anything," he quickly translated those works as well, and he promised to translate additional works by Monardes as soon as he got his hands on them.[18] Alone among Monardes's translators, Ziletti portrayed the Spanish texts as a work in progress rather than a stable source.

Ziletti also presented a strong voice in favor of translating faithfully from Monardes's original Spanish. He had read the Latin translation, and he was not impressed. The descriptions of materia medica had been "greatly altered" and were "truncated and imperfect," leaving the hapless reader to "disentangle" the various entries. It was fine to reduce Monardes for the sake of "brevity," Ziletti conceded, but it made no sense for the translator to then add "his own annotations." Although he never mentioned Clusius by name, the object of his criticism would have been obvious to his readers. This complete reworking, Ziletti complained, was nothing but an "extract" of Monardes's work and an "offense against the author," who had "wanted to write to make himself understood."[19]

[16] Garcia da Orta, *Aromatum, et simplicium aliquot medicamentorum apud Indos nascentium . . .*, trans. Carolus Clusius (Antwerp, 1574); Cristóbal Acosta, *Aromatum & medicamentorum in Orientali India nascentium liber* (Antwerp, 1582). See also Pardo-Tomás, "Two Glimpses" (cit. n. 3), 186–91.

[17] Originally based in Rome, Ziletti fled to Venice in the 1550s after the Inquisition accused him of printing forbidden books without permission; Katherine M. Bentz, "Ulisse Aldrovandi, Antiquities, and the Roman Inquisition," *Sixteenth Cent. J.* 43 (2012): 984–6. On the Italian translations of Monardes in general, see Pardo-Tomás, "Obras españolas" (cit. n. 5).

[18] Ziletti does not seem to have followed through with this intention; Nicolás Monardes, *Delle cose che vengono portate dall'Indie Occidentali pertinenti all'vso della medicina*, trans. Giodarno Ziletti (Venice: Giordano Ziletti, 1575), preface.

[19] Ibid., b1r–2r.

Ziletti thus blatantly accused Clusius of mistranslation, and he presented his own work as a contrast. He had decided to "change not a single point but to make you see [things] as they have been described by the . . . author." He expressed hope that Monardes would have no reason for annoyance with him, as he had merely "brought it from language into language [*di lingua in lingua*] in his honor, for the benefit of the world."[20] The process of translation, in his depiction, was an uncomplicated matter of semantics.

Yet Ziletti's criticism of Clusius did not arise from the botanist's elision of the archive of practice. He did not emphasize either experience or practice, and he completely ignored the topics of Amerindian knowledge or Spanish imperialism that foregrounded Monardes's work. Instead, his dedication to the Venetian nobleman Andrea Contarini addressed a community of erudite humanist Italians and trumpeted his overall aim of bringing printed books to "learned and virtuous men."[21] In that sense, he addressed the same readership as Clusius. His contrast between Clusius's "extract" and his own faithful rendition of Monardes may have been an attempt to give readers a reason to buy his books even if they already owned Clusius's Latin translation. While his full translation certainly would have transmitted more evidence of indigenous practice to his readers than Clusius's reworking, his paratext did not advertise this as an advantage.

The same pattern held true in a rival translation of Monardes from the original Spanish, published by the Neapolitan physician Annibale Briganti in 1576. This work translated both the *Dos libros* and Clusius's Latin edition of Garcia da Orta into Italian. Briganti's decision to translate Monardes directly from Spanish but to use Clusius's highly altered version of Orta suggests that he could not access Orta's work in the original Portuguese.[22] It also placed the East Indies and West Indies on similar footing, as underscored by the unwieldy title: *On the history of simples, spices, and other things, which are brought from the Oriental Indies for use in medicine by Don Garzia Dall'Horto . . . and two other books that likewise are brought from the Occidental Indies . . . by Nicolò Monardes*.[23] As would later become common in Monardes translations, Briganti brought together Monardes and da Orta under the banner of exotic drugs from afar.[24]

Unlike Ziletti, Briganti left few hints of the reasoning behind his translation choices, but like Ziletti, he disarticulated the archive of practice. His original 1576 edition

[20] Ibid., b1v–2r.

[21] Ibid., a2r.

[22] On da Orta's text and its circulation in Europe, see especially Palmira Fontes da Costa, ed., *Medicine, Trade and Empire: Garcia de Orta's Colloquies on the Simples and Drugs of India (1563) in Context* (Farnham, Surrey, UK: Ashgate, 2015); Pardo-Tomás, "East Indies, West Indies" (cit. n. 5); Hugh Cagle, *Assembling the Tropics: Science and Medicine in Portugal's Empire, 1450–1700* (Cambridge, UK: Cambridge Univ. Press, 2018), 101–32; and da Costa, "Geographical Expansion and the Reconfiguration of Medical Authority: Garcia de Orta's Colloquies on the Simples and Drugs of India (1563)," *Studies in History and Philosophy of Science Part A*, Reconsidering the Dynamics of Reason: A Symposium in Honour of Michael Friedman, 43 (March 2012): 74–81.

[23] Unlike Ziletti, Briganti only translated the *Dos libros*, not the *Segunda parte* or the treatise on snow. Briganti's first edition appeared with the Venetian printer's collective known as *Al segno della fontana*; Garcia da Orta and Nicolás Monardes, *Dell'historia de i semplici aromati, et altre cose che vengono portate dall'Indie Orientali pertinenti all'vso della medicina, di Don Garzia Dall'Horto . . . et due altri libri parimente di quelle che si portano dall'Indie Occidentali di Nicolò Monardes*, trans. Annibale Briganti (Venice: Nella stamperia di Giouanni Salis, 1576). See also Pardo-Tomás, "Obras españolas" (cit. n. 5), esp. 71–9.

[24] Pardo-Tómas notes that Clusius increasingly portrayed Monardes's text as exotic; Pardo-Tomás, "Two Glimpses" (cit. n. 3), 185.

contained no preface, and his 1582 edition (published by Giodarno Ziletti's nephew Francesco) focused more on the overall humanist project of medicine in Renaissance Italy than on the specific works at hand. In his dedication of the revised edition to Don Ferrante de Alarcon e Mendozza, he mentioned neither Monardes nor Orta by name but noted simply that "Two other divine authors have come to light in recent times, one writing in the Spanish-Castilian language, and the other in the language of his Portuguese nation." Because "our Italy" had been "without any light from these learned doctors," Briganti held, he decided to translate the two authors "from their foreign languages to our Italian." He acknowledged Clusius's intervention but put it in a positive light, noting that the text was "reduced, but with certain lovely annotations from Carolus Clusius."[25] At the same time, he did not differentiate between his translations of Orta (mediated through Clusius's Latin) and Monardes (translated directly from the original Spanish). As Dario Del Puppo has pointed out, Briganti did not even hint at the irony of combining works from bitter rivals Spain and Portugal.[26] Instead, he depicted Orta and Monardes as two sides of the same coin, as East Indian and West Indian counterparts. The overall point was access to information on drugs from afar. He gave no information at all about his practice, or the problems inherent in the transfer of knowledge from Amerindians to Europeans.

Although Ziletti and Briganti introduced their texts very differently, they both focused on Monardes's *intellectual* value to humanist readers. Both translators portrayed themselves as transmitters of crucial new knowledge that would benefit Italian readers, but neither evinced any particular interest in the archive of practice on which this knowledge was based. Eventually, these two separate Italian translations of Monardes were united. In 1589, Francesco Ziletti's heirs published a new edition that contained Briganti's translations of Orta and Monardes, followed by Giodarno Ziletti's translation of Monardes's *Segunda parte* and the treatise on snow.[27] The appeal of publishing works on the East and West Indies together appears to have triumphed over Ziletti's insistence on direct, word-for-word translation. Drugs gleaned from the "occidental" and "oriental" Indies were lumped into the overall humanist project of collecting and transmitting learned texts. The relatively direct translations propagated Monardes's original information on Amerindian practitioners and geography, but the translators completely ignored that practice in their framing.

JOYFULL NEWES

In contrast to the two Italian translations, the near-contemporary English version of Monardes focused less on high-minded notions of knowledge and more on the new

[25] Garcia da Orta and Nicolás Monardes, *Due libri dell'historia dei semplici, aromati, et altre cose; che vengono portate dall' Indie Orientali pertinenti all'vso della medicina . . . et due altri libri parimente di quelle che si portano dall'Indie Occidentali*, trans. Annibale Briganti (Venice: Francesco Ziletti, 1582).

[26] Del Puppo similarly mentions the appeal of this "two-for-one" deal; Dario Del Puppo, "All the World Is a Book: Italian Renaissance Printing in a Global Perspective," *Textual Culture* 6 (2011): 1–22.

[27] Garcia da Orta and Nicolás Monardes, *Dell'historia de i semplici aromati, et altre cose; che vengono portate dall'Indie Orientali pertinenti all'uso della medicina*, trans. Giordano Ziletti and Annibale Briganti (Venice: Li Heredi di Francesco Ziletti, 1589). Briganti's translation of the *Dos libros* continued to hold appeal in the seventeenth century, as evinced by a manuscript copy in the British Library, Sloane MS 253, fols. 1r-63v.

drugs' commercial and corporeal promise. We know far more about the background of this translation, John Frampton's *Joyfull Newes out of the Newe Founde Worlde* (1577), than any other vernacular Monardes translation. Frampton, a seafaring merchant, had spent the 1550s trading between Bristol and Andalucía before he was detained by the Spanish Inquisition in 1561, tortured, and imprisoned. He was later released but required to remain in Spain as a penitent. In the late 1560s, he escaped and found his way back to England. There he put his newfound Spanish language skills to practical use as a translator, "to pass the tyme to some benefite of my country, and to avoyde idleness."[28] His translation of Monardes was the first of six translations of Spanish books he completed between 1577 and 1581, all of which had themes of travel or trade.[29] The *Joyfull Newes* was his inaugural effort and his most popular. The original 1577 edition, which translated Monardes's *Dos libros*, had the mundane title of *Three Bookes written in the Spanishe tonge*, but the title was changed to the more evocative *Joyfull Newes* sometime in the early days of production.[30] A revised edition in 1580 added the full *Historia medicinal*, again directly from Monardes's Spanish.[31] That edition became the most widely distributed and was republished in 1596. It represented the only direct translation, in any language, of the full *Historia medicinal.*

As in the Italian translations, Frampton emphasized the importance of knowing about the new materia medica, but his emphasis was more firmly rooted in geopolitical competition. Like Ziletti, he presented a Monardes "truly and faithfully translated into English," but he focused on commerce rather than humanist knowledge.[32] He explained that he "tooke in hande to translate" Monardes's book not only because of its good repute in Spain and elsewhere, but also because "it might bring in tyme rare profite, to my Country folkes of Englande" through the improvement of their health. He boasted that the work held "much value" in the potential of health-giving drugs that could effect "wonderful cures of sundry great diseases, that otherwise . . . were incurable." Indeed, he noted, the new herbs had become "so precious a remedie for all manner of diseases" that they were supplanting "the olde order and manner of Phisicke."[33] England needed to become familiar with these marvelous remedies, he intimated, or its people's health would suffer in comparison with that of other countries. Frampton's

[28] Nicolás Monardes, *Joyfull Newes out of the Newfound World*, trans. John Frampton (London, 1580), *3r. On Frampton, see Lawrence C. Wroth, "An Elizabethan Merchant and Man of Letters," *Huntington Library Quarterly* 17 (1954): 299–314, https://doi.org/10.2307/3816498; Donald Beecher, "The Legacy of John Frampton: Elizabethan Trader and Translator," *Renaissance Studies* 20 (2006): 320–39, https://doi.org/10.1111/j.1477-4658.2006.00190.x; and Beecher, "Nicolás Monardes, John Frampton and the Medical Wonders of the New World" (cit. n. 5).

[29] His second most popular work was a translation of Marco Polo's travels, and his others related to seafaring and trade. One of his later works may even have been used by English pirates in the Caribbean: Beecher speculates that they may have viewed Frampton's *Briefe Description of the Portes, Creekes, Bayes, and Havens of the West Indias* (1580), now lost, as a useful resource; Beecher, "Legacy of John Frampton" (cit. n. 28), 321–9.

[30] Wroth, "Elizabethan Merchant and Man of Letters" (cit. n. 28), 307.

[31] The Spanish trade embargo of Britain from 1568 to 1573, although only partially effective, may have made it difficult for Frampton to receive the *Historia medicinal*; Pauline Croft, "Trading with the Enemy 1585–1604," *Hist. J.* 32 (1989): 283.

[32] Monardes, *Joyfull Newes* (cit. n. 28), *3v; Beecher, "Nicolás Monardes, John Frampton and the Medical Wonders of the New World" (cit. n. 5), 154–7.

[33] Monardes, *Joyfull Newes* (cit. n. 28), *3r-v.

use of words like "value" and "profite" insinuated that the economic benefits also would not be negligible.

The archive of practice formed an important backdrop to Frampton's argument, but like Monardes, he deemphasized indigenous practice. He noted that the remedies had been "thoroughly and effectuously prooved and experimented" through "greate experience made in Spayne, and other Countries," with no indication of whether the "other Countries" included American regions. Instead, he quickly moved his discussion back to commerce. These "effectuously prooved" drugs, he claimed, were being "brought out of the West Indias into Spaine, and from Spaine hither into England, by such as doe daily trafficke thither," making it imperative that the English learn how to use them.[34] As Donald Beecher has pointed out, Frampton's claim of "daily trafficke" between Spain and England was highly exaggerated.[35] Nevertheless, Frampton's argument about the profit to England (in terms of both health and wealth) rested on claims of the drugs' overwhelming success in practice, challenging the established order of medical authority. In his telling, however, they had been proven most prominently in Spain, not in the lands from whence they came.

Even though Frampton's *Joyfull Newes* focused on practical uses, it emphasized European commerce and competition rather than the Amerindian origins of Monardes's medicine. This concern about trading relationships provides an important reminder that access to the marvelous new substances that Monardes described was uneven. The Spanish trade monopoly on New World drugs, the intermittent trade embargoes, and the ongoing geopolitical tensions all influenced perceptions of the new materia medica.[36] At the same time, many of these drugs eventually did trickle into Europe, and translators increasingly chose to concentrate on a few specific substances they perceived as most relevant to their readership. After the *Joyfull Newes*, the next full translation of Monardes did not come until Antoine Colin's French translation in 1602. Instead, translators began to take a modular approach to the text: they pulled out the sections on the most popular drugs and published them separately as stand-alone treatises. As a result, the drugs themselves took center stage and Monardes as the author (and authority) faded from view. As we will see, however, deemphasizing Monardes did not generally mean a recovery of indigenous practice.

FRENCH MECHOACHAN

The first translator to take this modular approach was Jacques Gohory (1520–76), a Parisian physician, lawyer, and Paracelsian who founded an intellectual space called the *Lycium philosophal Sanmarcelli* (a forerunner to the Jardin des Plantes).[37] In 1572, Gohory published a small book with treatises on tobacco and Mechoachan root. Only the latter represented a translation from Monardes; the tobacco treatise

[34] Ibid., *3v.

[35] Beecher, "Legacy of John Frampton" (cit. n. 28), 327. Patrick Wallis's study of exotic drugs in England has suggested that there was only sporadic trade before 1620; Wallis, "Exotic Drugs and English Medicine: England's Drug Trade, 1550–1800," *Soc. Hist. Med.* 25 (2011): 20–46, on 25–6.

[36] For a general overview of trade amid rising Spanish-English tensions and the revolt in the Netherlands, see Croft, "Trading with the Enemy" (cit. n. 31), 281–302.

[37] W. H. Bowen, "The Earliest Treatise on Tobacco: Jacques Gohory's 'Instruction Sur l'herbe Petum,'" *Isis* 28 (1938): 349–63. On Gohory's Paracelsian leanings, see Didier Kahn, "Le paracelsisme de Jacques Gohory,"in "*Paracelse et les siens*," special issue, *Aries* 19 (1996): 81–130.

was Gohory's own, drawn largely from Charles Estienne and Jean Liebault's popular *Maison rustique* (1567).[38] Although Monardes's glowing description of tobacco would become one of the iconic accounts of New World drugs, the plant was already well known in Europe by the time that he published his *Segunda parte* in 1569.[39] The earliest writers on tobacco were French, and Estienne and Liebault's description, subsumed in a far larger work on the aristocratic manor house, became the most widely distributed. Their entry on tobacco drew largely on the accomplishments of the French diplomat Jean Nicot, whom they credited with bringing tobacco to France—a common misconception in the sixteenth century, as W. H. Bowen has noted.[40] Gohory drew liberally from the *Maison rustique*, but he also added his own interpretation and details from his direct experience with the drug.

Ignoring the topic of the New World entirely, Gohory situated his book within a French—or, rather, Franco-Italian—context.[41] He dedicated the treatise on tobacco to an Italian duke, Gianfrancesco Carafa d'Ariano, in gratitude for the support that he and "other great learned persons of your nation" had given to Gohory's Lycium.[42] In other places, he cited well-known French and Italian physicians, referring to "our Fernel" and "your Mattioli."[43] Although he called tobacco by its Brazilian name, *petum*, he gave no explanation of this term or mention of its origin in New Spain. Instead, he placed tobacco among "other rare garden simples" known since antiquity and noted that its effects had been proven in Portugal and, thanks to Nicot, in France. He also renamed tobacco "Medicean" (*Medicée*) in honor of the powerful, Italian-born Queen Mother, Catherine de'Medici, who believed, according to Gohory, that military successes paled in comparison to a medicine that could bring health to its people.[44] Throughout the treatise, Gohory consistently called tobacco *Medicée* alongside the more common *petum*. Spain's dominance over the trade in Amerindian drugs did not even appear as a background theme in this discussion, which was situated firmly in the intellectual and political world of Franco-Italian relations. The New World, along with the Amerindian archive of practice, all but disappeared.

Gohory took a very different approach in the second half of his book, a translation of Monardes's entry on Mechoachan root from the original Spanish. Both the Spanish conquest and the indigenous archive of practice were prominent in his brief introduction to

[38] Charles Estienne and Jean Liebault, *L'agriculture et Maison Rustique*, 2nd ed. (Paris, 1567), book 2, chap. 49.

[39] The Italian physician and botanist Pietro Andrea Mattioli included it in an updated edition of his bestselling herbal in 1565; Mattioli, *Commentarii in sex libros Pedacii Dioscoridis Anazarbei De medica materia* (Venice, 1565). See also Bowen, "Earliest Treatise on Tobacco" (cit. n. 37), 352; Sarah Dickson, *Panacea or Precious Bane: Tobacco in Sixteenth-Century Literature* (New York: Public Library, 1954); and Grace G. Stewart, "A History of the Medicinal Use of Tobacco 1492–1860," *Med. Hist.* 11 (1967): 228–68.

[40] The explorer André Thevet was the first Frenchman to describe tobacco, but Nicot brought it to the French court and nearly universally received credit for it in the sixteenth century; Bowen, "Earliest Treatise on Tobacco" (cit. n. 37), 351–2.

[41] Gohory spoke Italian and made several translations from Italian to French, including a well-known translation of Machiavelli; Kahn, "Le paracelsisme de Jacques Gohory" (cit. n. 37), 81; Bowen, "Earliest Treatise on Tobacco" (cit. n. 37), 349n1.

[42] Jacques Gohory, *Instruction sur l'herbe petum ditte en France l'herbe de la Royne ou Medicée: Et sur la racine Mechiocan principalement* (Paris, 1572), part 1, 3r, http://www.biusante.parisdescartes.fr/histoire/medica/resultats/index.php?cote=extbmlyon357581&do=chapitre.

[43] Ibid., part 1, 2r.

[44] Gohory claimed he had wanted to call the herb "Catherinaire," but Catherine herself had modestly signaled that she preferred a term honoring past Medicean princes; ibid., part 1, 3r.

the work, which drew from Monardes's first paragraphs on the root. He mentioned Cortez's conquest of Mechoachan province in 1524 and explained that the Spanish had named the root after the region, but that "her Indians" called it Chincicila. Following Monardes (without attribution), he maintained that along with the riches of gold and silver, the Mechoachan region also had "good air" that produced "herbs of great virtue in combatting diverse maladies." These herbs, he related, attracted "all of the nearby Indian peoples" because of their efficacy.[45] He opened the body of the treatise with Monardes's extensive anecdote depicting the transfer of knowledge from the Amerindians to the Spanish. When the guardian of a Franciscan monastery in Mechoachan province fell gravely ill, an Amerindian seigneur named Casique, lord of Caçoncin, brought "an Indian of his who was a doctor" to treat the Franciscan with powdered Mechoachan root. He recovered his health, and afterward "many Spanish who were sick" found the root so effective that they brought it with them to Mexico, and thence to Spain.[46] Clusius left this transmission narrative out of his Latin translation, and Gohory truncated Monardes's account of Mechoachan in many other places (for example, he omitted Monardes's long anecdote about the Spanish Franciscans' spread through Mechoachan province), so it is significant that he included this story. Gohory had depicted tobacco as entirely French, but he made it clear that Mechoachan originated with indigenous healers.

Monardes himself barely appeared. Gohory did not mention him by name on the book's frontispiece, and the title page to the Mechoachan treatise described it only as a work translated from Spanish to French by Gohory. A panegyric sonnet situated directly after the title page praised Gohory, with no mention of Monardes, and Gohory's preface only briefly mentioned "Dr. Monardis" without identifying him as the author. Monardes's role as author was noted only on the first page of the actual treatise.[47] Even the manner in which Gohory included Monardes's first-person anecdote about his successful use of Mechoachan on patients made it unclear whether the experience came from Monardes or Gohory himself. Unlike Ziletti, who specifically insisted on giving Monardes his due, Gohory nearly erased the author he was translating. He also diluted the Spaniard's authority. At the end of the treatise, he added an excerpt on Mechoachan from the Latin herbal of Mathias de l'Obel and Pierre Pena, the Flemish and French botanists, giving them the last word on the root.

Why highlight the indigenous archive of practice but downplay the physician who had introduced copies of that archive to Europe? Put together, Gohory's translation strategies suggest an attempt to deemphasize the Spanish influence over New World drugs in favor of that of the broader European humanist community. Hie emphasis on Mechoachan's Amerindian origins could plausibly be seen as part of that strategy, as it made the Spanish passive recipients in the discovery of the root's medicinal properties. Gohory clearly felt a loyalty to a broader European intellectual community that he depicted as centered on France and Italy, not Spain. In that sense, he plugged into the community of humanist readers that included Ziletti and Briganti. Yet unlike the Italians, he focused on the practical application of specific American substances, a purgative (Mechoachan) and a near cure-all (tobacco). Indigenous practice remained highly mediated in his account, and like Monardes, Gohory assimilated Mechoachan

[45] Ibid., part 2, 3–4.

[46] Monardes, *Historia medicinal* (cit. n. 11), fol. 29r-v; Gohory, *Instruction sur l'herbe petum* (cit. n. 42), part 2, 5–6.

[47] Gohory, *Instruction sur l'herbe petum* (cit. n. 42), part 2, 1–5.

into a European framework by calling it the "rhubarb of the Indies." He nevertheless saw value in identifying the Amerindian archive of practice, not the Spanish colonizers, as the source of the marvelous root.

DUTCH TOBACCO

Not all translators were so set on editing out the Spanish. In 1580, three separate modular translations of Monardes were completed in the Netherlands and the Holy Roman Empire, a sign of the growing popularity of specific New World drugs.[48] A Dutch version of Monardes's entry on tobacco, translated by Nicolaes Jansz vander Woudt and published by Jan van Waesberghe in Rotterdam, emphasized the "wondrous powers and strengths" of this drug from the "West Indies." Vander Woudt identified himself as a collector of "flowers and herbs" who had received Monardes's Spanish book and decided to translate it "from the Spanish language into our Netherlendish."[49] His motivation for translating, he noted, was the recent introduction of tobacco into the region.

Even more than the previous vernacular translations, this treatise appeared at a politically sensitive time and place. The city of Rotterdam had declared itself on the side of the Dutch Revolt against Spain in 1573, and by 1580 the northern provinces were close to breaking away from Spanish rule. The printer Jan van Waesberghe and his family had recently fled Antwerp for Rotterdam owing to their Reformed leanings.[50] Vander Woudt made explicit attempts to avoid discussing the obvious political undertones in his decision to translate a Spanish text. He called attention to the interest in tobacco among various European and Dutch intellectuals, and he insisted that his only motivation for publishing the treatise was to inform Dutch speakers about the valuable drug, for their own benefit. Without his intervention, he argued, he was afraid the drug unwittingly would be "rejected or even misused." The issue of use in practice stood at the forefront of his efforts.[51]

Whatever his motive, vander Woudt did not show any particular enmity against the Spaniards. In addition to including the lengthy title of the *Historia medicinal* in Spanish, he translated Monardes's original dedication to the Spanish king.[52] His brief preface mainly gave an overview of the various names for tobacco, including the Nahuatl name *picielt*, but Spanish imperialism directly informed this discussion. Unlike Gohory, he placed the credit for discovering tobacco squarely on the Spanish, emphasized it as a drug from New Spain, and placed Spain at the center of the European intellectual community. He named several Spanish physicians who had recommended tobacco and thanked a Doctor Aguilera for sending him the Monardes text. Despite the strong anti-Spanish

[48] Harold J. Cook, *Matters of Exchange: Commerce, Medicine, and Science in the Dutch Golden Age* (New Haven, CT: Yale Univ. Press, 2007); Cook and Dupré, *Translating Knowledge* (cit. n. 3); Sven Dupré and Christoph Lüthy, eds., *Silent Messengers: The Circulation of Material Objects of Knowledge in the Early Modern Low Countries*, Low Countries Studies on the Circulation of Natural Knowledge 1 (Münster: LIT Verlag, 2011).

[49] Nicolás Monardes, *Beschrijvinge van het heerlijcke en vermaerde kruydt, wassende in de West Indien aldaer ghenaemt Picielt*, trans. Nicolaes Jansz vander Woudt (Rotterdam: Jan van Waesberghe, 1580), A2r.

[50] *Lexikon des gesamten Buchwesens Online* (2017), s.v. "Waesberghe, van," accessed April 15, 2021, https://doi.org/10.1163/9789004337862__COM_230012.

[51] Monardes, *Beschrijvinge van het heerlijcke en vermaerde kruydt* (cit n. 49), A2v.

[52] Ibid., A1v.

sentiment of Rotterdam and the van Waesberghe press, vander Woudt depicted tobacco as a product of Spanish colonialism.

At the same time, vander Woudt put special emphasis on the use (and usefulness) of tobacco, and for that purpose, the archive of practice was crucial. His frontispiece contained the striking image of an African boy smoking a tobacco pipe, one of the earliest depictions of that practice and a hint of the ties between tobacco and the African slave trade (see Benjamin Breen's essay in this volume[53]). The opening lines of his translation included Monardes's note that tobacco was "well known among the Indians, particularly those of New Spain," who taught the Spanish how to use it for curing wounds and other ailments. He followed Monardes's text very closely, including the extensive information on how the Amerindians used tobacco. While most of these descriptions were accounts of healing, he also included an anecdote on the religio-spiritual use of tobacco to provoke visions. Monardes claimed these visions were inspired by "el Demonio," which vander Woudt translated as "Satan" (*den Sathan*).[54] Like Monardes, then, vander Woudt undermined the archive of practice, even as he gave it credence.

As has been well documented in previous historical scholarship, tobacco became a major focus for the growing Dutch East India Company (VOC) in the seventeenth century.[55] Vander Woudt's treatise has received little attention, but it hints that there was already significant Dutch interest in tobacco by the second half of the sixteenth century, even before Gilles Everaert published his influential treatise on tobacco, *De herba panacea* (1589). For vander Woudt, the use of tobacco among Amerindians and African slaves provided evidence of its benefits for Europeans.

GERMAN SASSAFRAS

If tobacco represented a particular interest of the Dutch, German proclivities centered on a different commodity: sassafras. The first printed German translation of Monardes appeared in a short, anonymous work on sassafras published in Vienna in 1580 by the imperial printer Michael Apfel. In his preface, Apfel explained he had received an account of sassafras "so poorly and unintelligibly translated from Spanish into German" that he had to ask a physician to correct it. The unnamed physician had agreed, with the caveat that it would be far better if he could have compared it to the (unavailable) Spanish original. In the end, demand for knowledge of sassafras was so great that Apfel decided to print the corrected treatise anyway and hope for the best.[56]

Apfel was invested in providing local access to valuable information and mostly ignored the broader geopolitics. His preface focused on expertise, and he situated the learned physician as the unquestionable authority. In so doing, he alluded to an archive of practice, but one based on empirical use by nonphysicians in Vienna rather

[53] Benjamin Breen,"Where There's Smoke, There's Fire," in *Osiris* 37.

[54] Monardes, *Historia medicinal* (cit. n. 11), 47v; Monardes, *Beschrijvinge van het heerlijcke en vermaerde kruydt* (cit. n. 49), B2v.

[55] See especially Marcy Norton, *Sacred Gifts, Profane Pleasures: A History of Tobacco and Chocolate in the Atlantic World* (Ithaca, NY: Cornell Univ. Press, 2008); and Dickson, *Panacea or Precious Bane* (cit. n. 39). A helpful summary of tobacco use in Europe, and in the Dutch Empire in particular, can be found in Ina Baghdiantz McCabe, *A History of Global Consumption: 1500–1800* (New York: Routledge, 2014), 70–7. On Dutch trade in the West Indies, see Cook, *Matters of Exchange* (cit. n. 48), esp. 210–25.

[56] Nicolás Monardes, *Description oder Beschreibung des Holtzes Sassafras* (Vienna: Michael Apfel, 1580), A2r-v.

than Amerindians. He praised sassafras as a "versatile and useful" drug, but he also warned his readers against overuse. Sassafras was not good for every illness, and it should not be used in the same way for every complaint. On the contrary, readers should tailor its use to their humoral complexion and to the illnesses for which it was best suited, with the advice of a learned physician.[57] These warnings suggest that sassafras was already widely available in Vienna, and that Apfel felt it necessary to stage an intervention on behalf of learned medicine. This translation was intended to address existing Viennese practices which, in his opinion, needed improvement.

In the body of the treatise, the anonymous physician prefaced his translation with a few paragraphs portraying Monardes as the true expert on sassafras. He addressed the European fascination with "herbs and plants brought from far away," asserting that "people claim a lot, and one knows little truthfully of them except from people who have experimented or experienced them and tried them with special care and diligence."[58] Those diligent people, he made clear, were learned physicians, not indigenous practitioners or European empirics. As Alix Cooper and Christine R. Johnson have shown, there was a general concern about foreign and exotic drugs in German-speaking regions in the sixteenth century.[59] The Vienna translator gave a nod toward this skepticism, but he argued that firsthand experience from trustworthy physicians—especially Monardes—could help identify and verify useful substances. He obliquely referenced the indigenous archive of practice but once again identified European physicians as the authorities of that archive.

Indeed, the translator situated Monardes as an expert mediator of knowledge from the New World, explaining that "Doctor Monardes" had received sassafras wood and heard many accounts of its success and had written down "what he had discovered and seen through experience."[60] Monardes's text (both the original and the translation) explained that the knowledge of sassafras had originally come from indigenous peoples in Florida, who had helped French invaders when they had fallen ill with fevers. When the Spanish drove out the French, some remaining Frenchmen had passed on the knowledge to the Spanish. The instructions handed on from the Amerindians, Monardes noted, were vague and did not give specific quantities, as "the Indians have no weight or measure . . . and do no more than throw a piece of wood (in the water)," a point that the Viennese translator emphasized as the "Indian way to use this wood."[61] Later in the text, Monardes gave more specific measurements and explained how people of every humoral complexion should use the wood—which, the translator editorialized, was the proper method. Like the Dutch tobacco treatise, the Vienna sassafras pamphlet emphasized use in practice, but the translator made it clear that, in his view, the original Amerindian knowledge was insufficient.

The Vienna translation was not the only attempt to bring Monardes's entry on sassafras into German. In the exact same year, 1580, the Heidelberg physician Johannes Strupp penned his own translation of Monardes's account of sassafras, titled "The

[57] Ibid., A4r-v.

[58] Ibid., B1r-v.

[59] Alix Cooper, *Inventing the Indigenous: Local Knowledge and Natural History in Early Modern Europe* (Cambridge, UK: Cambridge Univ. Press, 2007), chap. 1; Christine R. Johnson, *The German Discovery of the World: Renaissance Encounters with the Strange and Marvelous* (Charlottesville: Univ. of Virginia Press, 2008), chap. 4.

[60] Monardes, *Description oder Beschreibung des Holtzes Sassafras* (cit. n. 56), B1r-v.

[61] Monardes, *Historia medicinal* (cit. n. 11), 52–3; Monardes, *Description oder Beschreibung des Holtzes Sassafras* (cit. n. 56), B2r-v.

new Indian tree, *pauame*, called sassafras in Spanish and French," in a manuscript belonging to the Palatine princely library.[62] Strupp identified himself as the son of Joachim Strupp, the Palatine court physician, and explained that he had translated the text from Clusius's Latin "on the command of the Christian nobility," that is, the Palatine princes. There appears to be no relationship between Strupp's sassafras and the Vienna translation. Not only did the source texts differ, the focus also diverged. Strupp placed a far greater emphasis on the tree's Amerindian origin and use, and he gave a completely distorted view of Monardes, whom he identified as French. He began his treatise by noting, "There is a land called Florida in the New World, with an elevation of 25° latitude . . . from which a special kind of wood was brought into Spain and described by a Frenchman, Nicolao Monardis, D[octor]" who had pinpointed the wood's many "wondrous virtues and powers." These virtues, he noted had been "tested and used" by "the Frenchman himself, along with many of his countrymen."[63] Strupp's short translation emphasized the wood's American origin, the validation of European experimentation, and the many conditions for which it was useful.

Strupp presented an idyllic picture of the Floridian landscape, describing how sassafras groves emitted a sweet-smelling aroma. He noted that the French and Spanish called it *sassafras* instead of the Floridian name of the tree, *pauame*, but "the reason for this is still unknown to us." In Strupp's telling, French and Spanish soldiers had learned how to use sassafras after they became sick from "drinking raw water and sleeping under the heavens," and "the inhabitants of this land Florida" had alerted them to the usefulness of the wood.[64] Significantly, Strupp did not repeat the claim that Monardes had improved and quantified the Amerindian method. With a narrow audience of the Palatine counts, his treatise focused on the practical use of the new drug. He included a numbered list of thirteen conditions it could cure, followed by several sections on how to administer it. Those sections followed Clusius closely and placed sassafras firmly within a Galenic medical framework.[65] The initial nod to Amerindian knowledge soon disappeared, as did the figure of Monardes: instead, the focus was on the drug's potential at the Palatine court.

The emergence of two independent sassafras treatises in 1580 likely points to a broader interest in (and availability of) sassafras in the Holy Roman Empire, a trend that Holly Dugan has also noted for early modern England.[66] The independent translations by Apfel and Strupp, completed contemporaneously, suggest that German physicians had begun to realize the impact of sassafras and felt it prudent to assert the physician's expertise over the popular new substance. The archive of practice provided crucial evidence of efficacy, but physicians were already engaged in a longstanding struggle with European empirical practitioners and had little motivation to devolve authority to Amerindian healers. While Strupp's interventions were subtle, Apfel was more explicit about his choice to disarticulate the indigenous archive.

[62] The excerpt is bound with a separate translation of Monardes's treatise on poison antidotes; Universitätsbibliothek Heidelberg, Germany, Cod. Pal. germ. 501, fols. 65r-69v.

[63] Ibid., fols. 65r-v.

[64] Ibid., fols. 65v-66r.

[65] Ibid., fols. 67r-69v.

[66] Holly Dugan, *The Ephemeral History of Perfume: Scent and Sense in Early Modern England* (Baltimore, MD.: Johns Hopkins Univ. Press, 2011), chap. 3.

"THE INDIAN WAY AND OUR WAY"

The German physician Johann Wittich, in contrast, chose a different tactic. Wittich's two books from 1589, as mentioned in the introduction, presented divergent views of Monardes and the indigenous archive of practice, both from Clusius's Latin. The first, *On Guaiac Wood*, was a short pamphlet containing excerpts on New World medicinal trees, including guaiac, West Indian China, sarsaparilla, and sassafras. The second, *On the Wondrous Bezoar Stone*, was a larger treatise on poison antidotes and wonder drugs from the New World, Asia, and Europe, which incorporated accounts from Monardes along with those of many other authors, especially da Orta's work on the materia medica of the Indian subcontinent.[67] Both treatises became reasonably popular and went through several editions.[68] The two books, however, differed starkly in both their intended audience and their approach to Monardes's text.

The guaiac treatise drew mainly on Monardes (via Clusius), save for a brief excerpt from the French explorer André Thévet on the Brazilian hyuourahe tree. Like Gohory, however, Wittich did not mention Monardes until the beginning of the treatise itself. The title page explicitly listed the author as "Johann Wittich, Doctor of Arnstadt." Instead of a translation of Monardes, the book advertised itself as a treatise on exotic medicinal trees, authored by Wittich and intended for the use of "all surgeons and barbers."[69] He dedicated the treatise to seven surgeons and barbers residing in four closely clustered Thuringian cities, whom he characterized as "good friends and dear brothers."[70] Because few barbers and surgeons read Latin, the treatise translated relevant portions of Monardes for practitioners who would not otherwise have had access to his text.[71] Rather than portraying these drugs as generally useful substances, as in the two German sassafras treatises, Wittich described his pamphlet as a discourse on medicines "for curing the unchaste diseases," generally the purview of surgeons and barbers.[72]

Wittich particularly emphasized the bountiful global variety of materia medica. He boasted that one could find "many glorious herbs" in the "mountains, gardens, and places of the German lands," but he also noted that "there is such a wondrous variety of trees, shrubs, and herbs in all lands," which had substantial "usefulness for all sorts of illnesses." Moreover, he continued, "there is such a diversity of how the Indians, Italians, French and Germans prepare and use such [herbs] for all sorts of ailments." Because guaiac wood had become the standard remedy for the "unchaste pox" and had recently been joined by other "Indian" items such as China wood, sarsaparilla, and sassafras, he wanted to introduce the way that "the Indians . . . prepare these [things] . . . which accomplishes more than we Germans." And thus, he continued, "you can compare both the Indian way and our way of preparation side by side and judge from it what would

[67] Wittich, *Ligno guayaco* (cit. n. 1); Wittich, *Bericht von den wunderbaren bezoardischen Steinen* (cit. n. 1). On Wittich, see Klaus Hafemann, *Magister Johann Wittich (1537–1596)* (Würzburg: Universität Würzburg Georg-Sticker-Institut für Geschichte der Medizin, 1956).

[68] The *Ligno guayaco* was republished in 1592 and 1603, while the bezoar treatise went through new editions in 1592, 1601, and 1612.

[69] Wittich, *Ligno guayaco* (cit. n. 1), title page.

[70] Ibid., b1v.

[71] Wittich does not appear to have been familiar with either previous German treatise on sassafras.

[72] On the use of New World drugs against the French disease and the tensions between physicians and surgeons, see Claudia Stein, *Negotiating the French Pox in Early Modern Germany*, History of Medicine in Context (Farnham, Surrey, UK: Ashgate, 2009).

be the most useful for your patients."[73] Alone among the translators of Monardes, he put a spotlight on the Amerindian archive of practice and posited the indigenous preparation method as potentially more valuable than the European version.

Yet even this characterization was a façade. In the end, Wittich focused very little on indigenous methods and very much on German applications. He attempted to make the New World methods accessible to German barber-surgeons by translating all apothecary measurements into common German units of measure.[74] However, he referred generally to the "Indians" without giving any sense of specific geography, and he frequently overlooked the substantial differences between Amerindian and Spanish practices. Following a long explanation of the traditional way to prepare and use guaiac wood, for example, he noted that "this is the way the Indians and the Spanish thoroughly expel and drive away the French Disease." He then added a separate "common" method of preparation in Germany.[75] Wittich thus gave a vague sense of foreign and exotic practices, combining them into one and contrasting them with German methods.

Wittich also added his own anecdotes. Most strikingly, he included a very personal observation on Monardes's claim that sassafras worked wonderfully against gout, "which I, M[agister] Wittich, can truthfully attest and praise, for on the 17th of March in the year 1587, I was struck by a very painful gout, and I lay ill day and night for six weeks and could not move my entire body." After trying some other medicines, he said, "I had this decoction prepared, and drank it fifteen mornings in a row, nice and warm, from which I sweated and (God be praised) felt a definite improvement." Not only that, he continued, he regained a hearty appetite after feeling virtually no hunger during his illness, "and I also by nature have a weak, cold, and evil stomach, which is always full of wind." The sassafras decoction had "gently warmed the whole stomach, promoted digestion, violently awakened the appetite for food, stilled the gurgling, and also dried out other fluxes that were burdensome to me, so that apart from God, I have much to thank this wood for, and from my own personal view cannot praise it enough."[76] Even more than in the 1580 German works on sassafras, Wittich focused on the wood's use in practice locally, including on his own body. Despite opening the door to Amerindian practice, his most effusive evidence came from Monardes, backed up by his personal experience.

Accordingly, Wittich's main additions to the text were recipes: how Germans could practically make and administer the various New World cures. He answered the question of Germanness vs. foreignness with the practical use of the foreign drugs—which he essentially Germanized, despite his initial nod to the "Indian way." Although he framed his treatise as a friendly gesture toward barbers and surgeons, he thereby asserted his own authority over substances deriving from an indigenous archive of practice and intended for use by European empirical healers.

THE EXOTIC MONARDES

As the title itself suggested, Wittich's second Monardes translation, *On the Wondrous Bezoar Stone*, was a very different type of book. Rather than a small pamphlet aimed

[73] Wittich, *Ligno guayaco* (cit. n. 1), a3r-v. Anna Winterbottom has shown the very interesting ways that China root circulated as a syphilis drug in early modern Europe; Anna E. Winterbottom, "Of the China Root: A Case Study of the Early Modern Circulation of Materia Medica," *Soc. Hist. Med.* 28 (2015): 22–44, https://doi.org/10.1093/shm/hku068.

[74] See, e.g., a recipe for preparing China wood on c3v.

[75] Wittich, *Ligno guayaco* (cit. n. 1), c1r-v.

[76] Ibid., e4v.

at surgeons, it was a much larger work on a variety of near-miraculous poison antidotes aimed at a broader German reading public. The title page advertised descriptions of a variety of wondrous strange plants, animal products, and minerals from all the world. Although it contained several local German remedies, it especially highlighted drugs from the East and West Indies, and the title page advertised it as a translation of both Monardes and da Orta.[77] Wittich's main focus in this work, as the title indicates, was wonder, and he made little attempt to separate the various exotic places from one another, or from wonders that came from nearby. The dedicatory preface, addressed to three advisors to Elector Christian I of Saxony, noted that "many wonderfully powerful things have been brought to light in these past thirty years, about which the ancient writers knew little or nothing."[78] In contrast to his *On Guaiac Wood*, which specifically engaged with the question of German drugs and exotic imports, he framed this work as a contrast between the ancients and the moderns, a comparison in which the moderns had the definite advantage. A display of humble German herbs this was not.

If the focus was wondrous, these wonders came with the validation of experience. The very beginning of Wittich's dedicatory preface noted that powerful people needed "powerful and artful medicines that have been proven adequately in a trial [*proba*]."[79] His treatise strived to fill that gap: many of the remedies he listed included explanations of situations in which they had been tried. Unlike in the Vienna sassafras treatise, this experience was not restricted to learned physicians. Many of Wittich's examples came from Monardes's claims of observing the efficacy of a given New World drug, but he interspersed these instances with examples from other authors and from his own contacts' experience in practice. In writing about the use of jade as an amulet to help against kidney pain, he stated that Monardes had observed its success on a nobleman and that the duchess of Bavaria had set the stone in a golden armband and worn it every day for ten years against kidney pain. He added that the stone was hard to come by and was used mainly by princes and kings, but that Countess Catherine of Schwarzburg told him one could get it from the Portuguese in Arnstein. She had obtained one for her husband, and it had worked well.[80]

This pattern continues throughout the book—the explanation of a drug, often (but not always) from Monardes, followed by the description of its successes, in many cases giving specific names of people Wittich knew who could attest to it and, in some cases, where in Germany one could buy it. This wide-ranging discourse on experience included numerous references to indigenous practitioners. For example, the description of bezoar included a long description of "how the Indians use this stone" against poison, the bites of poisonous animals, and melancholic diseases, although Wittich did not indicate whether the "Indians" referred to inhabitants of the East Indies or West Indies.[81] Despite his inclusion of indigenous knowledge, his understanding of non-European places, peoples, and practices was hazy at best. He described Peru as an island, frequently mixed up the East and West Indies, and gave very little sense that the place from whence a drug came mattered. He only rarely cited his

[77] Clusius had printed an edition of da Orta in 1582; Garcia da Orta, *Aliquot notae in Garciae Aromatum Historiam*, trans. Carolus Clusius (Antwerp, 1582).
[78] Wittich, *Bericht* (cit. n. 1),)a(2r.
[79] Ibid.,)a(3r.
[80] Ibid., 23–35.
[81] Ibid., A3r.

sources, including information from Orta and Monardes interchangeably. In short, Wittich completely repackaged Monardes in ways that he felt could be useful to elite Germans, while putting the focus on the exotic origin and occult powers of various drugs. It was less a translation of Monardes than an interpretation of Clusius. While experience was central to Wittich's text, Amerindian practitioners blurred into an indistinct impression of the exotic.

Although his complete dismantling of Monardes was somewhat unusual, Wittich's portrayal of foreign drugs as amorphously exotic had been a common thread from the very first translations. Monardes himself often compared American drugs to substances from the East Indies, and translations of Monardes and da Orta frequently were bound together. The trend towards "exoticizing" Monardes appears to have only strengthened in the early seventeenth century. In 1602, the French apothecary Antoine Colin published the first complete translation of Monardes in several decades, from Clusius's Latin, as part of a larger work that also included Clusius's translations of Acosta and da Orta.[82] Clusius himself published a new version of Monardes in 1605, a large Latin volume titled *Exoticorum libri decem* (Ten books of exotica), which merged excerpts from Monardes, da Orta, and other authors under the general theme of exotic drugs. His approach was very much in the vein of Wittich's volume, but in Latin and interspersed with images. He also, however, included an index of decidedly non-exotic European plants that he found useful.[83] In contrast to the modular translations, which focused on the benefits of particular drugs in a particular setting, these works revealed an increasing tendency to put Monardes's works under the general heading of exotic drugs from afar, often in combination with other authors. The original texts and authors faded into the background, and the Amerindian origins of these substances barely surfaced. Instead, the drugs themselves took center stage.

CONCLUSION

In his essay on translation in early modern Europe, Michael Wintroub argues that "practices of translation oscillated between strategies for assimilation and strategies for domination and rule."[84] The Monardes translations were a little bit of both—certainly they aimed first and foremost at assimilating the knowledge coming from the New World, but the translators also alluded to broader trends in Europe: political rivalries, trade networks, questions of physicians' authority, the fascination with exotica, and the reality of some of these drugs entering the European pharmacopoeia. The practical interest in New World plants and trees that appeared in Northern European translations from the 1580s likely reflected the increasing availability of these substances at local pharmacies.[85] Monardes thus functioned on multiple levels, ranging

[82] Antoine Colin, *Histoire des drogues, espiceries, et de certains medicamens simples, qui naissent és Indes & en l'Amerique, divisé en deux parties. La premiere comprise en quatre livres*, 2nd ed. (Lyon, 1619).

[83] Florike Egmond, "Figuring Exotic Nature in Sixteenth-Century Europe: Garcia Da Orta and Carolus Clusius," in da Costa, *Medicine, Trade and Empire* (cit. n. 22), 167–94; Pardo-Tomás, "Two Glimpses" (cit. n. 3), 185.

[84] Michael Wintroub, "Translations: Words, Things, Going Native, and Staying True," *Amer. Hist. Rev.* 120 (2015): 1185–1217, on 1190.

[85] For example, in Wittenberg in 1599, sassafras cost 1½ gulden, sarsaparilla cost 1 gulden, and guaiac cost 2 dinari; *Taxa oder wiederung aller Materialien, so in den Apotheken zu Wittenberg verkaufft wurden* (Wittenberg, 1599).

from a demonstration of extensive new knowledge from the New World, to a representation of physicians' authority, to a useful guide for unfamiliar medicaments. Especially for vernacular translations, bringing Monardes into the local idiom became a highly charged process, not just an intellectual exercise. Translators frequently touted themselves as crucial figures in this hyperlocalizing of New World drugs, crowding out Monardes and obliterating Amerindian practices.

In this sense, the Monardes translations were very different from the pathways of another popular book on materia medica, the Italian physician Pietro Andrea Mattioli's *Commentaries on Dioscorides*. First published in Italian in 1544, Mattioli's work appeared in Latin, French, Czech, and German.[86] Like Monardes, Mattioli added pieces to his works over time, constantly updating them. Like Monardes, moreover, his work contained individual entries on various substances (including some New World plants like tobacco), which, in theory, could have been separated and republished on their own. But this did not happen. Unlike the many versions of Monardes's texts, the translations of Mattioli had very little variation. Nearly all of them were attempts to translate the author's text directly "from language into language," to use Ziletti's phrasing, and they all occurred with his encouragement and permission. "The Mattioli" became a paradigmatic text.[87] "Monardes," in contrast, became a pliable concept that could be molded to fit a specific linguistic context.

This difference underscores how central the indigenous archive of practice was to the trajectory of Monardes, even if translators often erased it. Mattioli's work rested on the ancient authority of Dioscorides, although he eventually added entries on New World drugs. In contrast, Monardes's books described novel commodities discovered through experience (and domination over indigenous Amerindians).[88] Translators accessed these new substances in variable ways, sometimes assimilating them into the humanist tradition, sometimes using them to assert professional authority, sometimes even explicitly trying (and failing) to access indigenous Amerindian knowledge. This malleability and modularity allowed translators to either emphasize or deemphasize Monardes as author, to foreground successful drug therapies, or to highlight commodities for their specific linguistic communities. Because the primary sources were Amerindian practice, Monardes himself became a vector for translators to use as they saw fit.

[86] Pietro Andrea Mattioli, *Di Pedacio Dioscoride Anazarbeo libri cinque della historia, & materia medicinale* (Venice, 1544); Mattioli, *Commentarii* (cit. n. 39); Mattioli, *Commentaires de M. Pierre André Matthiole . . . svr les six livres de Ped. Dioscoride* (Lyon, 1572); Mattioli, *New Kreüterbuch*, trans. Georg Handsch (Prague, 1563); Mattioli, *Herbarz: ginak Bylinář welmi vžitečný a Figůrami*, trans. Thaddeus Hajek (Prague, 1562).

[87] See especially Paula Findlen, "The Formation of a Scientific Community in Sixteenth-Century Italy," in *Natural Particulars: Nature and the Disciplines in Renaissance Europe*, ed. Anthony Grafton and Nancy G. Siraisi (Cambridge, MA: MIT Press, 1999), 369–400, esp. 375–7; and Sara Ferri, ed., *Pietro Andrea Mattioli: Siena, 1501–Trento, 1578:La vita, le opere: con l'identificazione delle piante* (Ponte San Giovanni, Perugia: Quattroemme, 1997).

[88] This was particularly true in the well-known case of Amato Lusitano; Findlen, "Formation of a Scientific Community" (cit. n. 87), 385–8.

When the Tallamys Met John French: Translating, Printing, and Reading *The Art of Distillation*

by Elaine Leong*

ABSTRACT

Centered on the life story of the Tallamy family's copy of John French's *The Art of Distillation* (London, 1651), this article explores translation, print, and medical reading in early modern England. It traces the adaptation and reuse of textual and practical knowledge across linguistic, geographical, gender, and spatial boundaries and shines light on the scientific labor of translators, technicians, and householders, historical actors who are so often hidden by structures of the archival record. By historically situating translation, reading, and writing practices, it joins recent calls to view each translation as an independent text shaped by new contextual settings. It concludes by offering the concept of "knowledge itineraries" as a framework for analyzing long-view connected histories of knowledge transfer across time and space.

In 1736, Rebecca Tallamy started a recipe collection. After inscribing her name and the title "Book of Stilling & Reccepts," she diligently gathered and wrote down medical and culinary recipes in the thick black leather-bound book. Like many householders of the time, Rebecca collected know-how from family and friends and took copious notes from contemporary printed medical books. And as was common practice, recipe collecting was a family affair.[1] However, unusually, the Tallamys did not follow

* Department of History, University College London, Gower Street, London, WC1E 6BT, UK; e.leong@ucl.ac.uk.

Situated between two book projects, this article gestated for an unusually long time. I am grateful to audiences at the AAHM, HSS, and RSA annual meetings and to seminar and workshop participants at Cambridge, Lund, New York, Paris, and Princeton for their questions and suggestions. I am especially indebted to Lauren Kassell and Pippa Carter for their challenging and helpful comments and to Laura Selle for research assistance. Finally, my deepest gratitude goes to members of the "Translating Medicine" project for their generosity, patience, and friendship. Research for this article was funded by the Max Planck Society and the Wellcome Trust (grant no. 209835/Z/17/Z).

[1] Wellcome Library, London, Western MS 4759 (all references in this article refer to the pencil foliation at the top left-hand corner of each recto page). Rebecca wrote "Rebecca Tallamy her Book of Receps" on fol. 2r and "Rebecca Tallamy Her Book 1738" on fol. 12r. She also wrote "Rebecca Tallamy her Book of Stilling & Reccepts 1736" on fol. 17r. Additional ownership notes by Rebecca can be found on fols. 40v and 72r. William and Patience Tallamy also signed their names on the title page. Additionally, there is a "Catalogue of Books p*er* WT: Divinity Books" dated "[17]26 July 29th" on fol. 155v. The (likely) unfinished list contains twelve entries, including, for example, Richard Sibbes, *A Heavenly Conference between Christ and Mary* (London 1654); John Flavel, *Sacramental Meditations upon Diverse*

Osiris, volume 37, 2022. © 2022 History of Science Society. All rights reserved. Published by The University of Chicago Press for the History of Science Society. https://doi.org/10.1086/719222.

the typical practice of storing their recipes in a notebook bought especially for this purpose; rather, they chose to build the family's collection in an eighty-year-old printed book: *The Art of Distillation*, written in 1651 by the physician John French (1616–57) (see fig. 1).

As advertised on the cover, *The Art of Distillation, or A Treatise of the Choisest Spagyricall Preparations Performed by Way of Distillation* contained the knowledge of "the most select Chymicall Authors of Severall Languages," know-how based on "the Authors manuall Experience," and descriptions of the "chiefest Furnaces and Vessels used by Ancient and Modern Chymists." Just in case that was not enough, it also included hundreds of recipes for various drugs and compound medicines, descriptions of diverse experiments and curiosities, anatomical knowledge, and instructions for the preparation of gold and silver.[2] The work drew heavily on French's previous experiences as a translator, particularly his work "Englishing" the *Furni novi philosophici*, a series of five German-language tracts published in Amsterdam in the 1640s. Written by the German chemist Johann Rudolf Glauber (1604–70), the *Furni novi philosophici* described a new alchemical furnace invented and sold by Glauber and offered relevant technical instruction and methods for making various iatrochemical substances. As outlined below, French's endeavors to adapt these tracts for English readers involved not just a linguistic translation but rather a reordering of the content and an expansion of the text. The result is a book organized around different kinds of medicines, much like other pharmaceutical texts and household recipe collections of the period.

With their copious notes, the Tallamy family tailored French's work to suit their needs, adapting knowledge designed to be used in an artisanal workshop to the eighteenth-century home. By personalizing the text with recipes gleaned from friends and family, they added new functions and layers of meaning to the object, utilizing it as an archive of family history and affording it social value. Yet the work of the Tallamys was not the first set of customization practices employed in the production of this object. *Those* occurred when the mid-seventeenth-century physician John French penned the *Art of Distillation* through his reading, translating, and compilation practices.

Books such as the Tallamys' handwritten compendia or French's printed *Art of Distillation* occupied a central place in the English early modern medical landscape. Seventeenth-century London saw a remarkable boom in vernacular medical printing, and book sellers stocked their shelves with books to fit every budget.[3] Titles addressed all branches of medicine, from physic to surgery to pharmacy, and were designed to aid readers from all walks of life with their everyday health practices. Householders

Places of Scripture (London, 1679) and *Touchstone of Sincerity* (London, 1679); and Nicholas Byfield, *The Marrow of the Oracles of God* (London, 1619). All these works appeared in multiple editions in the seventeenth and early eighteenth century, either as stand-alone entries or as part of omnibus editions.

[2] John French, *The Art of Distillation* (London, 1651), title page.

[3] See Mary F. Fissell, "The Marketplace of Print," in *The Medical Marketplace and Its Colonies c. 1450–c. 1850*, ed. Mark Jenner and Patrick Wallis (Basingstoke: Palgrave Macmillan, 2007), 108–32; Fissell, "Popular Medical Writing," in *The Oxford History of Popular Print Culture: Volume One: Cheap Print in Britain and Ireland to 1660*, ed. Joad Raymond (Oxford: Oxford Univ. Press, 2011), 417–30; Paul Slack, "Mirrors of Health and Treasures of Poor Men: The Uses of the Vernacular Medical Literature of Tudor England," in *Health, Medicine and Mortality in the Sixteenth Century*, ed. Charles Webster (Cambridge, UK: Cambridge Univ. Press, 1979), 237–73; and Elizabeth Lane Furdell, *Publishing and Medicine in Early Modern England* (Rochester, NY: Univ. of Rochester Press, 2002).

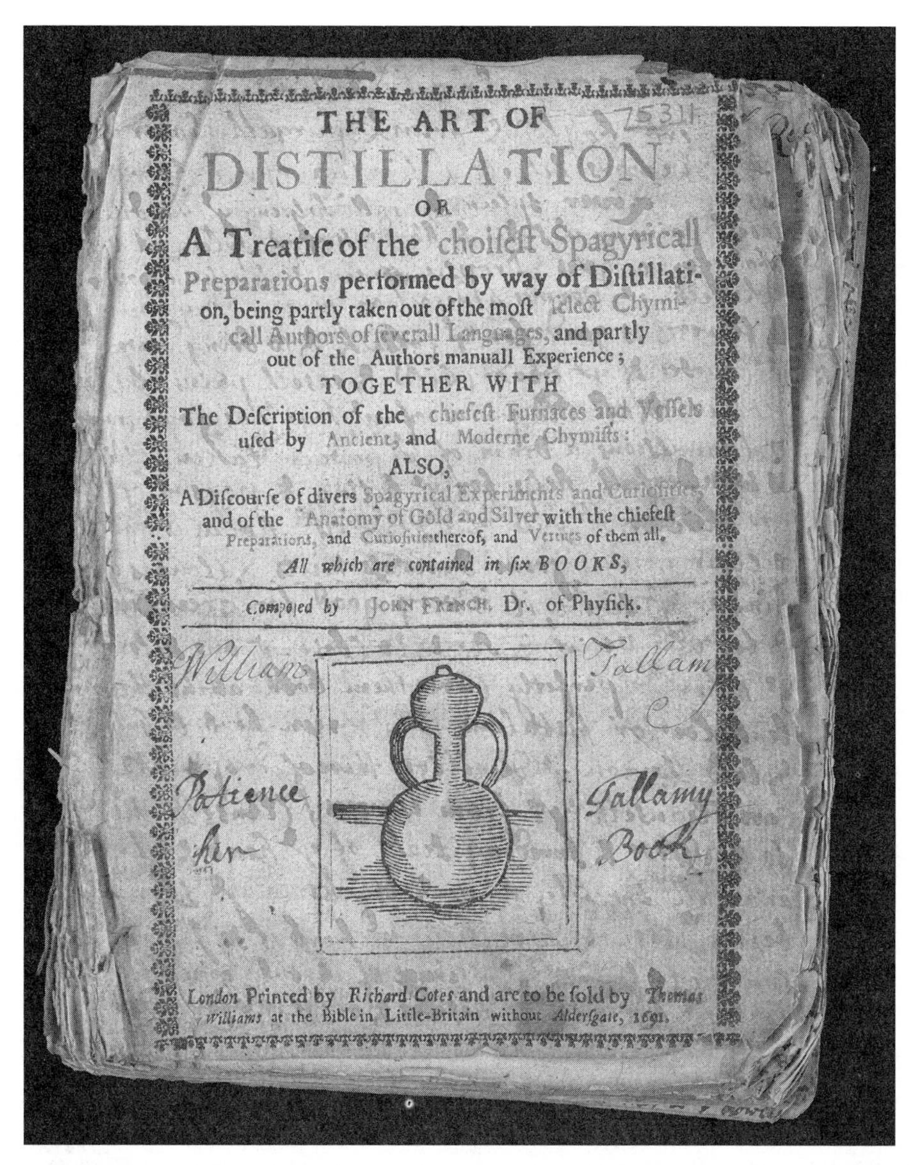

THE ART OF
DISTILLATION
OR
A Treatiſe of the choiſeſt Spagyricall Preparations performed by way of Diſtillation, being partly taken out of the moſt ſelect Chymicall Authors of ſeverall Languages, and partly out of the Authors manuall Experience;
TOGETHER WITH
The Deſcription of the chiefeſt Furnaces and Veſſels uſed by Ancient and Moderne Chymiſts:
ALSO,
A Diſcourſe of divers Spagyricall Experiments and Curioſities, and of the Anatomy of Gold and Silver with the chiefeſt Preparations, and Curioſities thereof, and Vertues of them all.
All which are contained in ſix BOOKS,
Compoſed by JOHN FRENCH, Dr. of Phyſick.

William Tallame
Patience her Tallamy Book

London Printed by *Richard Cotes* and are to be ſold by *Thomas Williams* at the Bible in Little-Britain without *Alderſgate*, 1651.

Figure 1. *Title page from the Wellcome Collection copy of John French's* Art of Distillation *(London: R. Cotes for T. Williams, 1651) with annotations written by members of the Tallamy family, including notes from works by Nicholas Culpeper. Wellcome Collection, MS 4759, fol. 1r.*

in particular avidly read the abundance of printed medical books available, and many left traces of their reading practices in margins and notebooks filled with handwritten notes. Householders' medical reading practices informed home-based medical practices *and* shaped decisions in medical encounters. Books were a crucial part of early modern medical economies.

While past studies have illuminated our understandings of medical book production, the intertwined textual practices at the core of this article still await further exploration.[4] Objects such as the Tallamy/French printed book/manuscript bring to the fore complex entanglements of translating, reading, and writing practices, shining light on the numerous changes that occur when a body of knowledge is in transit.[5] In many cases the boundaries between acts of translating, reading, and writing were flexible and continually changing.[6] Readers became translators, authors, and users, and through their own reading and hands-on practices extended the original text. By juxtaposing linguistic transfer against what we might consider appropriation or knowledge consumption, we open conversations about the utility of translation as an analytic and complicate notions of knowledge circulation.

Located in the intersection between histories of science and medicine and histories of the book and reading, this article traces the life story of the Tallamys' copy of *The Art of Distillation*. Through analysis of the knowledge practices evidenced in this one object, I shine light on the scientific labor of translators, technicians, and householders, historical actors who are so often hidden by structures of the archival record.[7] The focus on pharmaceutical processes and technologies offers an opportunity to examine the connections between translation and the practice of medicine production. As others have noted, linguistic translation aside, the transfer of practical knowledge often brings an additional layer of resistance.[8] Three points of knowledge transfer are examined in this essay. I begin by exploring the tensions, nitty-gritty practices, and multiple actors involved in producing *The Description of Philosophical Furnaces*, the English translation of Glauber's *Furni novi philosophici* that formed the basis of French's subsequent work, *The Art of Distillation*. I unpack John French's practices of compilation and assemblage in creating *The Art of Distillation* and then investigate how the Tallamy family customized a distillation manual for their home-based medical activities. My emphasis on historically situating translation, reading, and writing practices joins recent calls to view each translation as an independent text shaped by new contextual settings. Scrutinizing the practices of translation, reading, and writing in concert, I posit, enables us to better understand what "to English" meant to our historical actors. I conclude this essay by offering the concept of "knowledge itineraries"

[4] Early work on medical reading includes Mary E. Fissell, "Readers, Texts, and Contexts: Vernacular Medical Works in Early Modern England," in *The Popularization of Medicine, 1650–1850*, ed. Roy Porter (London: Routledge, 1992), 72–96; Peter Murray Jones, "Book Ownership and Lay Culture of Medicine in Tudor Cambridge," in *The Task of Healing: Medicine, Religion and Gender in Early Modern England and the Netherlands 1450–1800*, ed. Margaret Pelling and Hilary Marland (Rotterdam: Erasmus, 1996), 49–68; Jones, "Reading Medicine in Tudor Cambridge," in *The History of Medical Education in Britain*, ed. Vivian Nutton and Roy Porter (Amsterdam: Rodopi, 1995), 153–83.

[5] James A. Secord, "Knowledge in Transit," *Isis* 95 (2004): 654–72.

[6] Other essays in this volume also draw our attention to connections between translation, reading, and archive building; see Ahmed Ragab, "Translation and the Making of a Medical Archive"; Alisha Rankin, "New World Drugs and the Archive of Practice"; and Dror Weil, "Unveiling Nature"; all in *Osiris* 37.

[7] Recovering voices "lost" in our archival records is a theme running through essays in this volume; see Montserrat Cabré, "Female Authority in Translation"; Shireen Hamza, "Vernacular Languages"; and Pablo Gómez, "[Un]Muffled Histories"; all in *Osiris* 37.

[8] See, for example, Heinz Otto Sibum, "Reworking the Mechanical Value of Heat: Instruments of Precision and Gestures of Accuracy in Early Victorian England," *Studies in History and Philosophy of Science Part A* 26 (1995): 73–106; Pamela H. Smith, "In the Workshop of History: Making, Writing, and Meaning," *West 86th* 19 (2012): 4–31; Sven Dupré, "Doing It Wrong: The Translation of Artisanal Knowledge and the Codification of Error," in *The Structures of Practical Knowledge* (Cham, Switzerland: Springer, 2017), 167–88; and Thijs Hagendijk, "Learning a Craft from Books: Historical Re-Enactment of Functional Reading in Gold- and Silversmithing," *Nuncius* 33 (2018): 198–235.

as a framework for analyzing long-view connected histories of knowledge transfer across time and space.

TRANSLATING GLAUBER FOR ENGLISH READERS

In 1650, the Oxford-trained physician John French busied himself with a string of publications.[9] Within a little more than twelve months, he translated no fewer than four books on occult philosophy, alchemy, distillation, and iatrochemistry, including the works of Heinrich Cornelius Agrippa (1486–1535), Michael Sendivogius (1566–1636), Paracelsus (1493–1541), and the *Furni novi philosophici*, a series of five tracts in German by Johann Rudolf Glauber (1604–70) published in Amsterdam in the 1640s.[10] It was in this same period that French authored the work at the center of this study: *The Art of Distillation*.

As John French's list of publications demonstrates, translation from Latin and European vernaculars was a mainstay of his work as a book producer. In this, he was not alone, nor were his activities unusual. Early modern London was a sprawling metropolis and a vibrant multilingual community where French, Spanish, Italian, Dutch, German, and other immigrants rubbed shoulders along the narrow streets, exchanging ideas and knowledge.[11] This melting pot of cultures and languages fostered an active translation scene that rippled through different areas of the book world.[12] During the early years of English publishing most printed works consisted of texts translated and adapted from Latin, French, and other European vernaculars, a process often described as "to English" by contemporary book producers. As Anne Coldiron reminds us, the first book printed in English—the *Recuyell of the Hystoryes of Troye* (1473)—was a translation from Raoul Lefèvre's *Recoiel des histoires de Troie* (1473)

[9] For a biography, see Peter Elmer, "French, John (c. 1616–1657), Physician," in *Oxford Dictionary of National Biography* (Oxford University Press, 2008); online ed., January 3, 2008, https://doi.org/10.1093/ref:odnb/10164; and Charles Webster, *The Great Instauration: Science, Medicine and Reform, 1626–1660* (London: Duckworth, 1975), 279.

[10] French's translations included *Three Books of Occult Philosophy* by Cornelius Agrippa, which is a version of the *De occulta philosophia libri tres* first published in Paris in the 1530s; and *A New Light of Alchymie* by Michael Sendivogius, which was a version of his *Novum lumen chymicum* published in 1604. In English, the latter was often issued and bound with *Of the Nature of Things, Nine books*, reportedly by Paracelsus. *Of the Nature of Things*, as bibliographers have often shown, was based in part on *Dictionarium Theophrasti Paracelsi* by the German physician Gerhard Dorn and was first published in Frankfurt in the 1580s. French's translation of the *Furni novi philosophici* appeared as *A Description of New Philosophical Furnaces* (London, 1651).

[11] On multilingualism and language learning, see John Gallagher, *Learning Languages in Early Modern England* (Oxford: Oxford Univ. Press, 2019). On multilingual publishing, see Anne E. B. Coldiron, *Printers without Borders: Translation and Textuality in the Renaissance* (Cambridge, UK: Cambridge Univ. Press, 2014). On multilingualism in other European contexts, see, for example, Eric R. Dursteler, "Speaking in Tongues: Language and Communication in the Early Modern Mediterranean," *Past & Present* 217 (2012): 47–77.

[12] For translation of literature, see, for example, Peter France, ed., *The Oxford Guide to Literature in English Translation* (Oxford: Oxford Univ. Press, 2001); Fred Schurink, *Tudor Translation* (Basingstoke: Palgrave Macmillan, 2011); and S. K. Barker and Brenda M. Hosington, *Renaissance Cultural Crossroads: Translation, Print and Culture in Britain, 1473–1640* (Leiden: Brill, 2013). For history of science, see, for example, Bettina Dietz, ed., "Translating and Translations in the History of Science," special issue, *Annals of Science* 73 (2016): 117–21; Marwa Elshakry and Carla Nappi, "Translations," in *A Companion to the History of Science*, ed. Bernard Lightman (Chichester: Wiley-Blackwell, 2016); Sietske Fransen, Niall Hudson, and Karl E. Enenkel, eds., *Translating Early Modern Science* (Leiden: Brill, 2017); and Maeve Olohan, "History of Science and History of Translation: Disciplinary Commensurability?," *The Translator* 20 (2014): 9–25.

and was printed in Bruges by a bilingual printer-translator using continental printing technology, materials, and design.[13] The complexity of this "Englishing" process has been emphasized by literary scholars who argue that, oftentimes, these works were not solely translations but rather remakings of texts within specific contexts.[14] For Coldiron, to "English" in the fifteenth century involved "appropriative acculturation performed by means of verbal translation and material-textual mediation."[15] Guyda Armstrong similarly contends that the "translated book-object, as a historically situated 'container' of the text, carries its transmission history within itself."[16] The emphasis on translations as texts worthy of study in their own right has brought the crucial work of translators and book producers into the limelight, recovering the agency of the multiple actors involved in these practices.[17] These nuanced and multilayered interpretations of cultural translation offer helpful frameworks for understanding cases such as the Tallamys' reading of *The Art of Distillation*. Drawing on this rich historiography, this article takes the current narrative to the realm of medical publishing, extending our gaze to instructional texts and the various processes required to transfer technical know-how for drug production.

By the mid-seventeenth century, the bookshops near St. Paul's in London were stocking an astonishing array of English-language medical books and, crucially, many were translations from Latin or other European vernaculars.[18] For instance, one often reprinted and cited title, the *Praxis medicinae universalis; or A Generall Practise of Physicke* (London, 1598), was a translation of the Ausburg/Heidelberg physician and apothecary Christoph Wirsung's (c. 1500–71) popular *Artzney Buch* (Heidelberg, 1568). Another well-known example was John Frampton's translation of the Spanish

[13] Coldiron, *Printers without Borders* (cit. n. 11).

[14] Tania Demetriou and Rowan Cerys Tomlinson, eds., *The Culture of Translation in Early Modern England and France, 1500–1660* (Basingstoke: Palgrave Macmillan, 2015); Schurink, *Tudor Translation* (cit. n. 12).

[15] Coldiron, *Printers without Borders* (cit. n. 11), 1.

[16] Guyda Armstrong, "Translation Trajectories in Early Modern European Print Culture: The Case of Boccaccio," in *Translation and the Book Trade in Early Modern Europe*, ed. José María Pérez Fernández and Edward Wilson-Lee (Cambridge, UK: Cambridge Univ. Press, 2014), 126–44, on 126.

[17] Marie-Alice Belle and Brenda M. Hosington, "Translation, History and Print: A Model for the Study of Printed Translations in Early Modern Britain," *Translation Studies* 10 (2017): 2–21; Belle and Hosington, eds., *Thresholds of Translation: Paratexts, Print, and Cultural Exchange in Early Modern Britain (1473–1660)* (Basingstoke: Palgrave Macmillan, 2018). On translators, see, for example, Peter Burke, "Lost (and Found) in Translation: A Cultural History of Translators and Translating in Early Modern Europe," *European Review* 15 (2007): 83–94; and Andrea Rizzi, *Trust and Proof: Translators in Renaissance Print Culture* (Leiden: Brill, 2017).

[18] On medieval medical translation within the English context, see, for example, Faye M. Getz, *Healing and Society in Medieval England: A Middle English Translation of the Pharmaceutical Writings of Gilbertus Anglicus* (Madison: Univ. of Wisconsin Press, 1991); and Peter Murray Jones, "Four Middle English Translations of John of Arderne," in *Latin and Vernacular: Studies in Late-Medieval Texts and Manuscripts*, ed. A. J. Minnis, York Manuscripts Conferences, vol. 1 (Cambridge, UK: D. S. Brewer, 1989), 61–89. While Fissell, Furdell, and Slack all note the importance of translations in early modern English medical print, as yet, there are few detailed studies of medical translation in sixteenth- and seventeenth-century England; see note 4 of the present article. Recent works include Mary C. Erler, "The First English Printing of Galen: The Formation of the Company of Barber-Surgeons," *Huntington Library Quarterly* 48 (1985): 159–71; Isabelle Pantin, "John Hester's Translation of Leonardo Fiorvanti: The Literary Career of a London Distiller," in Barker and Hosington, *Renaissance Cultural Crossroads* (cit. n. 12), 159–84; and Elaine Leong, "Transformative Itineraries and Communities of Knowledge in Early Modern Europe: The Case of Lazare Rivière's The Practice of Physick," in *Civic Medicine: Physician, Polity, and Pen in Early Modern Europe*, ed. J. Andrew Mendelsohn, Annemarie Kinzelbach, and Ruth Schilling (Abingdon: Routledge, 2019), 257–79.

physician Nicolás Monardes's (1493–1588) *Historia medicinal de las cosas que se traen de nuestras Indias Occidentales* (1565) as *Joyfull Newes out of the Newe Founde Worlde* (1577), discussed in Alisha Rankin's essay in this volume.[19] By the 1650s, Wirsung's and Monardes's works sat next to the translated works of other European practitioners, from the Parisian physician Jean Fernel (1497–1558), to the German surgeon Fabricius Hildanus (1560–1634), to the French Royal apothecary Moise Charas (1619–98).[20]

Johann Rudolf Glauber, the German chemist whose *Furni novi philosophici* tracts were translated by French, first came into the purview of the English reading public through the work of the German émigré Samuel Hartlib (1600–62) and his circle of reformers. An intelligencer, reformer, and writer, Hartlib gathered around him a group of like-minded men and women who collected and made public useful knowledge as part of their schemes for the improvement of the Commonwealth.[21] Their considerable efforts to bring Glauber's works to England, as detailed below, were likely driven by these political aims. French's preface to *A Description of Philosophical Furnaces* outlines a commitment to opening access to knowledge in the name of public interest. Lamenting that it was a "pitty such useful and so learned writings should be obscured from the English Nation," French claimed that through reading his translation, "the poorest man may in a short time become very rich, the most sickly very healthy, and the basest truely honorable." And, thus, he vowed, "It shall be my practise as long as I live to be instrumental in promoting true knowledge, wheather by way of Translation,

[19] See Rankin, "New World Drugs" (cit. n. 6); and Antonio Barrera-Osorio, "Translating Facts: From Stories to Observations in the Work of Seventeenth-Century Dutch Translators of Spanish Books," in *Translating Knowledge in the Early Modern Low Countries*, ed. Harold John Cook and Sven Dupré (Zurich: LIT Verlag Münster, 2012), 317–32.

[20] Jean Fernel's *consilia* are included in Lazare Riviére's *The Practice of Physick* (London, 1658 edition onward), translated by Nicholas Culpeper and issued by Peter Cole. Hildanus's works most obviously appear as *Gulielm Fabricius Hildamus, His Experiments in Chyrurgerie* (London, 1642); and *Cista militaris, or, A Military Chest* (London, 1674). Moise Charas's *Pharmacopée royale galénique et chimique* was translated as *The Royal Pharmacopoea, Galenical and Chemical* (London, 1678). Also translated were his *Nouvelles expériences sur la vipère*, which appeared as *New Experiments upon Vipers* (London, 1670 and other editions). On Charas's arguments with Francesco Redi over experimentation, vipers, and poison, see Jutta Schickore, "Trying Again and Again: Multiple Repetitions in Early Modern Reports of Experiments on Snake Bites," *Early Sci. & Med.* 15 (2010): 567–617.

[21] On the Hartlib circle, see, for example, Webster, *Great Instauration* (cit. n. 9); Mark Greengrass, Michael Leslie, and Timothy Raylor, eds., *Samuel Hartlib and Universal Reformation: Studies in Intellectual Communication* (Cambridge, UK: Cambridge Univ. Press, 1994); Koji Yamamoto, "Reformation and the Distrust of the Projector in the Hartlib Circle," *Hist. J.* 55 (June 2012): 375–97; Paul Slack, *The Invention of Improvement: Information and Material Progress in Seventeenth-Century England* (Oxford: Oxford Univ. Press, 2015), chap. 4; Vera Keller and Leigh T. I. Penman, "From the Archives of Scientific Diplomacy: Science and the Shared Interests of Samuel Hartlib's London and Frederick Clodius's Gottorf," *Isis* 106 (2015): 17–42; Penman, "Omnium Exposita Rapinæ: The Afterlives of the Papers of Samuel Hartlib," *Book History* 19 (2017): 1–65; and Carol Pal, "The Early Modern Information Factory: How Samuel Hartlib Turned Correspondence into Knowledge," in *Empires of Knowledge: Scientific Networks in the Early Modern World*, ed. Paula Findlen (Routledge, 2018), 126–58. French's earlier forays into translation were commissioned by Hartlib, who recorded that he visited French at his lodgings at Warwick Court on November 30, 1652, receiving updates on the translation of "Erker" and Agricola and lending him copies of Glauber's tracts; Samuel Hartlib, Ephemerides 1652, Part 2, 1652 [7 October–31 December], Sheffield University Library, MS 61 28/2/37A-44B (28/2/42B), as published online by M. Greengrass, M. Leslie, and M. Hannon, *The Hartlib Papers*, The Digital Humanities Institute, University of Sheffield, 2013, http://www.dhi.ac.uk/hartlib (hereafter Hartlib Papers).

or any other way of making what is occult manifest."[22] As many scholars have noted, these kinds of sentiments were widely shared among members of the Hartlib circle.

Hartlib and his circle had strong interests in the potential of iatrochemistry, and it was Glauber's fame as a producer of medicines that caught their attention.[23] From 1644 onward, Henry Appelius (fl. 1640–58) and Johann Moriaen (c. 1592–1668) sent individual tracts of the *Furni novi philosophici* to Hartlib, along with descriptions of Glauber's Amsterdam laboratories, his inventions, and his whereabouts.[24] Throughout the late 1640s, various members of the circle, including Hartlib himself, tried their hand at translating Glauber's confusing prose. French acknowledges this work in *A Description of Philosophical Furnaces*, writing that "the greatest part of the treatise in private hands [was] already translated into English by a learned German."[25] However, the translation of Glauber's technical know-how required more than linguistic competence. The *Furni novi philosophici* tracts were likely written-down versions of Glauber's teachings. In August 1647, Appelius told Hartlib that Glauber "taught the furnaces et the mannour of distilling for monyes." However, despite the fact that direct instruction of his techniques constituted a source of income for him, Glauber planned to communicate these ideas to the "whole world" as soon as he could fund the publication.[26] In other words, from the start, because of the technical nature of these processes, Glauber took a multipronged approach to disseminating his expertise and know-how and to establishing his reputation and authority. The archive makes clear that Hartlib and others took a similarly ambitious approach to gaining information about Glauber's furnaces and techniques, with obtaining and translating the text of the *Furni novi philosophici* as just one path. This was crucial because the group quickly realized that although Glauber was happy for his printed tracts to be translated into French and Latin and considered them "no more his, but all mens," he was much more guarded when it came to divulging exact methods and recipes.[27] As Moriaen wrote, "he also wanted to

[22] John French, "Letter to the English Reader," in Johann Rudolf Glauber, *A Description of Philosophical Furnaces*, trans. French (London, 1651), sig. A4r-v.

[23] On chemical medicines in the 1640s and 1650s, see, for example, Antonio Clericuzio, "From van Helmont to Boyle: A Study of the Transmission of Helmontian Chemical and Medical Theories in Seventeenth-Century England," *Brit. J. Hist. Sci.* 26 (1993): 303–34; and Webster, *Great Instauration* (cit. n. 9), chap. 4. On the Hartlib circle and chemistry, see Stephen Clucas, "The Correspondence of a XVII-Century 'Chymicall Gentleman': Sir Cheney Culpeper and the Chemical Interests of the Hartlib Circle," *Ambix* 40 (1993): 147–70; and John T. Young, *Faith, Medical Alchemy, and Natural Philosophy: Johann Moriaen, Reformed Intelligencer and the Hartlib Circle* (Aldershot: Ashgate, 1998).

[24] The first mention of the *Furni novi philosophici* appears to be in a letter from Appelius to Hartlib sent in June 1644 in which the two discussed Hartlib's issues with kidney stones; Letter, Henry Appelius to Samuel Hartlib, 12 June 1644, Hartlib Papers 45/1/8A-B. Appelius and Moriaen wrote often about Glauber's movements. Appelius, for example, reported that Glauber had gone to Utrecht in September 1644 and to Arnheim in August 1647; Letter from Henry Appelius to Samuel Hartlib, 5 September 1644, Hartlib Papers 45/1/13A-B and 26 August 1647, 45/1/33A-34B.

[25] Glauber, *A Description of Philosophical Furnaces* (cit. n. 22), sig. A4r. On Glauber in London, see Young, *Faith, Medical Alchemy, and Natural Philosophy* (cit. n. 23); Pamela H. Smith, "Vital Spirits: Redemption, Artisanship, and the New Philosophy in Early Modern Europe," in *Rethinking the Scientific Revolution*, ed. Margaret J. Osler (Cambridge, UK: Cambridge Univ. Press, 2000), 119–36; and Stephen Clucas, "Correspondence," (cit. n. 23).

[26] Letter, Henry Appelius to Hartlib, 26 August 1647, Hartlib Papers 45/1/33A-34B.

[27] Ibid. On notions of openness and secrecy in craft and technical knowledge, see Pamela O. Long, *Openness, Secrecy, Authorship: Technical Arts and the Culture of Knowledge from Antiquity to the Renaissance* (Baltimore, MD: Johns Hopkins Univ. Press, 2001); and Elaine Leong and Alisha Rankin, eds., *Secrets and Knowledge in Medicine and Science, 1500–1800* (Farnham, UK: Ashgate, 2011), particularly the essay by Pamela Smith, "What is a Secret? Secrets and Craft Knowledge in Early Modern Europe," 47–66.

keep his thumb, as they say, in his hand and didn't reveal the secret."[28] Robert Child (1613–54) further exclaimed in relation to the recipe for the Alkahest: "I Cannot beleeve that Glauber will reveall it to any one, though perhaps they may get some particulars from him."[29] In fact, Glauber excelled at walking the fine line between desiring to communicate information freely and openly and protecting his own commercial interests.[30]

By the autumn of 1647, the group was eager to clarify matters and, as Glauber was reluctant to travel to England, they decided to send a member to gain firsthand knowledge of his processes and technologies, and assess their feasibility and utility.[31] Ideally, this person would possess skills "in Chymicall et Alchymisticall matters . . . [and] bee best able to judge of his Inventions." After all, as Henry Appelius reported, "[Glauber's] Operations are not so havy and long, they can better be tryed than disputed."[32] In early 1648, Benjamin Worsley (1618–77), a self-styled medical practitioner, traveled to the Netherlands to gather information on a range of topics including Glauber's furnaces.[33] While much can be written about Worsley's eventful time with Glauber, for the purposes of this article it suffices to note that despite sharing common skills and training in chemical operations, Worsley and Glauber found their time together challenging. The minutiae of everyday life intervened at every corner, and linguistic and technical issues abounded, for Worsley had no German, and Glauber, though able, was reluctant to communicate in Latin.[34] At various points, the two men brokering this knowledge exchange, Moriaen and Appelius, expressed doubts on whether Worsley could coax

[28] "Er hatt aber gleichwoll den daumen wie man sagt in der hand behalten vnd das secret nicht offenbahrt"; Letter from Johann Moriaen to Hartlib, July 1650, Hartlib Papers 37/163A-164B (37/163A). Translation mine. Unless otherwise noted, the translations in this article are mine.

[29] Letter from Robert Child to Samuel Hartlib, 2 February 1652, Hartlib Papers 15/5/18A-19B (15/5/18B).

[30] When asked if his book could be translated into Latin and French, Glauber answered that there was "no necessity to aske leave of him, seeing the book was no more his, but all mens"; Letter from Henry Appelius to Samuel Hartlib, 26 August 1647, Hartlib Papers 45/1/33A-34B (45/1/33B). See also Smith, "Vital Spirits" (cit. n. 25), 125; and Smith, *The Body of the Artisan: Art and Experience in the Scientific Revolution* (Chicago: Univ. of Chicago Press, 2004), on Glauber's efforts to protect his commercial interests.

[31] This episode is vividly described by John Young in *Faith, Medical Alchemy, and Natural Philosophy* (cit. n. 23), chap. 7. A letter from October 1647 goes into some detail on the logistics and costs of this arrangement. Appelius supposed that for "100^{lb} starling the friend may have of Glauber what hee desireth if not more"; Letter from Henry Appelius to Samuel Hartlib, 27 October 1647, Hartlib Papers 45/1/37A-B. On translators as mediators and cultural exchange, see Brenda M. Hosington, "Translation as a Currency of Cultural Exchange in Early Modern England," in *Early Modern Exchanges: Dialogues between Nations and Cultures, 1550–1750*, ed. Helen Hackett (Routledge, 2016), 27–54; and Peter Burke, "The Renaissance Translator as Go-Between," in *Renaissance Go-Betweens*, ed. Andreas Höfele and Werner von Koppenfels (Berlin: De Gruyter, 2005), 17–31.

[32] Letter from Henry Appelius to Samuel Hartlib, 26 August 1647, Hartlib Papers 45/1/33A-34B (45/1/33B).

[33] On Worsley, see Thomas Leng, *Benjamin Worsley* (Woodbridge, UK: Boydell Press 2008); and Charles Webster, "Benjamin Worsley: Engineering for Universal Reform from the Invisible College to the Navigation Act," in Greengrass et al., *Samuel Hartlib and Universal Reformation* (cit. n. 21), 213–35.

[34] Worsley's stay did not get off to a good start: he arrived at Glauber's lodgings only to be turned away as the latter's wife was in labor and the family had no desire to entertain an Englishman with no German in those circumstances; Letter from Johann Moriaen to Samuel Hartlib, 27 February 1648, Hartlib Papers 37/131A-132B. Hartlib had repeatedly asked Henry Appelius about Glauber's Latin skills and had received positive replies. See, for example, Letter from Henry Appelius to Samuel Hartlib, 27 October 1647, Hartlib Papers 45/1/37A-B (45/1/37A).

the required knowledge out of Glauber.[35] When Worsley returned to London in 1649, it was unclear whether this brief Dutch sojourn had yielded the hoped-for results.

Consequently, when John French took up the mantle to translate Glauber, he joined a years-long (albeit informal) collaborative operation. The efforts to bring the *Furni novi philosophici* to England went far beyond finding linguistic equivalence; rather, they became a lengthy and complex process requiring specialist expertise and first-hand experiential knowledge and involving assessment and trial of knowledge and techniques. The technical nature of the *Furni novi philosophici* and the artisanal/craft context within which it was created brought particular layers of resistance—the capture and transfer of tacit or gestural knowledge, the need to protect commercial interests, and issues of openness and secrecy.[36] Much of the translation, though now only archived on paper, took place within what historians of science have termed a "trading zone."[37] However, in this particular case, the exchange of knowledge was further encumbered by linguistic challenges. It is little wonder that *A Description of Philosophical Furnaces* turned out to be a wordy and cumbersome text, one not likely to induct newcomers to the trade.

CREATING *THE ART OF DISTILLATION*

Soon after his work translating the *Furni novi philosophici*, John French turned his attention to *The Art of Distillation*. Intended as a "generall treatise of Distillations," it offered "the choisest preparations of the selectest Authors both ancient, and moderne, and those of severall languages."[38] If the efforts to translate the *Furni novi philosophici* were collaborative, the work of remaking Glauber's technical tracts into a manual for general readers was conducted solely by French. And here, he did much more than just collate and complete the translations. French made clear that the knowledge contained within was gathered via his reading and translation practices and his "long, and manuall experience," extended by know-how he had "by way of exchange purchased out of the hands of private men, which they had monopolized as great secrets."[39] He thus fashioned himself as a compiler, a translator, an expert, and a maker. In producing what he considered a general guide to distillation, French articulated what he considered the most important texts and know-how in the field.

[35] In August 1648, for example, Appelius reported that Worsley's work was proceeding slowly, particularly as Glauber offered many compliments but was not forthcoming on exact methodologies and processes; Letter from Henry Appelius to Samuel Hartlib, 2 August 1648, Hartlib Papers 45/1/39A-40B. In September 1649, when he informed Hartlib of Worsley's return to England, Appelius stated plainly that the two men did not understand each other and that Glauber was hard on Worsley; Letter from Henry Appelius to Samuel Hartlib, 20 September 1649, Hartlib Papers 45/1/41A-B.

[36] On gestural knowledge, see Sibum, "Reworking the Mechanical Value of Heat" (cit. n. 8). On issues of openness and secrecy, see footnote 27 in the present article.

[37] The concept of "trading zones" was first developed by Peter Galison and was adapted and refined for early modern science (particularly artisanal science) by Pamela O. Long; see Long, *Artisan/Practitioners and the Rise of the New Sciences, 1400–1600* (Corvallis, OR: Oregon State Univ. Press, 2011); and Long, "Trading Zones in Early Modern Europe," *Isis* 106 (2015): 840–47.

[38] French, *Art of Distillation* (cit. n. 2), sig. *1r-v. More than twenty authors are named in the text, including contemporary or recent authors such as Michael Sendivogius, Paracelsus, Jean Baptista van Helmont (cited throughout the text), and Gregorius Agricola (177), and late medieval writers such as Albertus Magnus (178) and Thomas Aquinas (185). As with many early modern English recipe collections, there are also the usual references to recipes by hard-to-identify figures such as Dr. Burges (53) and Dr. Stephens (48).

[39] Ibid.

Within French's scheme for a general distillation guide, a central place was allotted to Glauber's inventions from the *Furni novi philosophici*. However, in order to create a general guide to distillation, French had to call upon common practices of textual compilation: extraction, reorganization, and embellishment. As befitting a publication selling a number of different furnaces and relevant technical know-how, each tract in the *Furni novi philosophici* is centered on a different kind of equipment. Aside from the opening chapter describing the distillation processes, *The Art of Distillation* is organized around types of medicines, with chapters on compound waters and on mineral and animal-based drugs. As a consequence, while *A Description of Philosophical Furnaces* and *The Art of Distillation* share common images and textual passages, these occur in different parts of the books and often have been significantly altered.

For example, the glass vessel in figure 1 was originally featured in the fifth tract of the *Furni novi philosophici*, offering advice for a range of processes from luting to the making of glassware and crucibles.[40] In *The Art of Distillation*, however, the illustration appears in the first book, in which French outlined the basics of the art, offering information on how to make instruments, build furnaces, and more. In fact, while almost all the images featured in the German and English version of *Furni novi philosophici* were included in *The Art of Distillation*, most appeared in the first book of *Distillation*, rather than dispersed across tracts dedicated to individual furnaces as per Glauber's original intention.

A Description of Philosophical Furnaces and *The Art of Distillation* were produced by the same printshop, and the images across the two works are almost identical, likely the result of the reuse of woodblocks. However, this was not a case of simple repurposing. As French moved the images from *A Description of Philosophical Furnaces* to *The Art of Distillation*, he made amendments and changes. For example, in the illustration of a glass vessel (see fig. 2), French added the label "D" with clear instructions on how to create a quicksilver (mercury) seal that would prevent any spirits stored within from escaping and preserve the glass.

Significant changes were also made to the text, and the recipe for aqua fortis is a good point of comparison. In *Furni novi philosophici*, the recipe can be found in the second tract, accompanying the description of a furnace designed for distillation. For reasons of protecting commercial interests and secrecy, recipes in *Furni novi philosophici* were often very brief. In this instance, the maker was simply told to mix vitriol and salt nitre in an equal or two-to-one ratio. In place of workable instructions, Glauber instead diverged into lengthy polemical discussions on the merits and faults of aqua fortis and salt nitre. This lengthy entry was reproduced largely unchanged in *A Description of Philosophical Furnaces*. However, when French featured the same recipe in *The Art of Distillation*, he not only repositioned it within the book but also significantly rewrote the instructions. Here, it sits in the section dedicated to recipes using "Minerals" alongside other instructions involving salt and vitriol. Omitting Glauber's discussion on aqua fortis and salt nitre entirely, French expanded the instructions, outlining the equipment required ("a glasse Retort coated, or earthen Retort that will endure the fire") and the production steps ("set them into the Furnace in an open fire, and then having fitted a large receiver distill it by degrees the space of 24 hours.")[41]

[40] Glauber, *Description of Philosophical Furnaces* (cit. n. 22), 293 ff.
[41] French, *Art of Distillation* (cit. n. 2), 70.

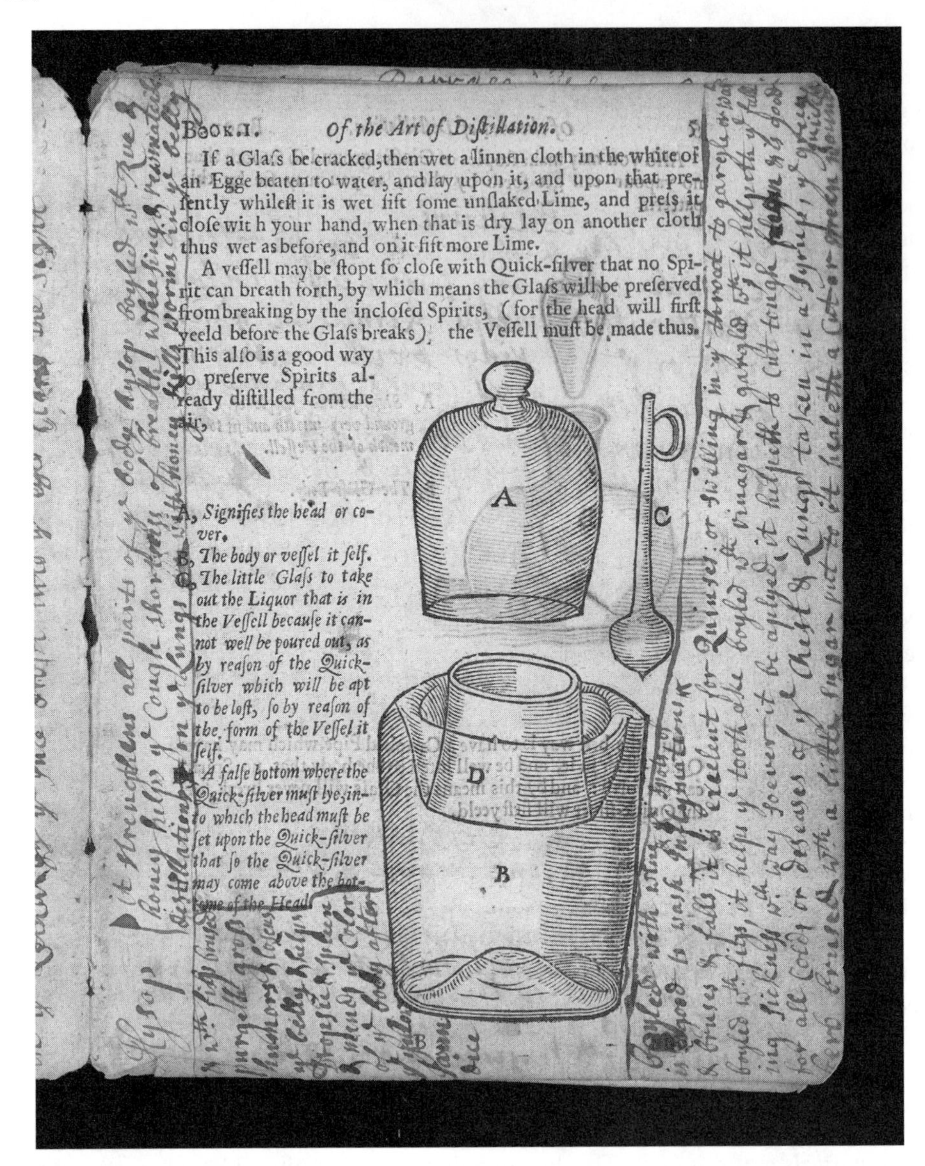

BOOK. I. *Of the Art of Diſtillation.* 5

If a Glaſs be cracked, then wet a linnen cloth in the white of an Egge beaten to water, and lay upon it, and upon that preſently whileſt it is wet ſift ſome unſlaked Lime, and preſs it cloſe wit h your hand, when that is dry lay on another cloth thus wet as before, and on it ſift more Lime.

A veſſell may be ſtopt ſo cloſe with Quick-ſilver that no Spirit can breath forth, by which means the Glaſs will be preſerved from breaking by the incloſed Spirits, (for the head will firſt yeeld before the Glaſs breaks) the Veſſell muſt be made thus.

This alſo is a good way to preſerve Spirits already diſtilled from the air.

A, *Signifies the head or cover.*

B, *The body or veſſel it ſelf.*

C, *The little Glaſs to take out the Liquor that is in the Veſſell becauſe it cannot well be poured out, as by reaſon of the Quick-ſilver which will be apt to be loſt, ſo by reaſon of the form of the Veſſel it ſelf.*

D, *A falſe bottom where the Quick-ſilver muſt lye, into which the head muſt be ſet upon the Quick-ſilver that ſo the Quick-ſilver may come above the bottome of the Head.*

Figure 2. *Page from the Wellcome Collection copy of John French's* Art of Distillation *(London: R. Cotes for T. Williams, 1651) showing diagram and instructions for creating a quicksilver (mercury) seal, with annotations written by members of the Tallamy family. Wellcome Collection, MS 4759, fol. 20r.*

If the *Furni novi philosophici* was part of Glauber's scheme to sell furnaces and medicines, French had other plans for *The Art of Distillation*, aiming to offer an accessible set of instructions. Indeed, many of the subtitles in *The Art of Distillation* resemble those of contemporary pharmaceutical tracts or books of medicinal recipes. This is not by chance, for if Glauber's *Furni novi philosophici* offered descriptions of alchemical devices accompanied by examples to illustrate their use, French's work is largely filled with recipes to make medicines. In that, it is closely related to one of the most popular medical genres of the day.

While both *A Description of Philosophical Furnaces* and *The Art of Distillation* were translations of Glauber's *Furni novi philosophici*, they represent two different paths to bringing continental vernacular works to English audiences, accentuating the many modes of translation adopted by early modern book producers as well as their differing receptions. French's first rendition of Glauber in English—*A Description of Philosophical Furnaces*—was never reprinted after French issued it in 1651. By contrast, French's subsequent reworking of Glauber—"Englished" in language as well as in cultural appeal—was well received, and *The Art of Distillation* remains his best known work, being issued four times with the final edition appearing in 1667. From the second 1653 edition onward, enterprising printers merged *The Art of Distillation* with *The Distiller of London*, a book of rules and directions issued by the Distillers' Company in 1639 in a bid to regulate practices.[42] For readers, this would have meant a bounty of additional recipes.

Yet, the story of Glauber in England did not end there. A second translation of Glauber, *The Works of the Highly Experienced and Famous Chymist, John Rudolph Glauber*, appeared in 1688, "Translated in *English* and Published for the Publick Good by the *Labour*, *Care* and *Charge*" of a physician named Christopher Packe (c. 1657–c. 1708). By then, there was enough interest in the work for it to be produced by subscription, with the list of all-male subscribers including gentlemen, physicians, surgeons, and apothecaries hailing from all around the country, from York to Somerset. One of the subscribers was Robert Boyle, who had also been involved in the efforts to translate Glauber the first time around.

Unlike his countrymen earlier, Packe was able to work from the recently available Latin translations of Glauber's works, and his publication was also a work of textual compilation. Packe took great pains to obtain the original Dutch copper plates for the images and hunted down twelve additional tracts "never printed in Latin, but in the German Tongue only" to ensure that he had as complete a set of works as possible. As were the earlier efforts by the Hartlib circle, this was a collaborative enterprise as the newly located German tracts were not translated by Packe but rather an anonymous man "well skill'd both in the High-Dutch, and also in Chymistry."[43] Following in Glauber's footsteps, Packe also paired the publication with a flourishing drug business, selling a number of Glauber's famed medicines at his house and laboratory next to the sign of the gun in Little Moorfields, London. Over time, Packe's translation became the standard edition of Glauber's works for English readers, including modern historians of science.[44] For many readers of Glauber, the collective efforts of the Hartlib circle and John French are largely forgotten, obscured by the mechanics of print and the ever-changing world of book production. In this case, print might have brought

[42] Company of Distillers of London, *The Distiller of London. Compiled and set Forth by the Speciall Licence and Command of the Kings Most Excellent Majesty: For the Sole Use of the Company of Distillers of London. And by Them to Bee Duly Observed and Practized* (London, 1639), sig. Bv. On the Distillers Company, see Webster, *Great Instauration* (cit. n. 9), 253–4.

[43] Johann Glauber, *The Works of the Highly Experienced and Famous Chymist, John Rudolph Glauber*, trans. Christopher Packe (London, 1688), preface, sig. A2r.

[44] J. R. Partington provides a long list of Glauber's publications in *A History of Chemistry*, but *A Description of Philosophical Furnaces* is not included; Partington *A History of Chemistry* (London: MacMillan, 1961), 341–61. Similarly, Packe was used as the main translation in other major English research on Glauber, including Kathleen Ahonen, "Johann Rudolph Glauber: A Study in Animism in Seventeenth-Century Chemistry" (PhD diss., Univ. of Michigan, 1972); and Anna Marie Roos, *The Salt of the Earth: Natural Philosophy, Medicine and Chymistry in England, 1650–1750* (Leiden: Brill, 2007).

Glauber's inventions to wider audiences, but it also flattened the complex sets of practices—textual and experiential—required to make this knowledge travel.

THE ART OF DISTILLATION IN THE EIGHTEENTH-CENTURY HOME

Sometime in or before the 1730s, a copy of *The Art of Distillation* fell into the hands of the Tallamys, a family likely from the port town of Bideford in Devon. While little is known about the Tallamys, the extant ownership notes suggest that the book once belonged to William, Patience, and Rebecca Tallamy.[45] While all three signed their name in the volume, Rebecca emerges as the most prominent owner and active annotator, signing her name multiple times over the course of 1736–8 and extending French's work with substantial notes. By the 1730s, when French's book reached the hands of the Tallamy family, it was almost eighty years old. The difficulties and tensions experienced by the Hartlib circle in obtaining Glauber's know-how were long forgotten, and *The Art of Distillation* was out of print. We can only speculate how this decades-old object became such a central part of the Tallamys' knowledge practices, but once it was in situ, the Tallamys customized their copy of the book, augmenting French's distillation guide with their own carefully gathered household knowledge. Running out of space in the margins, the Tallamys bound another 140 blank leaves to the book, filling it with a cornucopia of notes, including information on the medicinal virtues of herbs and hundreds of additional recipes.[46] Many of the additional entries contained information collated from friends and other printed books, including works by well-known medical authors such as Nicholas Culpeper and William Salmon.[47] Entries such as "Mrs Maines receipt from Liverpool to make currant wine" from 1806 indicate that the book continued to be used into the nineteenth century.[48] Clearly, for generations of the Tallamys, the object functioned as a treasured archive filled with everyday health knowledge tailored specifically for their family.

The Tallamys were not alone in their interest in pharmacy and medicines. The early modern home was a bustling site for a range of medical activities, from self-diagnosis and medication, to nursing and caring for the sick, to drug production, with women taking on key roles across this broad range of health practices. To further their knowledge about medicine and the body, householders accessed a wide variety of sources.

[45] The Tallamys have proved elusive to track down. A William Tallamy is mentioned in a deposition taken by the Commission on the King's Remembrancer side of the Exchequer in 1719. The deposition concerned the price of tobacco in the port town of Bideford in Devon. Additionally, in 1724, Hannah and Patience Tallamy described as "of Bideford" and "spinsters" were leased "Moiety of 2 messuages in Potters Lane" in Bideford by John Williams of Trewargey, Cornwall, and Lewis Stucley of Middle Temple. Finally, Katherine Allen has found mentions of the family name in eighteenth- and nineteenth-century records for Mortenhampstead, also in Devon; The National Archives, London, UK, E 134/9Geo1/Mich29; North Devon Record Office (South West Heritage Trust), B156/L/B/13/1; and Katherine Allen, "Hobby and Craft: Distilling Household Medicine in Eighteenth-Century England," *Early Modern Women* 11 (2016): 90–114, on 111n77.

[46] Wellcome Western MS 4759, fols. 120r-256v.

[47] One recipe is titled "A diet drink out of Culpeper"; ibid., fol. 232v. See below for further discussions of reading notes from Culpeper. A number of recipes are labelled with "Salmon" in the upper right-hand corner; ibid., fols 240r-241v. On how householders collected medical information and utilized their reading practices to build recipe collections, see, for example, Sara Pennell, "Perfecting Practice? Women, Manuscript Recipes and Knowledge in Early Modern England," in *Early Modern Women's Manuscript Writing: Selected Papers from the Trinity/Trent Colloquium*, ed. Jonathan Gibson and Victoria E. Burke (Aldershot: Ashgate, 2004), 237–58; and Elaine Leong, "'Herbals She Peruseth': Reading Medicine in Early Modern England," *Renaissance Studies* 28 (2014): 556–78.

[48] Wellcome Western MS 4759, fol. 183v.

While some turned to their family and friends for advice or conferred with medical practitioners of various sorts, many also consulted the rich offerings by contemporary book producers, leaving traces of their reading practices in book margins and manuscript notebooks. Know-how for drug production in particular was much sought after by householders. It was common to make medicines at home, and distillation was a production method used within many domestic spaces by both male and female actors. Household inventories list equipment such as glass stills, alembics, and water baths, and recipes for distilled waters are regularly found in recipe books.[49] As such, it is not surprising that the Tallamys had use for a distillation manual, and indeed, a number of the Tallamys' handwritten recipes required distillation, such as the instructions to make a good water for the stomach, Dr. Bate's medicine against consumption, and a range of other medicinal waters.[50] This is not to say that all of French's complex chemical procedures made their way into the Tallamys' everyday practices. Without greater knowledge of the Tallamys' circumstances and their wider reading habits, it is difficult to ascertain the exact role served by French's text and the book as a material object in their daily lives and knowledge practices. After all, we have few clues about whether they so heavily annotated all of their books, medical or otherwise, or whether or how they might have used this volume alongside other works on their bookshelves. As illustrated below, many of their annotations only engage passively with the content of French's work, and it is possible that they might have been primarily using the book pages as a space to record know-how on food and drug preparation. We might also view their interest in *The Art of Distillation* as aspirational—that is, they viewed the book as a trove of ambitious recipes they hoped to make one day rather than as a collection of know-how for use in everyday life.

The opening page of Book 1 in the *Art of Distillation* serves as a good example of the multiple ways in which the manuscript and printed books layered upon each other. As illustrated in figure 3, this page acts as a title-page of sorts for both the first chapter of the *Art of Distillation* and the Tallamys' recipe book. Rebecca Tallamy's ownership note, "Rebecca Tallamy her Book of Stilling and Reccepts 1736," and French's chapter title, "What Distillation is, and the kinds thereof," are both featured centrally. French's succinct explanation of distillation as an art appears directly under the chapter title; surrounding this block of printed text are Rebecca's handwritten notes on the herb madder copied out of Nicholas Culpeper's *The English Physitian Enlarged* (1653).

To signal that the herbal knowledge hails from another text, Rebecca turned the book sideways and wrote in the margins, inserting a boxed heading with the word "Madder" on the left (or the lower left-hand corner of French's page). The excerpt from Culpeper's entry on madder was then copied around the block of printed text, first filling in the left- and right-hand margins of French's page and then the space at the bottom. As a result, Rebecca's excerpt of Culpeper's entry on "madder" is superimposed onto French's printed text. Rebecca used the same layout for a number of entries taken from *The English Physitian Enlarged*, each with a boxed heading

[49] Elaine Leong, "Making Medicines in the Early Modern Household," *Bull. Hist. Med.* 82 (2008): 145–68; Allen, "Hobby and Craft" (cit. n. 45); Anne Stobart, *Household Medicine in Seventeenth-Century England* (London: Bloomsbury, 2016).

[50] Wellcome Western MS 4759, fols. 48v, 158v, and 124v (distilling fumitory); 165r (to make "Aqua Carminativa"); 222r (to make cordial water); 228r (to make a compound water of butter burrs and gentian water); 227v (surfeit water); 228v (cinnamon water); 253r (medicine for a hot and costive habit of body); and 253v (instructions to distill elder water and flowers).

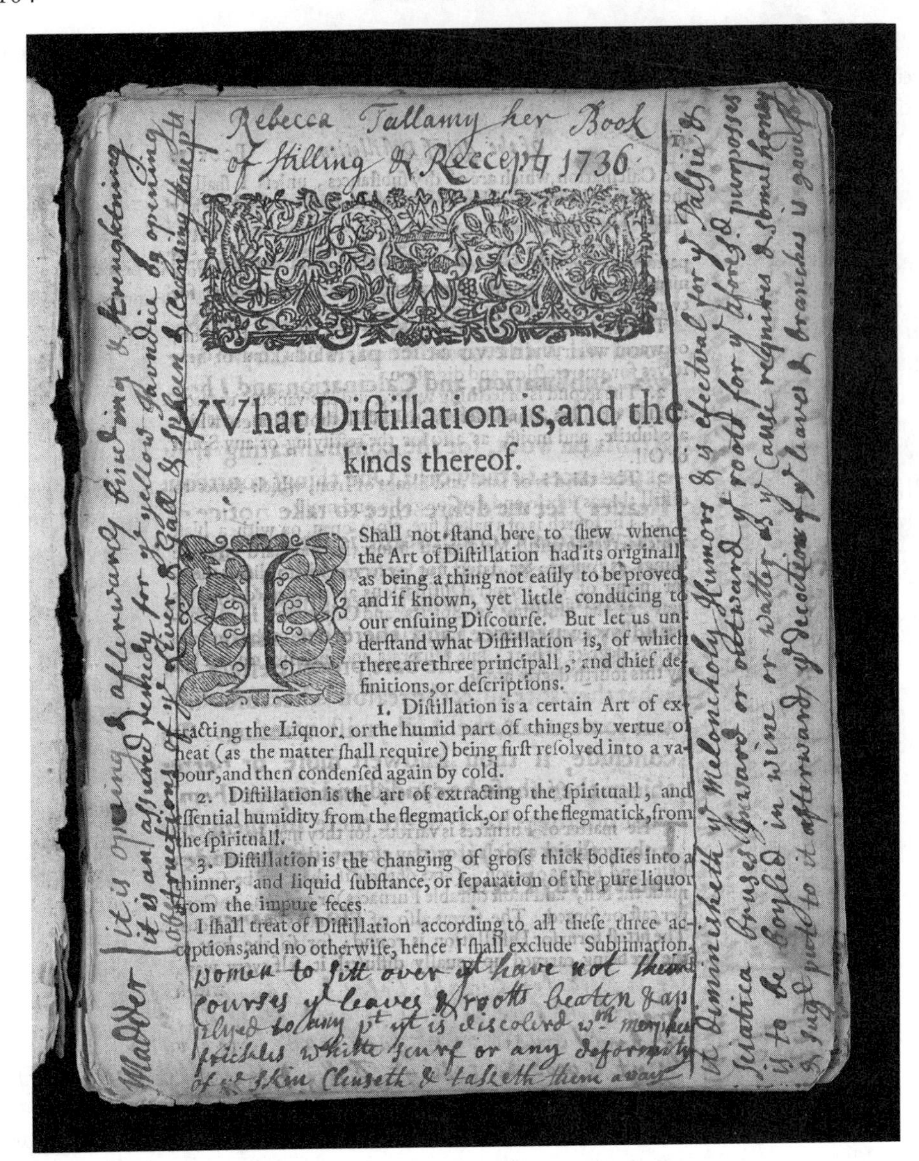

Figure 3. *Chapter opening from the Wellcome Collection copy of John French's* Art of Distillation *(London: R. Cotes for T. Williams, 1651) with annotations written by Rebecca Tallamy, including notes from works by Nicholas Culpeper. Wellcome Collection, MS 4759, fol.17r.*

placed in the corner. The eleven excerpts cover Culpeper's entries on sanicle, motherwort, mouse ear, tormentil, horehound, plantane, madder, nep or calmint, knot-grass, summer savory, and hyssop.[51] Directionality here is used to signal different kinds of knowledge, separating French's technical know-how from Culpeper's botanical

[51] Wellcome Western MS 4759, fols. 2v, 4v, 9v, 11v, 14v, 15v, 17r, 17v, 19r, 19v, and 20r. These entries are excerpted from Nicholas Culpeper, *The English Physitian Enlarged* (London,1653), 332, 164, 165, 359–60, 130–1, 301, 148, 171, 138, 334, and 128.

knowledge. Thus, when the book is orientated horizontally, Rebecca's consistently placed boxed headers work together to form a visual index of her herbal knowledge. In effect, Rebecca's canny mise-en-page enabled her to create a book within a book.

These were not the only excerpts Rebecca took from Culpeper's *The English Physitian Enlarged*. Another big batch of extracts, sometimes a full quarto-side long, can be found in the blank pages she bound with French's printed text.[52] Each entry in Culpeper's herbal offered information under four headings: description, place, time and government, and virtues. Rebecca's excerpts from the text were taken from the final part of each entry—government and virtues—and even in the long entries on madder and tormentil, her excerpts are selective. In choosing to record only the medicinal virtues and uses of herbs, Rebecca was following a fairly common practice at the time, particularly when the excerpts were combined with recipe knowledge in a single notebook.[53] It may be that the householders were seasoned foragers familiar with the appearance of common herbs, but more likely, they planned on buying their ingredients from apothecaries or herb women and did not see the need to acquire detailed knowledge or skills in botany.

Rebecca also turned to another one of Culpeper's popular works, *A Physicall Directory*, or *The London Dispensatory*, as it was titled from the second edition. This was a translation of *Pharmacopoeia Londinensis*—the official pharmacopoeia issued by the London College of Physicians.[54] Rebecca took numerous notes from this text and interspersed them throughout the French/Tallamy volume. For example, the notes on "roots" were written onto the recto side of a blank page inserted between pages two and three of French's printed work, where it is surrounded by passages taken from *The English Physitian Enlarged* on the two directly facing pages.[55] A comparison of Rebecca's excerpts and Culpeper's printed text demonstrates how this was not simply a copy but rather selected and amended passages pertinent to her own medical practices and needs. In other words, in her work of textual compilation, Rebecca interleaves excerpts from three different printed books, working across the print and manuscript medium to create her version of a household manual for health. Notably, each of these printed works—a herbal, a pharmacopoeia, a distillation manual—purported to offer a manual of specialist knowledge, and so, by bringing them into one, she also blurs the lines between different areas of medicine.

While Rebecca tended to write in the margins and blank spaces of the printed text, this was not always so. In one case, Rebecca's need to preserve or record information about materia medica overtook her need to retain French's explanations about distillation glasswork. Page six of French's text recommends particular types of glassware to preserve distilled spirits and contains the illustration of a glass and stopper, accompanied by explanatory labels (see fig. 4).[56] On this page, Rebecca added excerpts from

[52] The second run of entries includes information on burnett, butter bur, eyebright, featherfew, brown bugle, borrage and bugloss, liverwort and marigold, and much more.

[53] See Leong, "Herbals She Peruseth" (cit. n. 47) for examples.

[54] The publication history of the *Pharmacopoeia Londinensis* is complex. The notes in this article refer to the second English edition of the work, *Pharmacopoeia Londinensis, or, The London dispensatory further adorned by the studies and collections of the Fellows* (London, 1653).

[55] Wellcome Western MS 4759, fol. 18r. Folio 18v contains notes taken from the entry on juniper berries, and folios 17v and 19r contain excerpts on nep or calamint and knot-grass from *The English Physitian Enlarged*.

[56] Ibid., fol. 20v.

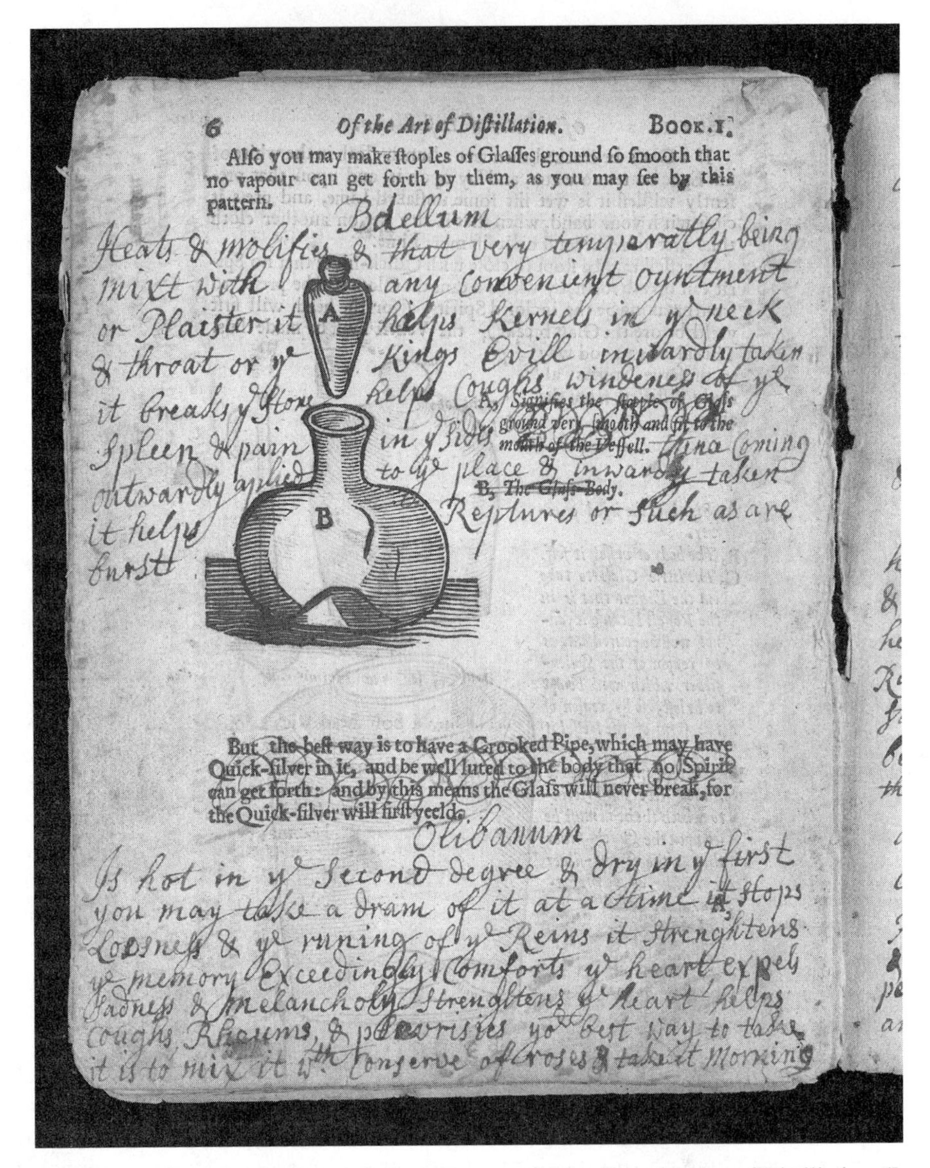

6 *Of the Art of Diſtillation.* BOOK. I.

Alſo you may make ſtoples of Glaſſes ground ſo ſmooth that no vapour can get forth by them, as you may ſee by this pattern.

Bdellium

Heats & molifies & that very temperatly being mixt with any convenient oyntment or Plaister it helps Kernels in ye neck & throat or ye Kings Evill inwardly taken it breaks ye Stone helps Coughs windeness of ye Spleen & pain in ye sides [illegible] coming to ye place & inwardly taken outwardly aplied it helps Reptures or such as are burst

A, Signifies the ſtopple of Glaſs ground very ſmooth and fit to the mouth of the Veſſell.

B, The Glaſs Body.

But the beſt way is to have a Crooked Pipe, which may have Quick-ſilver in it, and be well luted to the body that no Spirit can get forth: and by this means the Glaſs will never break, for the Quick-ſilver will firſt yeeld.

Olibanum

Is hot in ye Second degree & dry in ye first you may take a dram of it at a time it stops Loosness & ye runing of ye Reins it strenghtens ye memory Exceedingly comforts ye heart expels sadness & melancholy strenghtens ye heart helps coughs Rheums & pleurisies ye best way to take it is to mix it wth conserve of roses & take it morning

Figure 4. *Page from the Wellcome Collection copy of John French's* Art of Distillation *(London: R. Cotes for T. Williams, 1651), with annotations written by Rebecca Tallamy, including notes from works by Nicholas Culpeper. Wellcome Collection, MS 4759, fol. 20v.*

Culpeper's pharmacopoeia on two medical resins, bdellium and olibanum. Her entry on bdellium closely hugs the illustration of the glass and its explanatory labels, and in this instance, the handwritten excerpts overwhelm the printed text. To ensure that her notes are legible and clear, Rebecca crossed out most of the print on this page, including the explanatory labels for the illustration and French's recommendation for a second type of glassware, the crooked pipe. Rebecca's deliberate deletion of French's text suggests that she afforded more importance to her own reading notes than to French's technical knowledge, which she might not have found so useful. In these instances, the complicated compilation processes both enhanced and effaced French's original printed book.

Rebecca also customized information she gleaned from Culpeper's pharmacopoeia. For instance, page eight of *The Art of Distillation* offered detailed instructions on how to prepare the necessary equipment for distillation.[57] In this particular passage, French discusses the art of nipping, or sealing up a glass vessel. The maker is instructed to first heat the long neck of the vessel with pan coals and then cut off the excess glass with shears. Finally, in the step shown in the woodcut, the reader should pinch the neck closed with tongs. In the margins and blank spaces, Rebecca added information on two medical ingredients, "camphire" (camphor) and "styrax calamitis" (storax, a kind of natural resin).[58]

Both entries have been amended or extended. In the entry for camphor, for instance, Culpeper argued that it eased headaches coming from heat, but Rebecca thought it aided headaches stemming from cold. She also provided extra information on how to apply the medicine: "w*ith* oyle anoynt *the* temples easeth *the* head ake," undoubtedly useful information if you intended to use the medicine. The entry for *styrax calamitis* shows similar attention to practicalities, although in this case, the only addition made to Culpeper's text was the advice to "take ten grains made up in a pill." The focus on application methods in both these entries suggests that Rebecca relied on her personal experiences in administering the drug to extend the bookish knowledge offered by Culpeper.

Experiential knowledge also plays a key role elsewhere in the handwritten part of the French/Tallamy book. As mentioned earlier, not content with the blank spaces around the printed text, Rebecca extended the space available in the printed text by binding additional pages to the back of the book. While she continued to copy excerpts from other printed works into these pages, it is here that she (and other members of her family) transformed French's printed text into a family archive by merging social knowledge with natural knowledge. Scores of medical and culinary recipes, including instructions on baking cakes or making balsamic syrup and fumitory water, fill these pages. The recipes span a broad range of knowledge areas and were collected from a wide variety of sources, including recipe books belonging to relatives and newspapers. While only a few of the recipes are precisely dated, it is clear that they were gathered over a long time. Some of the earliest recipes in the book are connected with well known sixteenth- and seventeenth-century figures, such as "A Reccept of Metheglin made for Queen Elizabeth" and "A Cake ye way of ye Princes Elizabeth Daug to King Charles ye First."[59] And the latest recipes date from the early nineteenth century, including one "For a bad Mouth," dated 1805, and a recipe for raspberry wine by Mrs Newcomes, dated 1807.[60]

Like many recipe collectors, the Tallamys turned to a range of sources for health-related information. A long excerpt taken from the *Gentleman's Magazine* in 1802 recounts Sir Joseph Banks's trials with a sugar, milk, and ginger mixture for breakfast to ease gout symptoms, detailing his experiences with varying the amount and grind of

[57] Ibid., fol. 22v.

[58] *Oxford English Dictionary online*, s.v. "Camphor, n.," accessed August 6, 2018, http://www.oed.com/view/Entry/26800; ibid., s.v. "Storax, n.," accessed August 6, 2018, http://www.oed.com/view/Entry/190926.

[59] Wellcome Western MS 4579, fols. 201v and 129r.

[60] Ibid., fols. 196v and 184r.

the sugar/ginger mixture.[61] The Tallamys also excerpted from the handwritten recipe books of family and friends. For example, two entries—one headed "Doct Houards Syrup for a Consumtion" and another "A Red Powder to Expel any Disease from the heart"— are noted to be taken out of Unkle George Daveys or Davies Book."[62] Social visits with family and friends often resulted in access to prized recipe collections, and though undated, the proximity of the two excerpts within the notebook and the closeness of the addressed ailments indicate that these two recipes were likely collected during the same social visit.[63] Rebecca's uncle George was a regular source of recipes; other entries connected to him include instructions to make a remedy for a cough and short breath, a brown plaster, and an ague.[64] Other recipes, such as one "To Pickle Salmon the Newcastle way which will keep good twelve month," attributed to the Duke of Newcastle's Cooke, were both practical and brought social cachet.[65]

Many of the recipes were tried and tested, and sometimes rewritten accordingly.[66] The instructions to make gooseberry wine, for example, appeared with the endorsement "this I have had Experience of many years," written in Rebecca's distinctive handwriting.[67] Other recipes, such as the remedy "For a Nevoious Weakness of the Stomach," were simply marked out as "Tried."[68] The recipe for a plague water was originally titled "A Good Plague Watter," but the endorsement was heavily crossed out. As interlineal annotations in a different ink clarified the production process and added camomile to the ingredients, the rejection of this recipe likely happened after several trials (fig. 5).

The rewriting of recipes to reflect newly gained experience and knowledge was commonplace among recipe compilers and was a crucial step of this kind of textual compilation. At times, such notes—for example, "you may grate a little Lemon bread in it," written at the bottom of a recipe to make custards—are suggestive and adjustable to taste.[69] Other times, the changes reflect perfected trials and continual refinement of production methods. Interlineal notes on a seedcake recipe written in Rebecca's handwriting, for example, suggested the addition of brandy, doubled the amount of caraway seeds used, and advised makers to beat the butter with the eggs together first before adding dried sugar and flour (see fig. 6).[70] Open and malleable, recipe knowledge was continually updated and adjusted to suit the needs of the household.

Taken as a whole, the production history of *The Art of Distillation* offers new insights for the history of pharmacy. Recent studies have highlighted the complex set

[61] Ibid., fols. 189v-190r.

[62] Ibid., fols. 124r and 128r.

[63] For a detailed discussion of the sociability of recipe collecting, see Pennell, "Perfecting Practice" (cit. n. 47), 237–58; Michelle DiMeo, "Authorship and Medical Networks: Reading Attributions in Early Modern Manuscript Recipe Books," in *Reading and Writing Recipe Books, 1550–1800* (Manchester: Univ. of Manchester Press, 2013), 25–46; and Elaine Leong, *Recipes and Everyday Knowledge: Medicine, Science, and the Household in Early Modern England* (Chicago: Univ. of Chicago Press, 2018), chap. 1.

[64] These are all signed with the initials "GD"; Wellcome Western MS 4759, fol. 162v.

[65] Ibid., fol. 193v.

[66] On testing drugs, see essays in Elaine Leong and Alisha Rankin, eds., "Testing Drugs and Trying Cures," special issue, *Bull. Hist. Med.* 91 (2017). See also Leong, *Recipes and Everyday Knowledge* (cit. note 63), chaps. 5 and 6.

[67] Wellcome Western MS 4579, fol. 192r.

[68] Ibid., fol. 199r.

[69] Ibid., fol. 209v.

[70] Ibid., fol. 164v.

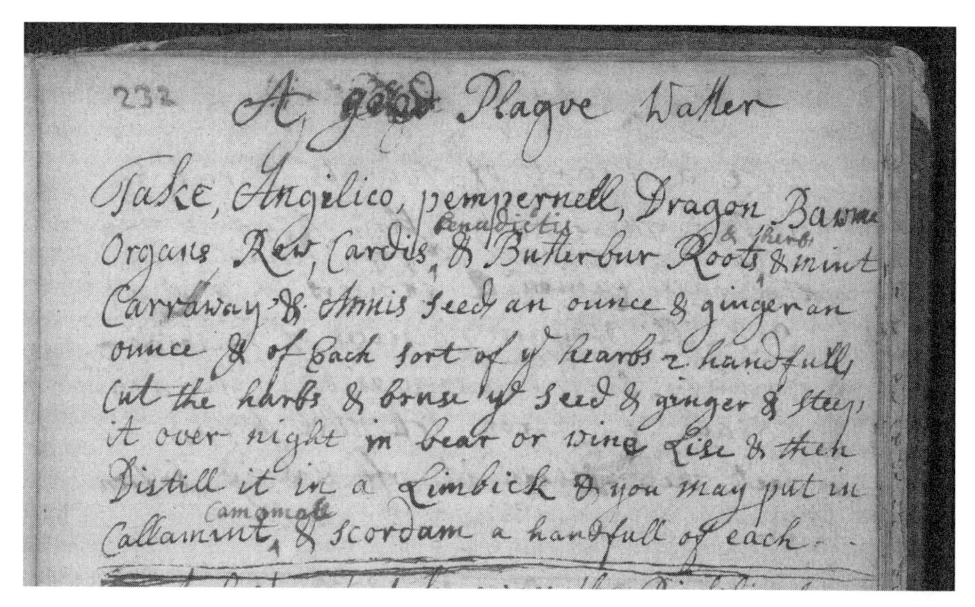

232
A ~~good~~ Plague Watter

Take, Angelico, pempernell, Dragon, Bawme
Organs, Rew, Cardus benedictis, & Butterbur Roots & herb & mint
Carraway & Annis seed an ounce & ginger an
ounce & of each sort of ye hearbs 2 handfull
Cut the harbs & bruse ye seed & ginger & steep
it over night in bear or wine Lise & then
Distill it in a Limbick & you may put in
Callamint camomile & scordam a handfull of each.

Figure 5. *Page showing recipe "A ~~good~~ Plague Watter" taken from the Wellcome Collection copy of John French's* Art of Distillation *(London: R. Cotes for T. Williams, 1651) with annotations written by members of the Tallamy family, including notes from works by Nicholas Culpeper. Wellcome Collection, MS 4759, fol. 232r.*

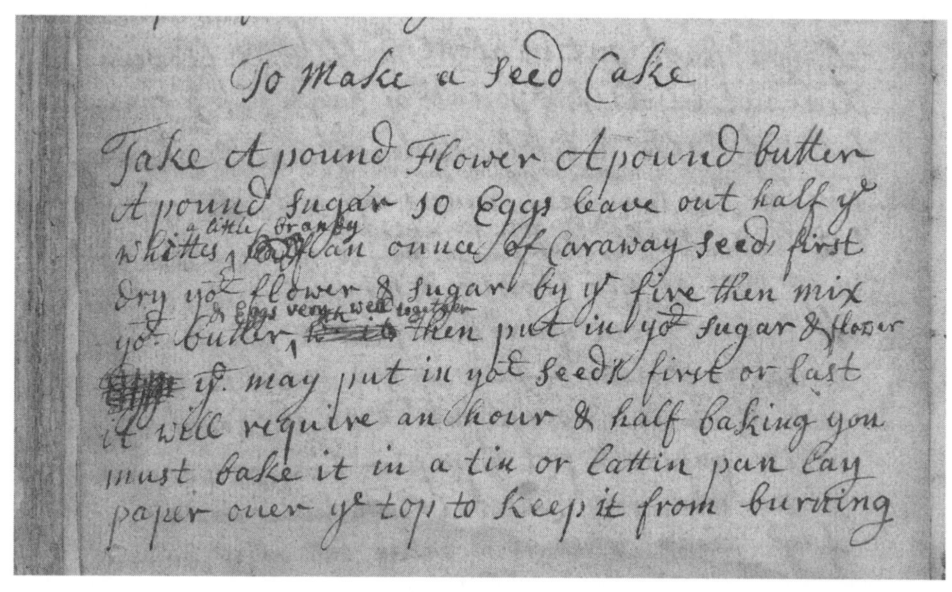

To Make a Seed Cake

Take A pound Flower A pound butter
A pound Sugar 10 Eggs leave out half ye
whites a little Brandy an ounce of Caraway seed first
dry yor flower & sugar by ye fire then mix
yor butter & Eggs very well together then put in yor sugar & flower
you may put in yor seeds first or last
it will require an hour & half baking you
must bake it in a tin or lattin pan lay
paper over ye top to keep it from burning

Figure 6. *Page showing the recipe "To make a seed cake" with interlineal annotations, from the Wellcome Collection copy of John French's* Art of Distillation *(London: R. Cotes for T. Williams, 1651) with annotations written by members of the Tallamy family, including notes from works by Nicholas Culpeper. Wellcome Collection, MS 4759, fol. 164v.*

of knowledge practices and power relations underpinning the movement of materia medica and botanical knowledge across the premodern world.[71] Complementing these studies, this story has focused on the transfer of the technologies and skills required for drug production. It has emphasized the resistance encountered in translating production processes that often could not be conveyed with mere words.[72] Concurrently, the Tallamys' annotations demonstrate that natural knowledge about materia media and technical knowledge and skills for drug production cannot be so easily separated.

CONCLUSION

From Glauber to French to Tallamy, the object, text, and body of knowledge now catalogued as Wellcome Western Manuscript 4759 traveled a long journey across national and linguistic borders, stopping at messy printers' workshops, the desks of London-based reformers and intelligencers, then moving all the way to the kitchens and stillrooms of an eighteenth-century household. This is a story about a printed book becoming a manuscript, and a story about how artisanal knowledge, touted for cash by a German inventor/chemist in Amsterdam in the 1640s, was read by householders in Devon in the 1730s. If the Hartlib circle's collaborative translation of Glauber's tracts tried to convey the latter's ideas somewhat faithfully, by the time John French read, extracted, and compiled from the tracts to make *The Art of Distillation*, the knowledge offered by Glauber was taken apart and reassembled. The Tallamys' additional notes further remake the text into a different kind of manual. Led by Rebecca, their extensive annotations and recipe writing brought the book into a different gendered space and intellectual milieu. The layers of handwritten notes and crossings-out reveal how the Tallamys confidently tested, and at times discarded, the knowledge offered by the printed text. The annotations also formed an additional layer, recording not only medical and technical know-how but also family history and social networks, and thus bringing this body of knowledge into new settings. As such, this story forcefully reminds us how one object can encode multiple layers of reading and writing practices, conducted over long time periods and across different knowledge communities. It encourages us to further investigate long-view histories of knowledge and to pay heed to how "knowledge in-use" responded to user needs and challenges as it moved from community to community.[73]

[71] See, for example, Matthew James Crawford, *The Andean Wonder Drug: Cinchona Bark and Imperial Science in the Spanish Atlantic, 1630–1800* (Pittsburgh, PA: Univ. of Pittsburgh Press, 2016); Crawford and Joseph M. Gabriel, eds., *Drugs on the Page: Pharmacopoeias and Healing Knowledge in the Early Modern Atlantic World* (Pittsburgh, PA: Univ. of Pittsburgh Press, 2019); Samir Boumediene, *La colonisation du savoir* (Vaulx-en-Velin: Des mondes à faire, 2016); Anna E. Winterbottom, "Of the China Root: A Case Study of the Early Modern Circulation of Materia Medica," *Soc. Hist. Med.* 28 (2015): 22–44; and Tara Alberts, "Curative Commodities between Europe and Southeast Asia, 1500–1700," in *Entangled Itineraries: Materials, Practices, and Knowledges across Eurasia*, ed. Pamela H. Smith (Pittsburgh, PA: Univ. of Pittsburgh Press, 2019), 79–98.

[72] On the transfer and "translation" of technologies across the premodern world, see the essay by Benjamin Breen in this volume, "Where There's Smoke, There's Fire: Pyric Technologies and Africa Pipes in the Early Modern World," in *Osiris* 37. On the transfer of medical skills, see Tara Alberts, "Translating Alchemy and Surgery between Europe and Southeast Asia," and Daniel Trambaiolo, "Translating the Inner Landscape," also in *Osiris* 37.

[73] This article is part of a larger project examining the notion of "knowledge maintenance" over long time periods. My interest in investigating the idea of "knowledge in-use" and maintenance is inspired by historians of technology such as David Edgerton, Lee Vinsel, and Andrew L. Russell; Edgerton, *The Shock of The Old: Technology and Global History since 1900* (London: Profile Books, 2006); Russell and Vinsel, "After Innovation, Turn to Maintenance," *Technology and Culture* 59 (2018): 1–25.

Reading and writing, translation and transformation are, of course, the processes by which text and knowledge traveled. The itineraries of these travels, I have shown, are meandering and complex and brought about significant epistemic consequences. Like other scholars, I am drawn to the term *itineraries* to emphasize nonlinearity. As a heuristic device, it encourages us to explore the seemingly never-ending and convoluted ways in which knowledge moved, pushing us to recognize the significance of each stop made along this journey rather than just the destination.[74] One does not have to assume that an itinerary has an end point or that it will complete a circuit or that we need to study a particular itinerary in its entirety. Itineraries can break off or connect or reconnect. We might think of each part of my story as a contact point for epistemic change; when joined together, they constitute a knowledge itinerary.[75] Notably, this particular itinerary encompassed a variety of contact points, from in-person exchange of know-how and ideas by Worsley and Glauber, to collective translation by the Hartlib circle, to textual engagement and enhancement by the Tallamy family. Some of these contact points, such as Worsley's stay with Glauber, are long and filled with linguistic, social, and cultural friction. Others, such as the translation of the Glauber tracts and French's assemblage of the *Art of Distillation*, are collaborative and involved deep entanglements of textual and experiential knowledge. Still others, such as the Tallamys' annotations in French's printed work, seem to be bare connections of text written upon text, knowledge added to knowledge. Paying attention to contact points, thus, pushes us to explore the many overlapping and interconnecting epistemic practices that occur at a particular juncture. It urges us to study practices across the traditional boundaries of print/manuscript, author/reader, and knowledge producer/consumer, and inspires us to recognize that bodies of knowledge are continually changing, responding to the needs and interests of different users, makers, and remakers.

This article used one particular material object to trace the itinerary of a body of knowledge—Glauber's ideas about the "philosophical furnace." The twisty itinerary through which this material object was created was deeply framed by contemporary social, political, and intellectual contexts. Building up thick descriptions of how these contexts shape instances of translation, reading, and writing—or contact points of epistemic change—allows us to better understand how vernacular medical knowledge was codified, transferred, and appropriated by a range of users and actors across early modern Europe. Moreover, narratives such as the story of the Tallamy family's copy

[74] Here, I join the salient call by Neil Safier and Pamela H. Smith to use "itinerary" as a heuristic term; Safier, *Measuring the New World: Enlightenment Science and South America* (Chicago: Univ. of Chicago Press, 2008); Safier, "Global Knowledge on the Move: Itineraries, Amerindian Narratives, and Deep Histories of Science," *Isis* 101 (2010): 133–45; Smith, *Entangled Itineraries* (cit. n. 71).

[75] Past models for book production and circulation have focused on historical actors or the material object of the book; by turning our attention to knowledge, we create a flexible system in which the impact of different contributing factors fluctuates as the body of knowledge journeys through its itinerary. See, for example, Robert Darnton's "communication circuit" and Thomas R. Adams and Nicholas Barker's revision of this model as the "socioeconomic conjuncture." Darnton's model, with clear roles assigned to authors, publishers, printers, suppliers, shippers, booksellers, and readers, beautifully highlights the role of human agency in book production and consumption. Adams and Barker's model, focusing on "events"—publication, manufacture, distribution, reception, and survival—shifts our attention to the book itself and the myriad processes the material object passes through at different stages of its life cycle; Darnton, "'What Is the History of Books?' Revisited," *Mod. Int. Hist.* 4 (2007): 495–508; Adams and Barker, "A New Model for the Study of the Book," in *A Potencie of Life: Books in Society: The Clark Lectures 1986–1987*, ed. Barker (London: British Library, 1993), 5–43.

of *The Art of Distillation* enable us to get a better sense of how "expert" and "popular" knowledge intersected and merged. Focusing on knowledge contact points and itineraries might also provide us with new ways of conceptualizing the local and the global. Finally, and most crucially for this volume, it might help us figure out the place of translation in histories of knowledge.

TRANSLATION BEYOND THE TEXTUAL

Vernacular Languages and Invisible Labor in Ṭibb

*by Shireen Hamza**

ABSTRACT

Glossaries providing the names of materia medica across languages are ubiquitous in medical manuscripts composed in the Islamic world. These anonymous medical glossaries were produced by physicians who sought out regional names for materia medica from nonliterate people, verifying them for local use in a process called *taḥqīq*. In early modern South Asia, glossary entries were Arabic or Persian terms, while translations were offered in Persian, Hindavi, or other vernacular Indian languages. Translation was a dynamic process, demonstrating the continuous use of multiple languages in *ṭibb*, or learned medicine, not often acknowledged by historians of science in Islam. I focus on four glossaries accompanying a Persian medical text composed by Shihāb Nāgaurī in 1388 in Western India. While affirming the hierarchy of certain languages over others, the glossaries offer us a glimpse into how the experiential knowledge of nonliterate people constituted *ṭibb*.

> In this region, the *ʿajam* (non-Arabic) language is common, but in the *Ikhtiyārāt-i Badīʿī*, the names of the medicines are written in Arabic. Thus, I have compiled this glossary with the name of each thing in both Persian and Arabic so that from this text, the names of each object in Persian and Arabic may be easily known, and Allah knows best what is right.
> —Anonymous, Index of Names of Medicines in the *Ikhtiyārāt-i Badīʿī* (*Fihrist-i Asāmī-i Adwiyya-i Ikhtiyārāt-i Badīʿī*)[1]

A physician enters the room and takes her place. While examining the patient, she understands the diagnostic sensations under her fingers through the formal language of medicine, but when it comes time to speak to the patient, to ask further questions, and to communicate her prognosis, she must reach for words in another language entirely. And when she leaves the patient's side in search of the many ingredients needed to prepare the remedy, ingredients even the pharmacist does not carry, in what language does she

* Department of the History of Science, Harvard University, Cambridge MA, 02138, USA; shireenhamza@g.havard.edu.

My gratitude to the Social Science Research Council for financial support, to all the archivists who helped me with this article, especially Dr. Tanwir Fatima Saheba of the Oriental Manuscript Library & Research Institute (Hyderabad), and to the many colleagues and mentors who gave me feedback on this article, especially the editors and participants of *Translating Medicine*.

[1] Salar Jung Museum Library, Hyderabad (hereafter cited as SJML), Tibb, Farsi 31, Acc 1661, fol. 117a; Bibliothèque Nationale de France (hereafter cited as BnF) Supplément Persan 335 fol. 1b. All translations in this article are my own.

Osiris, volume 37, 2022. © 2022 History of Science Society. All rights reserved. Published by The University of Chicago Press for the History of Science Society. https://doi.org/10.1086/719223.

communicate then?[2] Medical practice was and is multilingual, though formal texts and documents do not openly reflect on this. Oral translation remains an integral, everyday part of medicine.

In the premodern Islamic world, learned physicians wrote their texts in cosmopolitan registers of Arabic and Persian, but they used other languages and registers to communicate orally with patients and procurers of medicines. Their success in medicine depended on their ability to communicate with nonliterate people. However, historians immersed in the formal texts authored by physicians, the only sources surviving of their practice, may forget that physicians worked in vernacular languages.[3] This article explores the orality of *ṭibb* (learned medicine) through an overlooked archive—anonymous medical glossaries abundant in premodern medical manuscripts. These glossaries translated the names of medicinal plants and diseases from formal, scholarly language to local, vernacular names. I will focus on *ṭibb* in South Asia through a study of four glossaries that accompany the Persian medical text *Shifā' al-Maraẓ* (Illness's cure). The author, Shihāb Nāgaurī (fl. 1390s), lived in Rajasthan and Gujarat in Western India and composed this text in 1388 CE. I will refer to this text as the *Ṭibb-i Shihābī* (Shihāb's medicine), as it was often known, after its author. Written in rhyming, didactic verse (*rajaz*) and covering diseases and their treatments head to toe, this text was meant to be accessible to students. Within the text itself, the author includes a number of terms in Hindavi, a major vernacular language in North India. The text has remained in circulation in South Asia since its time of composition; the four glossaries accompanying this text were included in manuscripts copied from the seventeenth to nineteenth centuries.[4] Why might a physician capable of reading the *Ṭibb-i Shihābī*, a formal text of medicine written in a scholarly idiom, need these glossaries? With whom did physicians need to discuss the names of medicines in vernacular languages? And why did these glossaries rarely leave their marginal location at the end of manuscripts, to become a genre of their own?

I argue that much of the content in these glossaries was gained and verified through oral and experiential means, a continuous process of verification of the names of materia medica used among literate and nonliterate people in a given locale. Alongside formal terminology, physicians needed to know the most up-to-date names for materia medica in the places they practiced medicine. The formal names physicians studied in texts of *ṭibb* were not necessarily known by their patients, or the people who planted, collected, prepared, transported, and traded medicinal substances. By interrogating the content of these glossaries within the contexts of their production, we can better understand how the knowledge and labor of nonliterate people undergirded literate medicine. The changing nature of spoken languages necessitated the continuous production and verification of glossaries: as soon as a medical glossary was written down, it was in need of being verified again. This article is the first study of *taḥqīq* (verification) in the realm of practical knowledge, a term shown to have important implications for Islamic

[2] I have used feminine pronouns in this paragraph to call attention to gender. I acknowledge the possibility of women's participation in these interactions, but the premodern authors whose names I know are all men. I explore "medical masculinity" at length in my dissertation; Shireen Hamza, "Islam and Medicine in the Medieval Indian Ocean World" (PhD diss., Harvard University, forthcoming).

[3] For an exception to this trend, see Ahmed Ragab, "'In a Clear Arabic Tongue': Arabic and the Making of a Science-Language Regime," *Isis* 108 (2017): 612–20, at 614.

[4] My gratitude to the physician-scholar Hakim Syed Zillur Rahman for pointing out a reference to the *Shifā' al-Maraẓ* in his uncle's own twentieth-century text in verse.

philosophy, philology, and medicine.[5] After contextualizing multilingualism in *ṭibb* and analyzing four glossaries accompanying seventeenth- to nineteenth-century manuscripts of the *Ṭibb-i Shihābī*, I show how *taḥqīq* operates in this genre of medical texts. Then, through a discussion of the autobiography of Nāgaurī, I explore how medical education related to the development of glossaries. This article may encourage historians to understand oral translation as constitutive of medical practice. Centering the oral reveals the participation of people whose labor is made invisible by the conventions of scholarly language.

MEDICAL GLOSSARIES IN ṬIBB

Historians have often approached the history of *ṭibb* by studying formally titled and authored Arabic texts written in the medieval period. They have considered *ṭibb* to be a discipline studied, maintained, and furthered by male physicians, who learned it from experienced senior physicians. From the ninth century onward, these texts of *ṭibb* were written in a register of Arabic and Persian that would be legible to scholars and bureaucrats from "the Balkans to Bengal," and, by the fourteenth century, from Kedah (in what is now Malaysia) to Kanem Bornu (in what is now Chad and Nigeria).[6] Across these regions, medieval physicians were able to both write texts in cosmopolitan registers of Arabic, Persian, and Hebrew, and to speak a dazzling variety of vernacular languages, though they rarely referred to them in their texts.[7] This hierarchy between cosmopolitan and vernacular languages was also a norm among learned physicians using Sanskrit, Pali, Uighur, Tibetan, Armenian, Chinese, and Latin. Linguistic complexities are to be expected within a textual tradition of medicine stretching from Greek and Roman antiquity to the Byzantine Empire and into the Islamic world. The full geography of *ṭibb* shows us that there are many relevant languages to consider, beyond the cosmopolitan.[8] As Ahmed Ragab argues elsewhere in this volume, the Abbasid-era Translation Movement of texts from Greek to Syriac to Arabic, considered to be the birth of science in the Islamic world, also came about through multiple generations of physicians and astronomers orally translating aspects of their work into

[5] Kamran I. Karimullah, "The Emergence of Verification (taḥqīq) in Islamic Medicine: The Exegetical Legacy of Faḫr al-Dīn al-Rāzī's (d. 1210) Commentary on Avicenna's (d. 1037) Canon of Medicine," *Oriens* 47 (2019): 1–113; Matthew Melvin-Koushki, "Taḥqīq vs. Taqlīd in the Renaissances of Western Early Modernity," *Philological Encounters* 3 (2018): 193–249; Islam Dayeh, "From Taṣḥīḥ to Taḥqīq: Toward a History of the Arabic Critical Edition," *Philological Encounters* 4 (2019): 245–99.

[6] Shahab Ahmed, *What Is Islam? The Importance of Being Islamic* (Princeton, NJ: Princeton Univ. Press, 2015), 32–3.

[7] Sheldon Pollock, *The Language of the Gods in the World of Men: Sanskrit, Culture, and Power in Premodern India* (Berkeley: Univ. of California Press, 2006). To ascribe agency to vernacular actors in the premodern world, Pollock focuses on the literarization of vernacular languages, primarily in the disciplines of grammar, epic, and courtly poetry, referring to these literarized languages as the "cosmopolitan vernacular." In the medical glossaries under study in the present work, vernacular languages like Hindavi are literized, or written down, but not literarized. The agency of vernacular actors is minimized, relegated to the appendices of texts in prestige registers of Arabic and Persian. The work of Pollock and others carefully disaggregates the vernacular from the oral and the subaltern, which had been lumped together in historiography. But in this article, the vernacular does refer to oral communication and local knowledge. My gratitude to Eric Gurevitch and Anand Venkatkrishnan for helping me think through this.

[8] See Siti Marina Mohd. Maidin, ed., *Al-Tibb: Healing Traditions in Islamic Medical Manuscripts* (Kuala Lumpur: Islamic Arts Museum Malaysia, 2018); and Genie Yoo, "Clues in Recipes and Verses: Rethinking Gender in the Transmission of Malay-Language Books of Medicine and the Mediation of Natural Knowledge in the Dutch East Indies" (presented at the Harvard-Princeton Graduate Workshop on Early Modern History, Princeton University, February 15–16, 2019).

Arabic to engage their new patrons. Oral translation between the textual and spoken registers has been an integral part of *ṭibb* since its inception.[9]

Though anonymous medical glossaries may well have been produced throughout the medieval period, most surviving glossaries can be found in manuscripts produced from the seventeenth century onward.[10] While searching for vernacular languages in premodern science, we catch sight of a different geography of knowledge production than the one illuminated by authoritative texts. These short, anonymous glossaries often appear without introduction or title, in contrast to the formal introductions provided by authors of most medical texts, and thus cannot be easily cited by later writers. They are often appended to such formal texts at the end or beginning of a manuscript. From their marginal position within manuscripts, these glossaries can demonstrate a hierarchy between cosmopolitan and vernacular languages. Arabic medical terminology—heavily inflected by Greek and Syriac—maintains pride of place in the shifting hierarchy of languages in the sciences. Arabic terminology remained a part of *ṭibb* even in Persian texts composed in regions where Persian was the language of literature, governance, and science. We can better understand Arabic and Persian as languages of science and medicine by recognizing their place within a multilingual world.[11]

Medical glossaries range from a few dozen terms to a few thousand, from a single folio to fifty, and can be found appended to monographs or compendia. While they share a purpose, these glossaries demonstrate a great diversity of form, length, and language, composed as they were in vastly different geographies across the Islamic world. In Yemen, for example, the lemmas and translations in the glossaries were often from one register of Arabic to another, or took three steps, from transliterated Greek and Syriac to cosmopolitan Arabic and then to local Arabic.[12] This is not surprising; the regional specificity of Yemeni terms for materia medica is reflected in many texts.[13] In al-Andalūs, where Galenic medicine had been practiced in multiple languages even before it arrived as *ṭibb* in Arabic texts, "synonyms" for materia medica across languages were included as a column in tabular texts, alongside the qualities, dangers, uses of, and substitutes for these materials.[14] In the Maghreb, the Greek and Syriac transliterations and Arabic

[9] Ahmed Ragab, "Translation and the Making of a Medical Archive," in *Osiris* 37; Ismail Hussein Abdalla, *Islam, Medicine, and Practitioners in Northern Nigeria* (Lewiston, NY: E. Mellen Press, 1997).

[10] For medieval tables of pharmaceutical equivalences based on the visual shape of transliterated words, see Oliver Kahl, "The Pharmacological Tables of Rhazes," *Journal of Semitic Studies* 56 (2011): 367–99. Quid pro quo substitution lists, a long tradition, were sometimes similar in form to medical glossaries but played a different role by creating an equivalence between two different substances; see Alain Touwaide, "Quid Pro Quo," in *Herbs and Healers from the Ancient Mediterranean through the Medieval West*, ed. Anne Van Arndell and Timothy Graham (Farnham, Surrey, UK: Ashgate, 2012), 19–61.

[11] Dimitri Gutas, *Greek Thought, Arabic Culture: The Graeco-Arabic Translation Movement in Baghdad and Early 'Abbasid Society (2nd–4th/5th–10th c.)* (New York: Routledge, 1998).

[12] For example, see Biblioteca Ambrosiana, Milan, Italy, Arabi Nuovo Fondi H20, fols. 258a-259b.

[13] These range from a pharmacopeia produced locally by a sultan in the thirteenth century to the pharmaceutical text of the famous eleventh-century polymath al-Bīrūnī, who, in all his travels, had never been to Yemen. See al-Malik al-Muẓaffar, *al-Muʿtamid fī 'l-adwiyya al-mufrada* (Beirut: Dār al-Kutub al-'Ilmiyya, 2000); and Muḥammad ibn Aḥmad al-Bīrūnī, *al-Biruni's Book on Pharmacy and Materia Medica*, trans. Hakim Mohammad Said and Sami Khalaf Hamarneh (Karachi: Hamdard Academy, 1973).

[14] Emilie Savage-Smith, "Ibn Baklarish in the Arabic Tradition of Synonymatic Texts and Tabular Presentations," in *Ibn Baklarish's Book of Simples: Medical Remedies between Three Faiths in Twelfth-Century Spain*, ed. Charles Burnett (London, UK: Arcadian Library; New York, NY: Oxford Univ. Press, 2008), 113–39; Charles Burnett, "The *Synonyma* Literature in the Twelfth and Thirteenth Centuries," in *Globalization of Knowledge in the Post-Antique Mediterranean, 700–1500*, ed. Sonja Brentjes and Jürgen Renn (Oxford: Routledge. 2016), 131–41. See also Max Meyerhof, "Sur les noms ibero-portugais des drogues dans les manuscrits arabes medicaux, et supplément aux noms portugais dérivés de l'arabe," *Petrus Nonius*, fasc. 2 (1939): 85–96.

lemmas were paired with synonyms in Amazigh languages.[15] In West Africa, handbooks of Quranic medicine "include the names of local trees and plants (in Bambara, as well as other local languages) and instructions for their preparation and interaction with the verse in question."[16] And in Ottoman contexts, Arabic-Turkish glossaries were commonly included in compendia of medical texts.[17]

Though there are many parallels with these Middle Eastern and North African contexts, *ṭibb* in South Asia differed in several respects. From its spread in the thirteenth century, *ṭibb* remained one of multiple learned medical traditions in the subcontinent, alongside Ayurveda, for example. Glossaries translated lemmas or headings of transliterated Greek, Syriac, Arabic, and Persian into Persian and Hindavi words. Across these regions, several terms were used as generic titles for the glossaries, including *lughat* (language), *farhang* (glossary), *fihrist* (index), and *tafsīr* (elaboration/interpretation). Regardless of their language, these glossaries' lemmas likely consisted of strange words that a reader would never have heard spoken aloud; this is demonstrated by the diacritics found in some glossaries that instructed readers in pronounciation.[18] These terms were then translated into names the reader might more easily recognize. Hundreds of extant glossaries have not merited scholarly attention, perhaps owing to the hierarchy of the cosmopolitan over the vernacular at work within many archives of premodern science, including those of South Asia.[19] These glossaries were part of a broader process of negotiating the presence of vernacular languages, such as Hindavi, in Persian or Arabic texts, through marginalia and even occasionally as glosses within the body of authoritative medical texts.

The conventions of these medical glossaries differ from those of authoritative medical texts. They make visible what is hidden and unnamed in formal texts of *ṭibb*—the community of knowers beyond physician and patient, including people who made, used, and circulated material remedies of *ṭibb*, such as farmers, pharmacists, and yogis. These people are not cited conventionally, as the authors of texts of *ṭibb* would be. Rather, they are mentioned only in the third-person plural, as the people who name or speak of things (*yusammūn, gūyand*). At most, they are specified by their professional

[15] As in the glossary in Cambridge Univ. Library, Cambridge, UK, Or. 1021 (12), fols. 27–30. By contrast, most of the thirty "Berber" terms included by Ibn Baklarish in Spain may not match up to any spoken dialect; Ana Labarta, "Ibn Baklarish's Kitab al-Musta'ini: The Historical Context to the Discovery of a New Manuscript," in *Ibn Baklarish's Book of Simples* (cit. n. 14), at 25.

[16] Ali Diakite and Paul Naylor, "Medical Texts from Timbuktu—Local Pharmacological Remedies with Qur'anic Verses," *HMML Stories*, Hill Museum and Manuscript Library (HMML), April 15, 2021, https://hmml.org/stories/series-medicine-medical-texts-from-timbuktu/.

[17] For one such Ottoman glossary see Österreichische Nationalbibliothek, Vienna, Austria, Cod. Mixt. 944, fols. 119v-156v. Digitized by the Hill Museum and Manuscript Library (HMML), Saint John's University, Collegeville, MN.

[18] Most of the glossaries I examined did not include diacritics, unlike those examined in Vivek Gupta, "Images for Instruction: An Illustrated Multilingual Dictionary in Sultanate India (BL Or. 3299)," in *Muqarnas: An Annual on the Visual Cultures of the Islamic World* (forthcoming). Similarly, in Europe, sometimes the Latin synonyms of Arabic materia medica had accent marks "to assist in the oral delivery of the terms"; see Burnett, "*Synonyma* Literature" (cit. n. 14), 133.

[19] One exception is a formal text by Maimonides, *An Explanation of the Names of Drugs*, which goes into more detail than the glossaries but served largely the same purpose. Maimonides often included the local names for a given material according to "the common people of the Maghreb" or "the people of Egypt" specifically. See Max Meyerhof, *Sarh Asma' al-'Uqqar. (Explication des noms de drogues.) Un glossaire de matière médicale composé par Maimonide. Texte publié pour la prèmiere fois d'après le manuscrit unique avec traduction, commentaire et index*, Memoires presentes à l'Institut d'Egypte, vol. 41 (Cairo, 1940).

groups (*fallāḥīn*, *ʿaṭṭārīn*, *jūgīyān*).[20] The abundance of these glossaries makes it clear how important it was for physicians to know the names that these "speaking people" used for materia medica, and even certain illnesses. While historians have no doubt imagined the interactions between physicians, pharmacists, and the many kinds of people who cultivated, foraged, traded, and transported a vast number of medicinal substances across continents, we have done little to study these exchanges. The growers, foragers, and traders who supplied pharmacists and physicians have not left us texts of their making, so we catch glimpses of these people by reading against the grain of surviving texts. Between the sparse frame of lemma and gloss, lemma and gloss, the glossaries signified their usefulness to a world beyond the text. The reader would have to draw on tacit knowledge to understand when and how the glossaries could be relied on, since they included no statement of method. Unlike formal texts of *ṭibb*, in which the validity of principles and remedies were supported by the author's reputation and citations of authorities from previous generations of physicians, these glossaries employed a different strategy to be convincing and useful tools for physicians—a continuous process of communal verification of knowledge (*taḥqīq*). The reader would know that the author was drawing on both textual and oral sources and, importantly, verifying the knowledge with a number of people in their vicinity.

Because all sources remain unmarked in medical glossaries, besides the occasional references to groups such as farmers, it seems that the names of materia medica were recognized as local, contingent information.[21] The lack of title and anonymity of glossaries showed that they were not meant to be cited by later generations. Both lemma and gloss were written in the same Perso-Arabic script, so these glossaries may not appear at first to be multilingual. All of the lemmas in the glossary in figure 2, *Farhang-i Ṭibb-i Shihābī*, were marked with a red line on top. Glossaries were ordered alphabetically by lemmas, each new entry marked by a red line, but the text was sometimes written in a regular block, like any prose text. The red lines were thus necessary to highlight these terms and make the glossary easily searchable. All the languages were subsumed into a single Perso-Arabic script, even in the many glossaries written in tabular format, whether lined or unlined (as in fig. 1), whether demarcating entries through spacing or by writing the lemmas in red ink.[22] This continuity of script underscores the prestige of Arabic and Persian; texts of *ṭibb* in South Asia were exclusively composed and read in Arabic and Persian until the nineteenth century. But Persian especially functioned as a "composite expressive whole—a sort of hyperlanguage—made up of several different and recognizable languages," including Arabic, Turki, and Hindavi, as Stefano Pellò argues in his study of glossaries in fifteenth-century India.[23] Reflecting on the "almost complete

[20] See Biblioteca Ambrosiana, Arabi Nuovo Fondi H20, fol. 258b; and Telangana State Oriental Manuscript Library and Research Institute, Tibb 355, fol. 81a.

[21] Biblioteca Ambrosiana, Arabi Nuovo Fondi H20, fol. 258b. For example: "*Lisān al-ʿaṣfūr* is *al-ʿadārā* (sacred ficus), which happens to grow in farmland and has red flowers. It is known (*maʿrūf*) to farmers"; and "*Asbyūsh* is *barzqaṭunā* (psyllium), which is known (*maʿrūf*) to pharmacists."

[22] Or each lemma is written larger than the gloss, as in Staatsbibliothek zu Berlin, Glaser 98, fols. 41b-42a.

[23] Stefano Pellò, "Local Lexis? Provincializing Persian in Fifteenth-Century North India," in *After Timur Left: Culture and Circulation in Fifteenth-Century North India*, ed. Francesca Orsini and Samira Sheikh (New Delhi: Oxford Univ. Press, 2014), 166–85, at 176. According to the well-known expression of Amīr Khusrau (d. 1325), "Arabic is knowledge and learning, Turki is mastery, Persian is sweet."

Figure 1. *Page from opening chapter of* Farhang min Ṭibb-i Shihābī, *Tibb 355, fol. 81b-82a. Reprinted courtesy of Telangana State Oriental Manuscript Research Institute and Library.*

absence of Hindavi from Persian texts," Francesca Orsini uses the term *diglossia* to index the hierarchy between Persian and Hindavi, among other vernacular languages in medieval North India.[24] Thus, we realize Hindavi was relevant to the daily practice of *ṭibb* before the nineteenth century only through manuscripts in the Perso-Arabic script, rather than other contemporary scripts.

Across South Asia, the diglossia of Persian and Hindavi is also apparent in the location of these anonymous medical glossaries within manuscripts as mere appendages to properly authored Persian texts. However, Persian and Hindavi were in use at the same time, with oral Hindavi genres circulating among Indo-Persian audiences.[25] Persian-Hindavi glossaries were also included alongside popular erotological texts, like the fourteenth-century *Lazzat al-Nisā*ʾ (The pleasure of women) by Ḍiyāʾ al-Dīn

[24] Francesca Orsini, "Traces of a Multilingual World: Hindavi in Persian Texts," in Orsini and Shaikh, *After Timur Left* (cit. n. 23), 404–5. By contrast, in some manuscripts of the Mediterranean *synonyma* literature, different scripts were used to list the names of materia medica in Latin and Arabic, while names in vernacular *ʿajamiyya* (referring to Romance languages) were written in the Perso-Arabic script. Local words are taken from earlier generations of textual authorities, only occasionally drawn from the author's experiences with people locally. See note 14 of the present article.

[25] Orsini, "Traces of a Multilingual World" (cit. n. 24), 404–5.

Nakshābī (d. ca. 1350). This and other texts in the "Nakshābī tradition" drew heavily on the erotological *Kokaśāstra* tradition in Sanskrit and had a large readership beyond medical practitioners.[26] This is just one example among many of the formal translation of Sanskrit medical knowledge into Persian over seven centuries, a process which—while moving knowledge from one cosmopolitan language to another—often involved oral conversations in a shared vernacular among the learned translators.[27] But as informal medical glossaries show, multilingualism also shaped the more mundane practice of *ṭibb*, beyond formal translation.

Teasing out the relationship between textual and oral knowledge, and between cosmopolitan and vernacular languages, leads us to a better understanding of any science's full community of knowers. The glossary entry is the outer edge of the cosmopolitan, a juncture from where the relationships between languages in *ṭibb* and other sciences are visible. It is in medical glossaries that cosmopolitan and vernacular languages are calibrated with each other, and medicinal substances mediate this link or "articulation." The differences between these registers are softened through their linkage, but also reinforced. The lemma was made to be the formal term: its gloss was contingent, in need of verification and replacement.

As Orsini puts it, "The oral Hindavi world got transcribed in Persian, or else was left out of the archive."[28] However, the hierarchy of languages in the Persianate sphere was not representative of all South Asian languages. By the turn of the second millennium, practitioners of Ayurveda had to choose whether to compose their medical texts in cosmopolitan languages (Arabic, Persian, and Sanskrit) or in vernacular languages (like Bangla, Tamil, Kannada, and Hindavi). There was precedent for composition in both registers.[29] This choice would have been even more pronounced by the seventeenth century, when the glossary in figure 2 was appended to one manuscript of the *Ṭibb-i Shihābī*. Vernacular-language texts of Ayurveda proliferated by the early modern period, perhaps creating an environment in which multiple authors of *ṭibb* explicitly justified their choice to write in Persian over Hindavi.[30] Though they affirmed the language hierarchy, they acknowledged the vernacular as an (inferior) option. And as Persian texts about Ayurveda and Sanskrit texts about *ṭibb* demonstrate, the relationship between language and medical tradition was by no means fixed.[31] Medical glossaries

[26] For examples of this, see Österreichische Nationalbibliothek Cod. Mixt. 801, fols. 96b–136a, digitized by Hill Museum and Manuscript Library (HMML), Saint John's Univ., Collegeville, MN. See also Susanne Kurz, "Laḏḏat al-nisā'," Perso-Indica: An Analytical Survey of Persian Works on Indian Learned Traditions, ed. Fabrizio Speziale and Carl Ernst, November 5, 2018, http://www.perso-indica.net/work/kokasastra_%28laddat_al-nisa%29. For more on this text and the broader erotological tradition, see Sonia Wigh, "The Body of Words: A Social History of Sex and the Body in Early Modern India" (PhD diss., Univ. of Exeter, 2021).

[27] Fabrizio Speziale, *Culture Persane et médecine Ayurvédique en Asie du Sud* (Leiden: Brill, 2018), 36.

[28] Orsini, "Traces of a Multilingual World" (cit. n. 24), 406.

[29] Peter Friedlander, *A Descriptive Catalogue of the Hindi Manuscripts in the Library of the Wellcome Institute for the History of Medicine* (London: Wellcome Institute for the History of Medicine, 1996).

[30] Speziale, *Culture Persane* (cit. n. 27), 56.

[31] Jan Meulenbeld, "Mahādevadeva's Hikmatprakāśa—A Sanskrit Treatise on Yūnānī Medicine. Part I: Text and Commentary of Section I with an Annotated English Translation," *eJournal of Indian Medicine* 5 (2012): 93–133, https://indianmedicine.nl/article/view/24744.

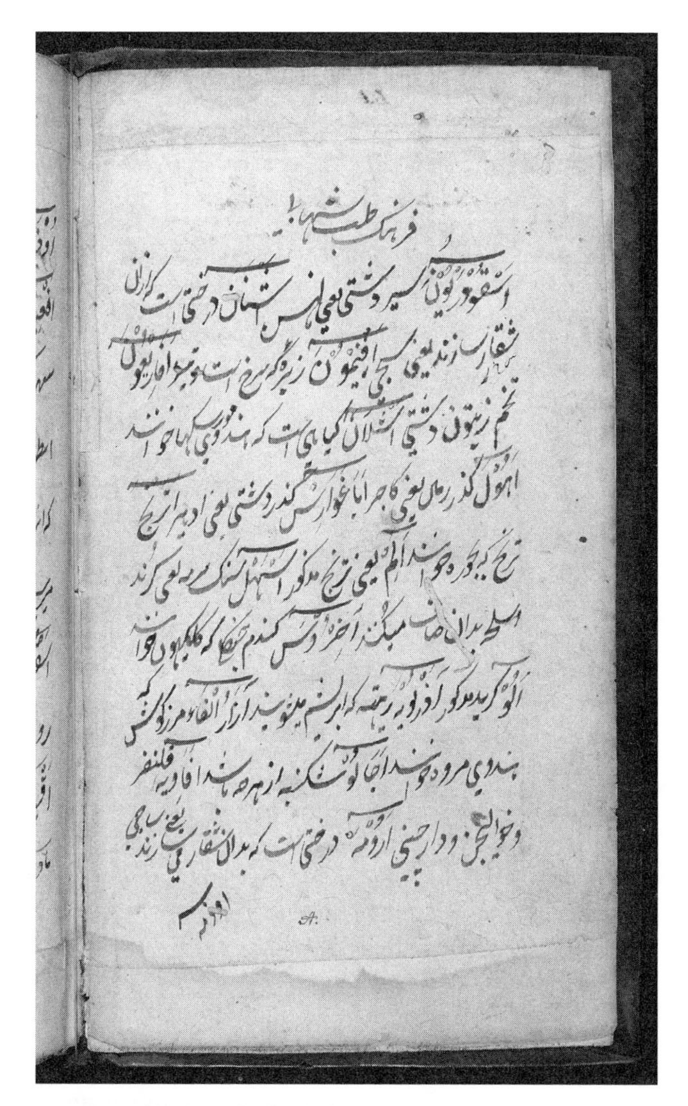
فرهنگ طب شهابی

Figure 2. *Opening page of* Farhang-i Ṭibb-i Shihābī, *The Oriental Collection, Pers. 23, fol. 28a. Reprinted courtesy of The Royal Danish Library under a Creative Commons License.*

enabled the continued use of cosmopolitan languages in *ṭibb* amid these dynamic shifts.

THE FOUR GLOSSARIES OF THE ṬIBB-I SHIHĀBĪ

While the creation and use of medical glossaries was a broader phenomenon across the Islamic world, I focus in this article on *ṭibb* in South Asia rather than the Middle East. We need to disaggregate the "Islamic world" and understand *ṭibb* in its regional specificities to comprehend a medical system used across a huge swathe of Eurasia and parts of Africa. The glossaries give us crucial insights into the particularities of their sites of composition—something that formal texts of *ṭibb* do not usually acknowledge

Table 1. *Four glossaries accompanying the* Ṭibb-i Shihābī

Name	*Farhang min Ṭibb-i Shihābī* (First glossary)	*Farhang-i Ṭibb-i Shihābī* (Second glossary)	*Farhang* (Third glossary)	*Farhang-i Ṭibb dar bayān-i lughāt-i mufradāt bar tartīb-i ḥurūf-i tahajjī* (Fourth glossary)
Manuscript location and accession number	Telangana Government Oriental Manuscript Library and Research Institute, Hyderabad, Tibb 355	The Royal Danish Library, Copenhagen, The Oriental Collection, Pers. 23	Salar Jung Museum Library, Hyderabad, Tibb 145	British Library, London, Indian Office Islamic 1735
Format	Tabular	Prose, lemmas marked by red	Tabular	Prose, lemmas marked by red
Date of composition	ca. 18th century	ca. 17th century	ca. 18th century	February 4, 1721
Number of folios	15 folios, with final folios missing (out of 96 total)	25 folios (out of 111 total)	16 folios (out of 99 total)	40 folios (out of 160 total)
Number of lemmas	1,947 (incomplete)	1,324	1,406	993

or emphasize. The four glossaries I focus on in this article were added to the beginning or end of manuscripts of the *Ṭibb-i Shihābī*, but they still differ in length, format, and content. These appended glossaries were written in manuscripts copied from the seventeenth to the nineteenth centuries. Of the thirteen surviving manuscripts of *Ṭibb-i Shihābī*, five have accompanying glossaries.[32] Often, medical glossaries are not titled, but as illustrated in table 1, the titles of the glossaries range from a single word, *Farhang* (Glossary), to descriptive, as in *Farhang-i Ṭibb dar Bayān-i lughāt-i mufradāt bar tartīb-i ḥurūf-i tahajjī* (Medical glossary explaining the language of simples in alphabetical order). Two of these glossaries are organized in tabular format and tend to be terse, providing a name of a drug for a name of a drug. The other two glossaries occasionally provide longer entries that list the available types of a substance, describe their differences, and state which is most efficacious. This style is facilitated by their prose format;

[32] I studied four manuscripts of the *Ṭibb-i Shihābī* that include glossaries: First glossary: Telangana Government Oriental Manuscript Library and Research Institute, Hyderabad (hereafter cited as TGOMLRI) Tibb 355; Second glossary: The Royal Danish Library, Copenhagen (hereafter cited as RDL), The Oriental Collection, Pers. 23, accessed online at http://www5.kb.dk/manus/ortsam/2009/okt/orientalia/object79867/en/; Third glossary: SJML Tibb 145; Fourth glossary: British Library, London, IO (hereafter cited as BL IO) Islamic 1735. The following manuscripts of the *Ṭibb-i Shihābī* do not include glossaries: Jamia Hamdard, New Delhi, Acc. 2683 (19th century); BL IO Islamic 1828 (18th century); BL IO Islamic 951 (18th century); Princeton University Library, Islamic Manuscripts Third Series no. 829 (early 17th century); TGOMLRI Tibb 292 (undated); Leiden University Library Ar. 4237 (18th century); Hyderabad, Osmania University Library Farsi 264; and SJML Tibb 146, 147, and 148. Fabrizio Speziale relies on two manuscripts, the Teheran, Kitābkhāna-yi Majlis, 14404, which I have not gained access to, and the RDL manuscript mentioned above. See Speziale, *Culture Persane* (cit. n. 27), 79n137.

the author would have had to compose or copy the glossary from beginning to end, since each term takes up a different amount of space.

Before delving further into the specificities of each glossary, it may be helpful to trace mentions of a single substance through each, to give a sense of how much individual glossaries diverge. Ajwain, a common seed used in medicine and cuisine in South Asia and beyond, appears in all the glossaries as the meaning for a variety of terms. In many Arabic and Persian texts of *ṭibb*, ajwain is called *nānkhwāh;* some Arabic texts explain that this is a Persian word meaning "desirer of bread."[33] They largely agree on its heating and drying qualities. Unlike any of the authoritative medical texts, three of the four glossaries offer ajwain as the gloss for the lemma *qurrat al-ʿayn*, a phrase that means "darling" (lit. "solace of the eye"). This lovely phrase occurs along with the more common *nānkhwāh*, though sometimes, *qurrat al-ʿayn* is described as a different type of ajwain from *nānkhwāh*. In the first glossary, further specifications are provided as to where *qurrat al-ʿayn* can be found. In the second glossary, this same phrase is translated as *ajwāin khurdaʾī*, or "edible ajwain." And in addition to these names for ajwain, two of the glossaries have several other lemmas that also translate to it. In the second glossary, the author provides *naqar khwān, walakhnīt*, and *wadd* as equivalents for *ajwain khurāsānī*. In the third glossary, *ajmūd* is the gloss for *ajwain khurāsānī* and *zanyān*, while *shanā* is the gloss for *chūrā ajwain* and *naghzkhūlān*. This profusion of terms for variants of a single material was not uncommon. While some of these names remain mysterious, others have come to refer to other plants: *qurrat al-ʿayn* for "watercress" in standard Arabic, *ajmūd* for "parsley" in Persian. The variation across glossaries points to the temporal instability and regional specificity of terminology, adding strength to Guy Attewell's argument that *ṭibb* is not one system across time. Calling *ṭibb* a system has its own history rooted in the professional concerns of practitioners at the turn of the twentieth century.[34]

The authors of glossaries were in conversation with a variety of people to obtain this profusion of names for materia medica, people whose role is obfuscated in formally authored texts. Many historians of science are now attending to the epistemologies of people who were efficacious in manipulating the natural world, working primarily in the oral and experiential rather than the textual.[35] The glossaries point to how generative their methodologies will be for the history of science and medicine in the Islamic world. A focus on this line of inquiry would lead to more sources being considered as relevant to the history of medieval science. For example, recent work on the medieval history of the language of the peripatetic Banū Sāsān people shows that among speakers of this language were medical practitioners—but not physicians. Speakers of this "language of *ghurabāʾ* (strangers)," as Banū Sāsān's language was often called, included bloodletters, cuppers, "sellers of unguents and medicines,"

[33] al-Muẓaffar, *al-Muʿtamid* (cit. n. 13), 372.
[34] Guy N. A. Attewell, *Refiguring Unani Tibb: Plural Healing in Late Colonial India* (New Delhi: Orient Longman, 2007), 21–9.
[35] Pamela Smith, "Nodes of Convergence, Material Complexes, and Entangled Itineraries," in *Entangled Itineraries: Materials, Practices, and Knowledges across Eurasia*, ed. Smith (Pittsburgh, PA: Univ. of Pittsburgh Press, 2019), 5–24; Pablo Gómez, *The Experiential Caribbean: Creating Knowledge and Healing in the Early Modern Atlantic* (Chapel Hill: Univ. of North Carolina Press, 2017); Marcy Norton, "Subaltern Technologies and Early Modernity in the Atlantic World," *Col. Latin Amer. Rev.* 26 (2017): 18–38.

and practitioners of eye medicine, alongside beggars, astrologers, and others.[36] Training in literary Arabic and Persian was largely accessible to particular classes and genders of people in a social hierarchy. Looking at other languages may reveal more of premodern medicine beyond male, literate physicians, in whose work historians of medicine have primarily been interested. No doubt, there is still much work to be done in that vein, considering the number of Arabic and Persian *ṭibb* manuscripts in libraries all over the world that have not yet been studied.[37] Persian texts—and thus the history of sciences in Persia, Central Asia, and South Asia—have received less attention in European language scholarship than Arabic texts, largely because Persian texts were not as directly relevant to the history of European science as Arabic texts composed in the Mediterranean littoral.[38] This paucity should soon be remedied. But these glossaries may help to put the textualized knowledge of physicians in broader context. As Alberts, Fransen, and Leong argue in their introduction to this volume, the "articulation" of translation is contingent and context-dependent; the "linkage" between words in a glossary is, as Stuart Hall notes, "not necessary, determined, absolute or essential for all time."[39] The format, anonymity, and marginality of the glossaries show just how contingent the authors of these translations into the vernacular considered them to be.

GLOSSARIES BEYOND MEDICINE

The glossaries appended to the ends of texts of *ṭibb* were mostly concerned with correctly identifying plant-based materia medica. Comprehensive coverage of medical terminology was taken on by hefty dictionaries like the *Baḥr al-Jawāhir fī taḥqīq al-muṣtalaḥāt al-ṭibbiyya* (The Sea of gems in the verification of medical terminology), composed in 1518, mentioned below. The short multilingual glossaries never became an autonomous genre of their own—they aimed to assist readers in medical encounters.[40] Their content was also less consistent than that of formal herbals or pharmacopeias, which described and classified the properties of materia medica at length, though pharmacopeias sometimes included local names for a given substance, specifying the region in which a given name was used. A genre of dictionaries in *ṭibb* about "Indian medicines" became prominent only at the turn of the nineteenth century, though glossaries translating from Sanskrit to vernacular languages in Ayurveda already constituted their own genre.[41]

[36] Kristina Richardson, "Tracing a Gypsy Mixed Language through Medieval and Early Modern Arabic and Persian Literature," *Islam-Zeitschrift für Geschichte und Kultur des Islamischen Orients* 94 (2017): 115–57, at 124, 128, and 139.

[37] Let alone texts in Judeo-Arabic, Judeo-Persian, Bengali, Tamil, Dakhani, Telugu, and Malay (Jawi)! Attewell gives examples of Tamil and Telegu texts printed in the early twentieth century, including "a glossary of Hindustani and Telegu names of drugs"; Attewell, *Refiguring Unani Tibb* (cit. n. 34), 19–20.

[38] Greco-Arabo-Persian medicine began not long after the inception of Greco-Arabic medicine; see L. Richter-Bernburg and H. M. Said, "Medical and Veterinary Sciences," *Differentiation* 750 (1996): 305–30, at 310–1.

[39] Stuart Hall and Lawrence Grossman, "On Postmodernism and Articulation," *Essential Essays, Volume 1: Foundations of Cultural Studies* (Durham: Duke University Press, 2018), 53, cited in Alberts, Fransen, and Leong, "Translating Medicine, ca. 800–1900" (*Osiris* 37).

[40] Speziale, *Culture Persane* (cit. n. 27), 78–80. For a few rare exceptions, see the following: BL IO Islamic 88; Dār al-Kutub, Cairo, Tibb 291; TGOMLRI Tibb 282; and BL Add. 17948.

[41] Speziale, *Culture Persane* (cit. n. 27), 49 and 67. Thank you to Eric Gurevitch for bringing these freestanding glossaries, *nighantus*, to my attention. For more on *nighantus* in colonial contexts, see Minakshi Menon, "What Is Indian Spikenard?," *South Asian History and Culture* (forthcoming).

Medical glossaries were more concise and practical than literary glossaries designed to aid in the understanding of canonical authors and cultivate literary expertise, but they borrowed the titles of *farhang* or *lughat* from this long literary tradition. Lexicography, called *ʿilm al-lugha*, or "the science of language," was a discipline of growing importance from the eighth century onward in the Islamic world, as Arabic became a language of governance in addition to literature and the religious and natural sciences.[42] As the use of Arabic as a language of religio-political significance spread in the eighth century, these dictionaries, written by and for poets, bureaucrats, and ulema, set standards of aesthetic excellence. The earliest bilingual Arabic-Persian glossaries were compiled in the twelfth and thirteenth centuries to help readers of Persian navigate literary and religious texts. Within this broader literary context, the reader of a medical glossary may have understood how to use the text, though medical glossaries were meant to help a reader navigate oral rather than literary interactions.

The dictionary in the Persian tradition, often known as a *farhang*, was "an ancillary discipline to poetry," meant to help students of Persian read epic poetry by glossing the difficult regional or archaic vocabulary of works like the *Shāhnāma* (Book of kings) by Firdausī (d. 1020 CE).[43] Though these *farhangs* were "monolingual" Persian literary glossaries, compiled from the ninth to the eleventh centuries and organized topically or alphabetically, they worked across registers to translate words unknown outside a given province. While early *farhangs* in the ninth century generally contained one to two thousand entries, the *farhangs* of the fourteenth century swelled to tens of thousands of words.[44] Readers of literary *farhangs* needed access to language that was increasingly removed in time and space from the original site of composition of texts such as the *Shāhnāma*. This was the key to writing and speaking a refined Persian for people living in Persia as well as those in the broader "Persianate" realms of Central and South Asia, where Persian was the language of governance, literature, and science but few people's mother tongue.[45] India became the primary site for the compilation of Persian *farhangs* between the thirteenth and eighteenth centuries, which were "explicitly comparative enterprises that sought to mediate between the desire for cosmopolitan coherence, on the one hand, and the interplay of local vernacularity on the other."[46] The challenge of maintaining a literary tradition is reminiscent of the problem faced by physicians, who tried to keep medical terminology in *ṭibb* consistent across many languages. The variation in medical glossaries show this was not always possible in the vernacular.

Like the medical glossaries, lexicographical texts were all written in the same Perso-Arabic script, whether they were in Arabic, Persian, or Arabic-Persian. Visually and

[42] J. A. Haywood, D. N. MacKenzie, and J. Eckmann, "Ḳāmūs," in *Encyclopaedia of Islam*, 2nd ed., ed. P. Bearman, Th. Bianquis, C. E. Bosworth, E. van Donzel, and W. P. Heinrichs (Leiden: Brill, 2000); online ed., 2012, http://dx.doi.org.ezp-rod1.hul.harvard.edu/10.1163/1573-3912_islam_COM_0434.

[43] Pellò, "Local Lexis?" (cit. n. 23), 173–4.

[44] Solomon I. Baevskii, *Early Persian Lexicography: Farhangs of the Eleventh to the Fifteenth Centuries*, trans. N. Killian, rev. John R. Perry (Kent, UK: Global Oriental, Ltd., 2007), 29–116.

[45] Nile Green, ed., *The Persianate World:The Frontiers of a Eurasian Lingua Franca* (Berkeley: Univ. of California Press, 2019).

[46] Rajeev Kinra, "This Noble Science: Indo-Persian Comparative Philology, c. 1000–1800 CE," in *South Asian Texts in History: Critical Engagements with Sheldon Pollock*, ed. Yigal Bronner, Whitney Cox, and Lawrence McCrea (Ann Arbor, MI: Association for Asian Studies, 2011), 359–85, at 361.

textually, these literary glossaries absorbed regional variants and vernacular languages into the cosmopolitan language. Some scholars have argued that "using an Arabic and a [Hindavi] lexical equivalent in the same entry and at the same level would have given the latter language a certain prestige," and thus have played some role in the literarization of vernacular languages in South Asia.[47] A lexicography completed by Badr al-Dīn Ibrāhīm in 1433, the *Farhang-i zafāngūyā u jahānpūyā* (Lexicon of the polyglot and experienced world-traveler), likely received the patronage of the Ghurid dynasty in Malwa, Western India.[48] At least half of the ninety-two Hindavi equivalents to Persian words in this text refer to medicines and medicinal plants; clearly, there was a desire to know vernacular names of materia medica among a reading public well beyond physicians.[49] As the *farhang* tradition grew, it included more terms relating to everyday life, making these texts valuable sources of social history and local knowledge.[50] Alongside these massive lexicographies, humble medical glossaries exemplify this untapped historical potential.

GLOSSARIES AS LOCAL TEXTS

Anonymous medical glossaries circulated in local communities. Those writing the glossaries had expectations about what their readers would and would not be familiar with. Many substances within the glossaries are listed as "well-known" (*maʿrūf*) or "famous" (*mashhūr*), or are marked with a mīm, the first letter of these two words.[51] In the second glossary (mentioned in table 1), for example, we find "*hawhīrah*: a famous medicine (*adwiyya mashhūra*)"—the term is thus left undefined.[52] In the third glossary, *zaytūn* (olives), *rayḥān* (aromatic plants), and *kūknār* (poppy seed) are all described as famous (*mashhūr*).[53] Similarly, *rūdantī* is simply glossed as "a famous plant" in the fourth glossary.[54] The *maʿrūf* convention was also common in the literary *farhang* tradition, but it takes on a new meaning in the context of glossaries aiding physicians in the acquisition of medicinal substances or in communicating diagnoses and treatments with patients in a multilingual environment. While certain substances listed as "known" had to be learned in the cosmopolitan language, with no vernacular equivalent, the convention also gestures toward the destructive aspect of translation, as one term wins out over multiple, local names of a substance or disease, and all of their conceptual specificities. As Alberts, Fransen, and Leong argue in their introduction to

[47] Pellò, "Local Lexis?" (cit. n. 23), 182.

[48] While these texts may have contributed to the reputation of patrons, these translations were not institutionally produced, unlike the bilingual glossaries in multiple scripts created by the imperial Ming's Translators' College; Carla Nappi, "Full. Empty. Stop. Go. Translating Miscellany in Early Modern China," in *Early Modern Cultures of Translation*, ed. Karen Newman and Jane Tylus (Philadelphia: Univ. of Pennsylvania Press, 2015), 206–20.

[49] This text also included six hundred "Turki" words, a name that "refers to Chaghatai . . . a Turkic literary language of Central Asia of the fourteenth to sixteenth centuries"; Dilorom Karomat, "Turki and Hindavi in the World of Persian: Fourteenth- and Fifteenth-Century Dictionaries," in Orsini and Shaikh, *After Timur Left* (cit. n. 23), 130–65, at 153.

[50] Baevskii, *Early Persian Lexicography* (cit. n. 44), 177–206; Roy Mottahedeh, "Medieval Lexicography on Arabic and Persian Terms for City and Countryside," *Eurasian Studies* 16 (2018): 465–78.

[51] See BnF Supplément Persan 335, fols. 1b-2a, for an example of the mīm convention.

[52] RDL, The Oriental Collection, Pers. 23, fol. 27a.

[53] SJML Tibb 145, fols. 9b, 10b, and 14a.

[54] BL IO Islamic 1735, fol. 16b.

this volume, this process can be accidental and reluctant, or intentional and, in some contexts, violent, with the newly dominant term considered to be an "improvement" over the one it replaces.[55] This intentionality of improvement is reminiscent of the perspective of physicians seeking to promote and normalize the use of their discipline's terminologies, shuffling the glossaries with vernacular terms off into the margins.

Though local language was less valued by physicians than cosmopolitan terminology, it is a boon for historians. The glossaries sometimes specify where medicinal substances can be obtained or where the best variety can be found. The lemma for *marjān*, or "coral," in the fourth glossary explains, "They call it *bisidra* as it resembles a tree from the middle of the sea. They bring it out from a mountain."[56] Some descriptions specify the kind of environment where a given plant grows: one example explains that "wild *fawdaj* is *naghḍ*, known to grow close to rivers"; another describes the three types of *zift*, or "tar," as of the sea, desert, and mountains (*daryā'ī, ṣaḥrā'ī, kūhī*).[57] This may imply that the glossaries were meant to circulate in a relatively immediate geography, where the market for locally cultivated (*bustānī*), wild (*barrī*), and imported versions of each plant would have been consistent. Since the availability of substances varied widely across the vast geographies in which texts of *ṭibb* circulated and informed practice, they would hardly have had relevance throughout the entire region.[58] In specifying the location where plants were best or most easily obtained, the glossaries often index the broader world of Indian Ocean trade.[59]

Except for in the fourth glossary, the languages of both lemma and gloss are not marked; the reader would be expected to figure out which word was in which language, a task for which they would need familiarity with *ṭibb*. Most entries are words in transliterated Greek or Syriac, as well as more obscure Arabic and Persian words.[60] The hierarchy of languages is further affirmed by the alphabetical organization by lemmas in Arabic and Persian, rather than by the glosses in Persian and Hindavi.[61] These sometimes include categories rather than specific objects, like aromatic plant (*rayḥān* in Arabic, glossed as *sabzī* in Persian), wind (*rīh* in Arabic, glossed as *bād* in Persian), and flower (*ward* in Arabic, glossed as *gul* in Persian).[62] Because many Persian lemmas in the glossaries were actually loan words from Arabic, these glossaries also demonstrate how the boundary between Arabic and Persian terminologies

[55] Leong, Fransen, and Alberts, "Translating Medicine, ca. 800–1900" (cit. n. 39).

[56] BL IO Islamic 1735, fol. 34a. Another example can be found in the Yemeni glossary composed in 1720, Biblioteca Ambrosiana, Arabi Nuovo Fondi H20, fol. 258a: "*[J]āwshīr* is a red gum which is brought from Fārs (Persia)."

[57] BL IO Islamic 1735, fol. 17b.

[58] Sometimes, the place names added to plants are unique to a locale, as when many plants are called "*al-hindī*," or from India, in the Yemeni glossary mentioned in n. 56. These are common substances, not known as "Indian" except in Yemen: for example, the well-known black seed of prophetic medicine (Biblioteca Ambrosiana, Arabi Nuovo Fondi H20, fol. 258b).

[59] Some names come from origin stories about the substance, like the story for *al-ṭabāshīr* in the eighteenth-century Yemeni glossary mentioned in footnotes 12 and 56–8 (Biblioteca Ambrosiana, Arabi Nuovo Fondi H20, fol. 258b). The same story can also be found in Ibn Sīnā's Canon.

[60] A few of these terms were created when Sanskrit knowledge was appropriated in the ninth century Translation Movement; one such example is *aṭrīfal*, the Arabization of *triphalā*, an electuary of myrobalans in Ayurveda. *Aṭrīfal* is a lemma in almost every glossary; Speziale, *Culture Persane* (cit. n. 27), 57.

[61] BL IO Islamic 1735, fol. 38b; "*nārjīl pārsī jawz hindī*." The lemma is in Persian and it was included under the first letter of the lemma rather than the first letter of the gloss in Hindavi.

[62] RDL, The Oriental Collection, Pers. 23, fols. 13b, 13b, and 26b, respectively.

was shifting in medical contexts, as were the norms for medical instruction. Persian words also needed to be glossed in some of these entries, as is shown for the word *walnut* (*bakhgale* in Persian, *jawz* in Arabic, *akhrūt* in Hindavi), though Persian lemmas with Arabic glosses are rare.[63]

Further, the glossaries demonstrate that *ṭibb* had vocabularies and norms unique to local geographies, and that there was no one vocabulary consistent within so-called Islamic medicine. Some lemmas have generic meanings in Arabic but seem to refer to the specific names of illnesses in local contexts: *wajʿ*, which means "pain" in Arabic, is translated in the second glossary as *dard-i chashm*, or "eye pain," in Persian. Similarly, *yabūs*, which just means "dry" in Arabic, is translated as *khushgī-yi duhn*, or "dryness of the mouth," in Persian.[64] Bizarrely, *ruṭūbat*, or "moisture" in Arabic, is glossed in the fourth glossary as *sard-u sardī*, the Persian words for "cold," and in the first glossary as *tarī balgham*, the Persian words for "wet phlegm."[65] These specific meanings for general Arabic terms may have become colloquialisms that did not exist in other regions—they carried these meanings only for the people who compiled these glossaries in India. As the field turns toward more regional studies of science in the Islamic world, the agency of nonliterate people may become more easily visible.

LEARNING ṬIBB IN INDIA

As these examples demonstrate, the glossaries give us insight into the process of learning *ṭibb* in a multilingual context like South Asia. The question of medical education is crucial to understanding the origin of this genre. Shihāb Nāgaurī, the author of the text to which these glossaries were appended, grew up in an intensely multilingual environment. All that we know of his life comes from what he tells us in his two surviving texts, both written in Persian. He likely lived in Rajasthan before moving to Gujarat at the end of the fourteenth century to seek the patronage of the governor, Ẓafar Khān. There, he may have also witnessed multilingual inscriptions in multiple scripts, some combination of Sanskrit, Old Gujarati, Arabic, and Persian.[66] Other inscriptions, though bilingual in Arabic and Persian, would have been written in the same script. His inclusion of Hindavi words in the Persian script would not have seemed unusual, though he would likely have been aware that Hindavi had already been written in other scripts. Throughout his text, he gives the local names of ailments (using the verbs *khwānand* and *gūyand*) either as a part of the chapter title or in the first verse of a chapter.[67]

Nāgaurī did not start out practicing medicine but decided to undertake its study at a later stage of life, which perhaps prompted him to write the *Ṭibb-i Shihābī* in didactic verse: a concise and accessible text for students. He included an autobiographical chapter at the end of this book, which is unusual.[68] He tells us of his family's history

[63] Ibid., fol. 26b.

[64] Ibid., fol. 27a.

[65] BL IO Islamic 1735, fol. 16b; TGOMLR Tibb 355, fol. 89a.

[66] Samira Sheikh, "Languages of Public Piety: Bilingual Inscriptions from Sultanate Gujarat, c. 1390–1538," in Orsini and Shaikh, *After Timur Left* (cit. n. 23), 187–210.

[67] In the first verse of chapter forty-eight on *kulfah*, he says, "they call it *chahāʾī* in India" (*bi khwānand chahāʾī bi hindūstān)*; see Jamia Hamdard Acc. 2683, fol. 17b.

[68] Shireen Hamza, "A Hakim's Tale: A Physician's Reflections from Medieval India," *Asian Medicine* 15 (2020): 63–82. Some surviving manuscripts omit this chapter.

of migration from Ghazni to India several generations before he was born. They were appointed governors of Nagaur, Rajasthan—a position they then lost. The most recent generation of his ancestors thus worked in the administrations of other rulers, and he followed suit. Nāgaurī and his father became more devoted to Islam at the behest of a charismatic sheikh and decided to leave their positions for more humble work, which led Nāgaurī to medicine. Thus, as an Indian-born Muslim, he certainly spoke vernacular Indian languages like Hindavi, in addition to being trained in a refined Persian.[69] He would have orally translated between the cosmopolitan and vernacular on a daily basis, but he also made the decision to include some translations of Persian terms in Hindavi in the *Ṭibb-i Shihābī*.

Two aspects of Nāgaurī's autobiographical chapter are relevant to the present discussion. First, Nāgaurī stressed the place of orality in medicine. He believed that students of medicine needed teachers if they were to properly access the knowledge in texts, thus ensuring that learning would have an indispensable oral component. This was true in Nāgaurī's own education: he learned Galenic medicine from a *ḥakīm* for a month with the aid of a Hippocratic work and other books, and studied Ayurveda in the presence of a yogi.[70] Nāgaurī refers to medicine as both *ṭibb* and *ḥikmat*, but in addition to these more common names, he calls his treatment of patients "*guftan-i adwiyya*" and "*dawā gūyī*," literally, to speak medicine.[71] Nāgaurī writes about using multiple diagnostic processes, such as pulse diagnosis and urine analysis, in treating his patients, but "speaking medicine" suggests that perhaps he considered the conversations between practitioner and patient to be a diagnostic method as well. Nāgaurī cannot be made the sole representative of *ṭibb* in his time and period, or in the "medieval Islamic world." Other authors in fourteenth- and fifteenth-century Gujarat took different approaches to the inclusion of Indic medicine and language in their texts, reminding us that *ṭibb* was not a uniform system. Nāgaurī's biography shows that the oral was a lively realm of medical education and practice with unique, local qualities.

TEXTUALITY AND ORALITY

In the glossaries composed centuries after Nāgaurī's death, it is clear that the process of compiling a glossary was textual as well as oral. When I counted the number of terms glossed per chapter, I tracked which chapters had the most and least terms. The chapters of terms beginning with the letters *ḍād* and *dhāl*, letters with unique pronunciations in Arabic and which are used less commonly in Persian, had less than 1.5 percent of the total terms in the glossary. This ratio was true across all the glossaries and suggests that those writing the glossaries were often adapting their lists of lemmas from older glossaries, though they felt no need to acknowledge sources for either lemmas or glosses.

[69] Fabrizio Speziale has highlighted the creativity of Nāgaurī's synthesis of Ayurvedic and Galenic physiology; he incorporated wind (*vāta*, an Ayurvedic *dosa*) into a new four-humor framework; see Speziale, "A 14th Century Revision of the Avicennian and Ayurvedic Humoral Pathology: The Hybrid Model by Šihāb al-Dīn Nāgawrī," *Oriens* 42 (2014): 514–32; and Speziale, *Culture Persane* (cit. n. 27), 90–101 and 168.

[70] He also emphasizes orality when describing the effectiveness of the arguments (*ḥujjat*) of his sheikh.

[71] Hamza, "A Hakim's Tale" (cit. n. 68), 69. Nāgaurī also uses the words for physician derived from *ṭibb* (*ṭabīb*) and *ḥikmat* (*ḥakīm*) for practitioners of both Galenic and Ayurvedic medicine.

But the importance of the oral to the creation of these glossaries is most evident when they mention other groups of people, some nonliterate, whom physicians must have interacted with in an oral register. Most glossaries communicated equivalence between lemma and gloss through a simple juxtaposition, though some, like the second and fourth glossaries, include commentary in certain entries. Sometimes, longer entries specify the communities in which medicinal substances are known by a certain name.[72] These imply that the plants in question would be available from farmers or pharmacists who were knowledgeable about them. Thus, for a physician to obtain them directly from someone in that professional group, he would need to know the names in use among these professional communities. The first glossary includes many such entries: "*Astarak* is a plant which in India is called (*gūyand*) *makhnā mūl*. The yogis (*jūgīyān*) dig up its roots."[73] These group names remind us that physicians may have had strong incentives to build relationships with a range of nonliterate people to succeed in their practice.

Occasionally, the translations are qualified by verbs. As mentioned above, this usually takes the form of the verbs *gūyand*, "they say," and *khwānand*, "they read/recite," which generally means "as they say" or "which they call." For example, in the second glossary, we find "*lāghiyya*, a plant that they call (*gūyand*) *jhalat*."[74] The meaning of these verbs is significant and suggests an environment of orality in which the compiler and reader of the glossary seek to match what people call a plant or a disease with its name in authoritative terminology. Because most of the glossaries do not explain which word is in which language and expect the reader to be able to discern this, these glossaries are meant to aid those with enough prior knowledge of textual *ṭibb* to recognize the lemmas. At the least, the reader should be able to read the Perso-Arabic script and recognize the glosses provided. Still, because the text is organized alphabetically by the lemmas, it would be most useful to those already familiar with the specialized terminology of *ṭibb*. But the omnipresence of *gūyand* makes one wonder: who exactly is naming and speaking about these plants in the oral register?

These passive verbs, or verbs with vague subjects, emphasize the hierarchy of textual knowledge in *ṭibb*. Though some fluency in the vernacular languages of a region would have been useful, and necessary, to a practicing physician like Nāgaurī, this was not emphasized or valued as something that one should learn within authoritative texts of *ṭibb*. The language used by nonlearned people was considered tacit knowledge, something physicians would rely on while treating or trading with them. Although most physicians relied on the people who grew and traded in medicinal substances, these individuals are not often explicitly recognized within texts of *ṭibb*.[75] While texts of *ṭibb* are not known for their meticulous citation, it is a strategy used when a physician needed to marshal the authority of a source; for example, when entering a debate on a contentious physiological matter or emphasizing the efficacy of an unusual recipe. The

[72] Biblioteca Ambrosiana, Arabi Nuovo Fondi H20, fol. 258b. For an example, see note 21 of this article.

[73] TOGMLRI, Tibb 355, fol. 81a.

[74] RDL, The Oriental Collection, Pers. 23, fol. 28a. In this case, I was unable to identify either lemma or gloss.

[75] For a citation of a nonliterate man in a pharmacological text ("Ṭāriq al-Shajjār [the tree specialist] here in Cordova"), see Zohar Amar, Efraim Lev, and Yaron Serri, "On Ibn Juljul and the Meaning and Importance of the List of Medicinal Substances Not Mentioned by Dioscorides," *Journal of the Royal Asiatic Society* 24 (2014): 529–55, at 552.

conspicuous absence of even a single textual citation, the kind of citation more common in regular medical texts, indicates that authority functions differently in the glossary than in more formal genres of medical texts. The glossary contains useful and communally determined knowledge verified by the compiler and is not meant to have the lasting authority of the knowledge textualized by physicians in their tomes.

As the primary purpose of these glossaries was to facilitate the accessibility of materia medica and to correlate available medicinal substances and textual knowledge, the authors sometimes needed to mention specific communities. No individual is ever credited with helping the physicians identify these plants. No name is given as a source to make this knowledge more legitimate. All nonliterate people who are mentioned are anonymous, and their presence blurs together through the passive verb. While this may at first seem elitist, this third-person passive verb was a convention in the *farhang* tradition, as mentioned earlier. It could also demonstrate a sense of the communal origins and ownership of knowledge. Perhaps including individual names in such a sparse genre would have been counterproductive, especially the names of people who had not authored works and thus whose knowledge could not be located and accessed, even if cited.

As with the vague mention of "yogis" above, there are also minimal references in *ṭibb* to specific people who are learned in other traditions of medicine, even if those other traditions are also textual.[76] As Fabrizio Speziale has suggested about the premodern process of translating Ayurvedic texts from Sanskrit into Persian, this method involved a team of learned practitioners of both Ayurveda (*vaidyas*) and *ṭibb* (*ḥakīms*).[77] However, though many of these translations were done by respected scholars and sponsored by rulers in elite court contexts, most of those involved in the translation process are not named, and the Hindavi intermediary translations they almost certainly relied on have not been preserved.[78] If oral sources of knowledge are not cited within these august scholarly circles of textual translation, it is no surprise that the communities of farmers, foragers, and traders involved in more everyday translation processes were not acknowledged by the conventions of textual citation.

TRANSLATION AS TAḤQĪQ (VERIFICATION)

The authors of glossaries may have thought of this process of everyday translation as *taḥqīq*, as they sought to ensure the accuracy of the translations and equivalences they were drawing between the names of plants. Thus, none of the glossaries are identical—even those that bear the same name are unique in content. Glossaries were open and fluid texts, meant to be altered and emended by each physician who made one anew. The physician would verify this knowledge, ensuring it was useful in the region and period in which the glossary was produced. It is clear from these glossaries that a dynamic process of translation was a crucial part of medicine across the medieval Islamic

[76] Dror Weil's essay in this volume offers a fascinating counterpoint to the kind of translations discussed here, both in terms of how some translators maintained the link to original texts and authors, and in the seventeenth-century vernacular-language summaries of oral discourses explaining Arabo-Persian texts; Weil, "Unveiling Nature," in *Osiris* vol. 37.

[77] Speziale, *Culture Persane* (cit. n. 27), 72–4 and 162–3.

[78] Ibid., 72. The exceptions are the names of the authors of the source text and the translation, such as Vāgbhaṭa and Muḥammad ibn Ismā'īl al-Aṣāwilī respectively, when the *Aṣṭāṅgahṛdaya* was translated as the *Ṭibb-i-Maḥmūd Shāhī* in fifteenth-century Gujarat. See Hamza, "Islam and Medicine" (cit. n. 2).

world. Physicians became accustomed to creating equivalences between terms across a cosmopolitan-vernacular axis by matching older, textual terminology to the local names for materia medica, and sometimes ailments too. Because of the dynamism of spoken language and the variation of plant life across time and region, none of these translations could be taken for granted as stable: they needed to be continuously updated, trimmed, corrected—in short, "verified"—through a process both oral and textual.

But *taḥqīq* was not an individual critical mode; physicians relied on a community closer to their region and time than the remote authorities of *ṭibb*. The imperative of *taḥqīq* drove physicians practicing *ṭibb* in premodern South Asia to bring the vernacular into their Arabic and Persian texts. The belief that the cosmopolitan is more valuable than the vernacular as a language of science, in epistemologies both pre-colonial and Orientalist, has played too large a role in the history of medicine. The *taḥqīq* of physicians willing to acknowledge this reality closed the distance across languages kept apart in the hierarchy of diglossia even though they were in use at the same time.

Most scholars studying *taḥqīq* have focused on logic, philosophy, and theoretical medicine. Several historians of the Islamic world have recently written about the importance of *taḥqīq* as verification. Within medico-philosophical debates, *taḥqīq* was refined by Fakhr al-Dīn al-Rāzī (d. 1219), Ibn al-Nafīs (d. 1288), and other "post-classical" authors into a precise analytical and exegetical method, deployed by subsequent generations of writers debating theoretical topics within medical commentaries.[79] In the context of new movements in seventeenth-century philosophy, some scholars defined *taḥqīq* as the "independent logical demonstration of the truth" of a scholar's views, in contrast to *taqlīd*, the "uncritical acceptance of received philosophical views."[80] Some have gone so far as to characterize all of early modern "Afro-Eurasian" intellectual history in terms of *taḥqīq* and *taqlīd*, including that of early modern Europe.[81] Turning to medicine and other natural sciences will reveal the other facets of *taḥqīq* by acknowledging other critical methodologies besides logical demonstration—in this case, experiential and oral knowledge about material objects.

In the context of medical glossaries and dictionaries, an author's *taḥqīq* is a critical assessment of received wisdom by confirmation of the validity of equivalences across languages. *Taḥqīq* is both an actor's category and an analytical tool for understanding the epistemic work involved in translating medicine across multiple languages simultaneously. Despite Islamic authors' stated reverence for, and frequent references to, the earliest authorities of *ṭibb*, the continuing validity of previous information could not be assumed by each new generation of physicians—especially regarding materia medica. In other words, physicians recognized that there was a dynamic, local element to this knowledge.

•

[79] Karimullah, "Emergence of Verification" (cit. n. 5); Nahyan Fancy, "Verification and Utility in the Arabic Commentaries on the Canon of Medicine: Examples from the Works of Fakhr al-Dīn al-Rāzī (d. 1210) and Ibn al-Nafīs (d. 1288)," *J. Hist. Med. Allied Sci.* 75 (2020): 361–82. Fancy argues in pages 375–81 that Ibn al-Nafīs draws on his practical experiences and observations to intervene in theoretical debates. Though his process was more philosophically oriented than what we find in the medical glossaries, Ibn al-Nafīs' reliance on observation and experimentation lends strength to the notion that the authors of glossaries may have considered their experiential work to be a form of verification.

[80] Khaled El-Rouayheb, *Islamic Intellectual History in the Seventeenth Century: Scholarly Currents in the Ottoman Empire and the Maghreb* (Cambridge, UK: Cambridge Univ. Press, 2015), 28.

[81] Melvin-Koushki, "Taḥqīq vs. Taqlīd" (cit. n. 5).

Physicians were not alone in constituting *ṭibb*. Their dependence on material substances that they did not grow, gather, or transport themselves necessitated an interplay between the oral and the textual in fixing the names of medical substances. Of the six hundred plants found in Dioscorides's (d. ~90 CE) *De Materia Medica* or the hundreds more added by authors in subsequent centuries, some were no longer available by the ninth century, or were considered unidentifiable—knowledge lost to time. Many herbals, like the one written in the twelfth century by al-Ghāfiqī, included glossaries that traced the origin of plant names in use at the time, and provided equivalents across several languages, including Arabic, Greek, Syriac, Latin, Amazigh, Romance, and Persian.[82] Perhaps the diglossia of multiple cosmopolitan and vernacular languages at one end of the Islamic world, in al-Andalūs, may be productively compared with the other end of Eurasia, as studied by Dror Weil in this volume.[83] However, these long lists of names of materia medica across languages did not make it into most practical texts of *ṭibb*, let alone descriptions or illustrations of whole plants. Standard medical texts helped physicians identify dried seeds, roots, gums, and other commodities in the marketplace and would not have been useful to those foraging for whole plants. Even in an unusually practical pharmacist's manual composed by the Cairene al-Kūhin al-ʻAṭṭār in 1260, the processes of identifying and testing the quality of materia medica were undertaken at the shop, not in the field.[84] And while physicians and pharmacists are mentioned in biographical dictionaries of the time, other laborers, such as foragers, are not.[85] The division of labor in *ṭibb* is reflected in the archive. Though the few physicians who did forage themselves are praised, the prestige of their experiential knowledge was tied to their ability to contribute to a long textual tradition of botany, from Dioscorides onward.[86] The foragers, farmers, and yogis referenced in medical glossaries who did not participate in this textual tradition could not become named authorities.

The multilingual medical glossaries could have perhaps enabled literate, male physicians to enter the domain of growers and foragers. Glossaries may have shown physicians how to access materia medica directly instead of having to rely on the pharmacist.[87] The author of the fourth glossary introduces the text with this very suggestion about its usefulness:

> Know this—medicines are of two types. First, those which can be found in the pharmacist's shop. As for [the second], a perfect man should put what is known to good use![88]

[82] Christina Álvarez Millán, "The Historical, Scientific, and Literary Contexts of al-Ghāfiqī's Herbal," in *The Herbal of al-Ghāfiqī: A Facsimile Edition of MS 7508 in the Osler Library of the History of Medicine, McGill University, with Critical Essays*, ed. Faith Wallis and Jamil Ragep (Montreal: McGill-Queen's University Press, 2014), 51–71.

[83] Weil, "Unveiling Nature" (cit. n. 76).

[84] Leigh Chipman, *The World of Pharmacy and Pharmacists in Mamlūk Cairo* (Leiden: Brill, 2010), 96–101 and 271–9.

[85] Ibid., 128.

[86] In Ibn Abī Usaibiʿa's biographical dictionary of physicians, only two are described as "*al-nabātī*," or botanists. Rashīd al-Dīn ibn al-Ṣūrī is described as one who would go to "Mount Lebanon and other spots where particular plants were found," along with a painter whom he employed to illustrate young, full-grown, and withered plants; see E. Savage-Smith, S. Swain, and G. J. van Gelder, eds., *A Literary History of Medicine* (Leiden: Brill, 2020), https://doi.org/10.1163/37704_0668IbnAbiUsaibia.Tabaqatalatibba.lhom-ed-ara1.

[87] Without a history of pharmacy in premodern South Asia, it is unclear how comparable these pharmacies were to the "marketplace pharmacies" in the medieval Middle East as described by Leigh Chipman.

[88] BL IO Islamic 1735, fol. 1b.

The author implies that someone who has attained learning and improved himself, the "perfect man" (*mard-i kāmil*), should be able to move past the mediation of the pharmacist. He should put the local names in this glossary to good use and acquire the materia medica himself. Considering how useful it would have been to physicians to know these vernacular language names, it may be surprising that this material was relegated to glossaries and not often integrated into medical texts. The exclusion can be explained by the temporal and regional limits of local plant names, a limit that did not apply in the same way to physiology or etiology, as well as by the hierarchy inherent in the diglossia of textual and oral knowledge. As their ubiquity in medical manuscripts attests, these glossaries were an important tool for physicians.

This hierarchy was reinforced by the anonymity of the authors of these glossaries. It is also apparent in the work of authors of formal medical dictionaries, who relied on oral knowledge as a source for their process of authorship and *taḥqīq*. For example, Muḥammad ibn Yūsuf al-Harawī (fl. 1492–1518) wrote this introduction[89] to one of his two alphabetical medical dictionaries, *Baḥr al-Jawāhir*:

> All that I have reported here from trusted authorities of both kinds of science—that of bodies (*abdān*) and that of religions (*adyān*)—may aid [students] in their task. In working on this, I required the verification (*taḥqīq*) of that which is found within [this knowledge] about the human body as a whole, as well as its parts, and about medicines and foods, simple and compound, together with their dispositions and degrees, and some of their beneficial properties, as well as some ailments, both names and definitions, and other words in use. I found no compilation of all such words, so I compiled them from well-regarded books . . . and that which I did not find there, I became satisfied [to know] *by listening to learned physicians and experienced scholars*."[90]

Those to whom al-Harawī listened were learned and experienced physicians, but he still believed that his oral learning from them was epistemically inferior to what he learned from authoritative texts. Further, al-Harawī himself subjected this knowledge to *taḥqīq* through his own experiential and textual learning. Perhaps he considered the knowledge of authoritative texts to have been verified by each successive generation of physicians, and their ongoing use to be proof of a critical consensus. The way al-Harawī describes *taḥqīq* goes beyond the process physicians used to verify the names of materials in glossaries, applying more broadly to all aspects of knowing and treating the body in *ṭibb*.

Glossaries drew on and organized oral knowledge about the local names of medical terms according to the priorities of the broader tradition, creating equivalences between the cosmopolitan and vernacular registers that would augment the broader tradition, leaving its norms unchanged.[91] They did not unsettle the epistemic hierarchy of *ṭibb*, which valued textual learning over oral, cosmopolitan languages over vernacular. As Alisha Rankin says of the experiential knowledge of indigenous and European medical practitioners in Monardes's work, "the archive of practice faded behind a

[89] He begins the text praising Allah, the "knower, who gave people of understanding verification (*taḥqīq*) and the loftiest of languages, Arabic."

[90] Bethesda, National Library of Medicine (USA) MS A 6, item 1, fol. 2a (emphasis and translation my own).

[91] Medical marginalia were another means for verification across region and period; see Deborah Schlein, "Medicine without Borders: Tibb and the Asbab Tradition in Mughal and Colonial India" (PhD dissertation, Princeton Univ., 2019).

new authority"—the authority of the translator.[92] However, in this case, the anonymous translators never asserted themselves as authors. They remained unnamed verifiers of local, communal knowledge. Those encountering the glossaries were not looking for authors and sources, because they understood names of materia medica to be shared between regional communities of knowers.

***

What if medical glossaries were used in South Asia by those with primary facility in vernacular Indian languages and only minimal knowledge of Persian? Presumably, such a reader would be seeking specific Arabic or Persian medical terminology for a vernacular word they already knew. Walter Hakala suggests this possibility in his analysis of the *Qaṣīda dar Lughat-i Hind* (Poem on Indian terminology), written by Yūsufī in the sixteenth century, after he migrated to Agra.[93] This metered, rhyming poem contains Persian and Hindavi equivalents of the names of farm animals, foods, luxury commodities, travel provisions, medicinal plants, metals, familial relationships, times of day, and weather. These are listed one after another, with the Hindavi gloss following the Persian lemma; sometimes verses included the verbs *gūyand* and *khwānand* as well.[94] Hakala places the poem within a didactic genre called *niṣāb al-ṣibyān* (children's curriculum).[95] Persian language-learning in South Asia, evinced by *niṣābs* in Persian-Pashto, Persian-Punjabi, Persian-Urdu, and others, was a site for the translation of scientific and medical terminologies in literary contexts.[96]

The fourth glossary under consideration in this article also gestures at a broader readership, because it specifies the language of lemma and gloss.[97] The author knew that some readers were students of *ṭibb* for whom both Arabic and Persian terminologies were new. They would have been learning terms like *pain*, *eye*, and *cumin* in Persian medical texts but not necessarily known which were loanwords from Arabic, having learned them first in a vernacular like Hindavi or Dakhani. In this glossary, one finds Hindavi referred to as either *hindī*, *hindavī*, or *ahl-i hind gūyand* (as the people of India say).[98] Another tactic the author uses is to specify the language of a given entry with its name and the verb *gūyand*.[99] Occasionally, he spells it out as a full sentence:

[92] Alisha Rankin, "New World Drugs and the Archive of Practice," in *Osiris* vol. 37.

[93] Walter N. Hakala, "On Equal Terms: The Equivocal Origins of an Early Mughal Indo-Persian Vocabulary," *Journal of the Royal Asiatic Society* 25 (2015): 209–27, at 210. The physician known as Yūsufī was the son of the author of the *Baḥr al-Jawāhir*.

[94] Ibid., 212.

[95] This genre goes back to the beginning of Persian in India, inspired by the *Khāliq Bārī*, an Arabic-Persian-Hindavī poem attributed to Amīr Khusrau (d. 1325).

[96] Muzaffar Alam, *Languages of Political Islam in India, 1200–1800* (Chicago: Univ. of Chicago Press, 2004).

[97] BL IO Islamic 1735. In the *Fihrist-i Asāmī-yi Adwiyya-yi Ikhtiyārāt-i Badī'ī* (Index of names of medicines in the *Ikhtiyārāt-i Badī'ī*), terms are marked with a *fā* and an '*ayn*— the first letters of Persian (*fārsī*) and Arabic ('*arabī*); see SJML Tibb Farsi 31 Acc 1661.

[98] Because the author was likely among the "people of India" himself, it is strange to read this phrase in the third person. However, use of this phrase was a convention of the farhang genre and does not necessarily imply that the author excluded himself. For example, in the entry for *jībāl*, some properties of the plant are prefaced with "*ḥukamā gufte-and*," or "as the physicians have said," though the author was likely a physician. BL IO Islamic 1735, fol. 10b.

[99] For example, in the entry on bamboo sugar: "*ṭabāshīr hindavī nabs lūchan gūyand";* BL IO Islamic 1735, fol. 23a.

"the language of *ḍarb* is Arabic (*lughat-i ḍarb ʿarabīst*)."[100] The *niṣāb* and the glossary were treasured tools for students in early modern South Asia, who would encounter many Arabic as well as transliterated Greek and Syriac words in Persian texts. These short texts furthered the accessibility of *ṭibb* in the vernacular before the era of print.

By the end of the nineteenth century, *ṭibb* came to be known in South Asia as Unani ṭibb. Printed texts in both cosmopolitan and vernacular languages became a means to broaden the realm of authority and practice in *ṭibb*. Because of the efforts of practitioners (*ḥakīms*), Unani ṭibb was institutionalized and standardized by the colonial government—unlike in the Middle East, where colonial and postcolonial governments alike considered it an impediment to the spread of modern medicine.[101] Since the founding of Unani schools in South Asia at the turn of the twentieth century, *ṭibb* has been taught in modern Urdu, a decision closely tied to British colonial administrators' shifting ideas about education and language politics.[102] While scholars have worked on translation and print technology in the vernacularization of *ṭibb* and *ṭibb nabawī*, or prophetic medicine, few have considered the role of the vernacular in medicine before the eighteenth century. Perhaps between general lexicographies and medical glossaries, literate patients or traders of medicinal substances gained access to specialist medical terminologies.

As demonstrated by the study of medical glossaries, investigating the relationship between cosmopolitan and vernacular languages in medical practice suggests new modes of inquiry about the history of nonelite people in the premodern Islamic world. Though this article is a preliminary study of four glossaries, a tiny share of the medical glossaries still to be found in tens of thousands of extant medical manuscripts, I hope it is sufficient to open new directions in the study of *ṭibb*. The authors of these glossaries were using methods of *taḥqīq* (verification) to create these sparse but useful texts for other physicians—bridges between the lofty cosmopolitan language of authoritative texts and the everyday vernacular languages with which physicians acquired medicinal substances and communicated with patients. Throughout the early modern period, one in which vernacular languages in South Asia were gaining increasing traction as literary languages, many physicians continued to uphold the language hierarchy in *ṭibb*. This continuity shows that the epistemic ruptures of the nineteenth century were long and varied processes.

Attention to local specificities of *ṭibb* and the appearance of vernacular languages within cosmopolitan texts will lead to a disaggregation of generalizations about so-called Islamic medicine. Attending to the vernacular in our sources may help us to render visible the labor and experiential knowledge disappeared by authoritative texts of *ṭibb*. Medical glossaries maintain the primacy of textual and cosmopolitan language, but they show us that authoritative texts of medicine float on a sea of vernacular languages.

[100] Ibid.

[101] Attewell, *Refiguring Unani Tibb* (cit. n. 34); Seema Alavi, *Islam and Healing: Loss and Recovery of an Indo-Muslim Medical Tradition, 1600–1900* (Ranikhet: Palgrave Macmillan, 2008), 154–204.

[102] Alavi, *Islam and Healing* (cit. n. 101), 306–33.

Where There's Smoke, There's Fire:
Pyric Technologies and African Pipes in the Early Modern World

by Benjamin Breen*

ABSTRACT

The globalization of pipes and smoking in the early modern world is often thought of as a linear movement from the Americas to Africa and Eurasia. While this is true of tobacco smoking, other early modern cultures of smoking (such as the use of cannabis pipes) diffused from sub-Saharan Africa and South Asia, not from the Americas. This article traces the technological, linguistic, and cultural translations of smoking in the seventeenth and eighteenth centuries, with a focus on the Atlantic world. These movements provoked significant new questions. Early modern smokers (and their critics) grappled with the question of how pipes and other "pyric technologies" of elemental transformation interacted with the body and mind—and with debates about racialized theories of health, long-distance travel, the African slave trade, and the translatability of knowledge and habits.

> *Nor yet could Oedipus e're understand,*
> *How to turne Land to smoake, and smoake to Land.*
> *For by the meanes of this bewitching smother,*
> *One Element is turn'd into another.*
> —John Taylor, "The Praise of Hemp-Seed," in *All the workes of John Taylor the water-poet* (London: printed for James Bolwer, 1630), 60.

Among the words for *pipe* in the major European languages, there is little variation to be found. The French smoke *pipes*; the Spanish smoke *pipas*; the Dutch, *pijpen*. But in the midst of this flock of words that resemble the sound made by birds (this being the apparent inspiration for the words' Latin forebear, *pipare*) there is one significant exception.[1] Portuguese-speakers smoke *cachimbos*.

The word derives from a Kimbundu word, *kixima*, meaning "water well." Kimbundu is a Bantu language that was used extensively as a lingua franca along the West Central African coast during the early modern period. This fact, and the timing of this loanword's entry into Portuguese during the mid-seventeenth century, make it clear that

* History Department, University of California, Santa Cruz, 1156 High Street, Santa Cruz, CA 95064, USA; bebreen@ucsc.edu.

[1] Michel Desfayes, *The Origin of English Names of European Birds and Mammals* (Lugano, Switzerland: Musée Cantonal d'Histoire Naturelle, 2008), 51.

Osiris, volume 37, 2022. © 2022 History of Science Society. All rights reserved. Published by The University of Chicago Press for the History of Science Society. https://doi.org/10.1086/719224.

the story of the word *cachimbo* is also a story of the Atlantic slave trade.[2] This is not, in itself, surprising: tobacco was one of the major items of exchange in the Luso-Brazilian slaving centers of West Central Africa, and several other Kimbundu words (including the inspiration for the word *zombie*) made their way into Portuguese vocabulary in the same era.[3] The story of the word *cachimbo* becomes more distinctive, however, when we reflect on what the word's origin can tell us about the history of smoking as a constellation of material technologies and cultural practices which has had—and continues to have—an enormous impact on global society.[4]

Historians have often assumed that the transatlantic movement of tobacco from the New World to the Old during the Columbian Exchange was the sole route for the globalization of the pipe as a medical technology—and, relatedly, that when early modern accounts describe pipes, they can be assumed to be describing *tobacco* pipes.[5] Over the past few decades, however, geographers and archaeologists have been uncovering a different story.[6] Since the 1970s, it has become clear that smoking technologies *did* exist in the Old World before the Columbian Exchange, even if tobacco did not.[7] Leaving aside tentative identifications (like a bone cylinder pipe from circa 750 CE found in present-day Botswana), pipes with identifiable traces of cannabis residue appear to have been in use in Ethiopia from circa 1325 to 1400 CE onward.[8] A recent book on the subject by the geographer Chris Duvall proposes that what we might call a "cannabis-smoking complex"—the material technologies of smoking, the cultivation of the drug itself, and a set of medical and spiritual beliefs about it—was already widespread across much of the eastern half of sub-Saharan Africa in the two or three centuries after around 1000 CE.[9]

[2] On the role of Kimbundu in early modern West Central Africa, see Linda Heywood, "Portuguese into African: The Eighteenth-Century Central African Background to Atlantic Creole Cultures," in *Central Africans and Cultural Transformation in the American Diaspora*, ed. Heywood (Cambridge, UK: Cambridge Univ. Press, 2002), 103–4.

[3] Leonardo Dallacqua de Carvalho and Wesley Dartagnan Salles, "Varíola, Tabaco e Sistemos Atlânticos: As causas da ascensão da Costa da Mina e queda de Angola no comércio negreiro na segunda metade do século XVII," *Revista Brasileira do Caribe* 17 (2016): 249–80; William W. Megenney, "Common Words of African Origin Used in Latin America," *Hispania* 66 (1983): 1–10.

[4] Carol Benedict, *Golden-Silk Smoke: A History of Tobacco in China, 1550–2010* (Berkeley: Univ. of California Press, 2011); Sander L. Gilman and Zhou Xun, eds., *Smoke: A Global History of Smoking* (London: Reaktion, 2004). The literature on smoking as a factor in public health is vast, but one notable recent contribution is Sarah Milov, *The Cigarette: A Political History* (Cambridge, MA: Harvard Univ. Press, 2019).

[5] For instance, Timothy Brooks writes that "wherever tobacco showed up [in the early modern period], a culture that did not smoke became a culture that did," attributing the globalization of smoking to the dissemination of American tobacco by European traders; Brooks, *Vermeer's Hat: The Seventeenth Century and the Dawn of the Global World* (London: Bloomsbury, 2010), 126.

[6] Chris S. Duvall, *The African Roots of Marijuana* (Durham, NC: Duke Univ. Press, 2019); John Edward Phillips, "African Smoking and Pipes," *Journal of African History* 24 (1983): 303–19; Allen F. Roberts, "Smoking in Sub-Saharan Africa," in Gilman and Xun, *Smoke* (cit. n. 4), 46–57. See also Elizabeth A. Bollwerk and Shannon Tushingham, eds., *Perspectives on the Archaeology of Pipes, Tobacco and Other Smoke Plants in the Ancient Americas* (London: Springer, 2016).

[7] Even this is not quite accurate—a little-studied species of tobacco, *Nicotiana africana*, seems to be native to Namibia and has been found to contain nicotine; Danica Marlin, Susan W. Nicolson, Abdullahi A. Yusuf, Philip C. Stevenson, Heino M. Heyman, and Kerstin Krüger, "The Only African Wild Tobacco, *Nicotiana africana*: Alkaloid Content and the Effect of Herbivory," *PloS One* 9 (2014): e102661, https://doi.org/10.1371/journal.pone.0102661.

[8] Nikolaas J. van der Merwe, "Antiquity of the Smoking Habit in Africa," *Transactions of the Royal Society of South Africa* 60 (2015): 147–50.

[9] Duvall, *African Roots of Marijuana* (cit. n. 6), chap. 1.

Likewise, there is an ancient and well-attested tradition of cannabis smoking in South Asia.[10] Many early artifacts associated with African and South Asian cultures of smoking were "water pipes," utilizing a reservoir of water through which smoke passed. This may explain the unusual etymology of *cachimbo*, offering a clue that the first pipes that the Portuguese encountered may not have been the "dry" pipes typically used for tobacco smoking in the indigenous Americas, but the water pipes of a longstanding sub-Saharan African cannabis culture. Portuguese sailors, after all, were present in sub-Saharan Africa *before* the Columbian Exchange, with sustained contact between Portugal and the Kingdom of Kongo from the 1480s onward. This produced a number of linguistic borrowings and syncretic spiritual traditions. For instance, although the term *feitiço* (progenitor of the English word *fetish*) emerged from the mingled Christian, Muslim, and Jewish magical traditions of medieval Iberia, the word was soon repurposed to describe the practices of African healers and spiritual practitioners; by the seventeenth century, it was being used widely by Africans in contexts far afield from its medieval Iberian origins.[11] It was from this Luso-African milieu that the concept of "fetishes" and those who wield them (*feiticeiros* or "fetisheers") became widespread in languages like French, Dutch, and English.

The history of the word *cachimbo* offers a glimpse not just into an unusual etymology but into a larger process of cultural, medical, and technological translation. Histories of smoking often start from the premise that smoking originated in the pre-Columbian Americas and then spread to Africa, Asia, and Europe via that favorite (and vague) descriptor used by historians of the Columbian Exchange: "Iberian mariners."[12] This article offers a complementary, yet distinctly different account. It is my hope that it will be relevant not just to historians of early modern medicine and textual translation, but to those interested in translation across alternative hierarchies of knowledge and across different regimes of technological practice. It may even speak to the "translation" between material objects (in this case, the physical substance of smokable drugs) and immaterial subjectivities (their transformation into vapor and, ultimately, into a disembodied set of bodily or mental "virtues" experienced solely by the consumer).[13]

As we'll see, early modern medical authorities employed smoking technologies in a surprisingly capacious way. Patients in seventeenth-century Portugal, for instance,

[10] P. Ram Manohar, "Smoking and Ayurvedic Medicine in India," in Gilman and Xun, *Smoke* (cit. n. 4), 68–76.

[11] Roger Sansi, "Feitiço e fetiche no Atlântico moderno," *Revista de Antropologia* (2008): 123–53; Sansi and Luis Nicolau Parés, eds., *Sorcery in the Black Atlantic* (Chicago: Univ. of Chicago Press, 2011); James H. Sweet, *Domingos Álvares, African Healing, and the Intellectual History of the Atlantic World* (Chapel Hill: Univ. of North Carolina Press, 2011), 74–5, 99.

[12] Jerome Edmund Brooks, *The Mighty Leaf: Tobacco through the Centuries* (Boston: Little, Brown, 1952), 32 ("[T]he habits of smoking cigars and pipes spread, in the ports of Spain . . . to the harbors of Portugal"); Benedict, *Golden-Silk Smoke* (cit. n. 4), 1–2 ("Iberian mariners . . . carried tobacco to ports around the world"). Likewise, Beverly Lemire describes the rise of "the smoking phenomenon" in West Africa as "likely the work of Iberian mariners"; Lemire, *Global Trade and the Transformation of Consumer Cultures: The Material World Remade, c.1500–1820* (Cambridge, UK: Cambridge Univ. Press, 2018), 202.

[13] On the ways that "the 'materiality' of early modern substances does not always collapse into concrete forms," see Jennifer Rampling, "Transmuting Sericon: Alchemy as 'Practical Exegesis' in Early Modern England," in *Osiris* 29 (2014): 19–34, on 34. See also Timothy Carroll, "Of Smoke and Unguents: Health Affordances of Sacred Materiality," in *Medical Materialities: Toward a Material Culture of Medical Anthropology*, ed. Aaron Parkhurst and Timothy Carroll (London: Routledge, 2019); and Viktoria von Hoffman, *From Gluttony to Enlightenment: The World of Taste in Early Modern Europe* (Champaign: Univ. of Illinois Press, 2016), chap. 4.

might be advised by their doctor to use a cachimbo to combust not only tobacco but arsenic, marjoram, cannabis seeds, cloves, and various minerals and resins. Additionally, smoking interacted with a larger natural philosophical interest in the transformation of matter from a solid into vapor, which invoked new questions about how techniques like distillation or smoking—"pyric technologies"—could alchemically "sublimate" the stuff of the mind itself.

Smoking is a technology of mediation between the material and immaterial. It is fitting, then, that smoking's history plays a similar role. The spread of smoking in the early modern period depended on both the diffusion of a *material culture* of smoking implements and social formations (such as the pipe as a marker of authority or status, as seen in fig. 1) and an underlying *epistemology* of smoking that offered medical or

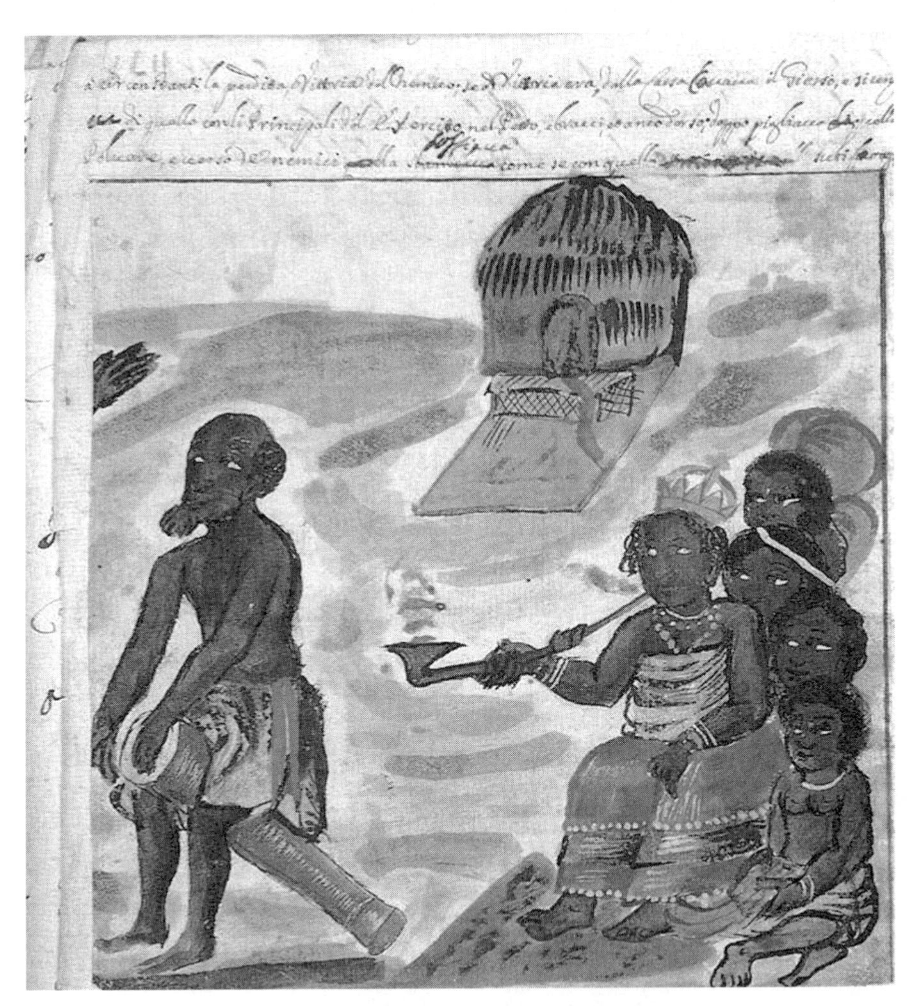

Figure 1. *Queen Nzinga of Ndongo and Matamba, one of the Portuguese empire's most committed opponents in West Central Africa, smoking a distinctive* cachimbo *as a mark of her authority, as depicted in the Araldi Manuscript (watercolor, ink, and gouache on paper, completed in 1671 by Giovanni Antonio Cavazzi). The image is reproduced from Ezio Bassani, ed.,* Un Cappuccino nell'Africa nera del seicento: I disegni dei Manoscritti Araldi del Padre Giovanni Antonio Cavazzi da Montecuccolo *(Milan: Quaderni Poro, no. 4, 1987).*

spiritual justifications for the practice. This history bids us to consider the ways that textual acts of translating medical concepts depended on the physical movements of objects and the habits—or addictions—that attended them. As such, it contributes to a growing body of scholarship on efforts to *translate technologies*, not just from the perspective of textual authority, but as an assemblage of material objects, vernacular techniques, and embodied practices.[14]

THE GLOBALIZATION OF SMOKING AS A MEDICAL TECHNOLOGY

Writing of the fears of Jean de Léry (1536–1613) about his fellow Frenchmen in Brazil who had supposedly become "traitors" because of their immersion in the practices of indigenous Brazilians, Michael Wintroub notes, "A translator who could fluently cross the borders of culture, language, and trust might also be in possession of powerful secrets and confidences." Such a translator "knew about customs, dispositions, desires, and intentions, as well as matters of prime tactical importance, such as the location and number of armed men, and the availability of food, water, and other valuables. Betrayal, it seems, was always lurking just around the corner of translation."[15]

For sixteenth-century Europeans, the practice of smoking epitomized this dual role of the translator as both expert and traitor. Throughout the final decades of the sixteenth century, in numerous languages and social environments, Europeans identified tobacco smoking as a quintessentially foreign practice. *A Counterblaste to Tobacco* (1604), an anonymous pamphlet authored by King James I of England (1566–1625), articulated this view in strikingly vivid language. In addition to attacking tobacco's "stinking suffumigation" and "venomous" nature, James obsessively meditated on the dangers of its foreign origins.[16] The plant was, he said, an "invention" driven by a "childish affectation of Noveltie" among a distant and poorly understood people from "beyond the Seas."[17] To have adopted such a custom was thus, in a sense, a certification of one's expertise in an unfamiliar new world, but it also raised the question of physical and mental corruption via the adoption of supposedly "savage" customs.[18] Because smoking is so intimately physical—involving, from a humoral perspective, the incorporation of smokables into one's constitutional makeup—it brought this question into high relief.

To be sure, European observers of smoking did have *some* earlier reference points for making sense of the habit. The practice of intentionally inhaling medicinal or psychoactive smoke is a widespread and ancient one. Reports of fumigated tents, or of breathing smoke from an open-air chimney or brazier, can be found in sources ranging

[14] Pablo F. Gómez, "[Un]Muffled Histories," in the present volume (*Osiris* 37); Jaime Marroquin Arredondo and Ralph Bauer, *Translating Nature: Cross-Cultural Histories of Early Modern Science* (Philadelphia: Univ. of Pennsylvania Press, 2019); Thomas S. Mullaney, *The Chinese Typewriter: A History* (Cambridge, MA: MIT Press, 2017); Maeve Olohan, "History of Science and History of Translation: Disciplinary Commensurability?," *The Translator* 20 (2014): 9–25; Carla Nappi, "The Global and Beyond: Adventures in the Local Historiographies of Science" *Isis* 104 (2013): 102–10.

[15] Michael Wintroub, "Translations: Words, Things, Going Native, and Staying True," *Amer. Hist. Rev.* 120 (2015): 1185–217.

[16] [King James I of England], *A Counterblaste to Tobacco* (R[obert] B[arker]: London, 1604), 20.

[17] Ibid., 8, 15.

[18] Jorge Cañizares-Esguerra, "New World, New Stars: Patriotic Astrology and the Invention of Indian and Creole Bodies in Colonial Spanish America, 1600–1650," *Amer. Hist. Rev.* 104 (1999): 33–68; Christopher Parsons, *A Not-So-New World Empire and Environment in French Colonial North America* (Philadelphia: Univ. of Pennsylvania Press, 2018).

from Greek accounts of Scythians to Tibetan Buddhist practices.[19] The gynecological use of fumigation had origins in ancient Mesopotamia and is well attested in medieval and early modern European medicine.[20] There are occasional accounts of medical breathing of smoke using personalized inhalation devices, if not quite pipes, from classical and medieval texts, although in Europe and the Mediterranean the practice does not appear to have been widespread. Pliny the Elder, for example, noted that "the smoke of dried cow-dung . . . is remarkably good for *phthisis* [tuberculosis], inhaled through a reed."[21] (One suspects that there might be good reasons why this practice was not quite as popular as tobacco-smoking was later to become.)

Pipes differed from most of these smoke-related practices for several reasons. First, the invention of pipes—not just makeshift reeds, but dedicated objects for smoke inhalation consisting of a bowl, a stem, and, optionally, a "carb" of some kind—offered a fully *individualized* route of administration. Only one person inhaled a targeted "unit" of smoke, rather than an entire room inhaling fragrance or smoke in a relatively uncontrolled manner. Although gynecological fumigation used in midwifery was individually targeted, it differed from pipes in that the practice typically required attendants, and the "dosage" of smoke could not be easily regulated. Pipes, by contrast, allowed relatively precise doses of smokable substances to be measured by the user. Pipes were also highly portable in a way that a large brazier or a fumigated tent or chamber could not be. Owing to these factors, the psychoactive or medicinal properties of combustibles native to the Old World (like cannabis, opium, "wild" dagga [*Leonotis leonurus*], and datura) are far more apparent than when inhaled from the ambient environment. All of this made pipes a novelty that was more difficult to translate into European medical vernacular—both as a word and as an assemblage of knowledge and bodily practices—than the mere act of employing smoke for medical reasons.

Little is known for sure about how *cachimbo* made its way into Portuguese from Kimbundu, but it is clear that Portugal's participation in the Atlantic slave trade in the sixteenth and seventeenth centuries created crossovers between the cannabis-smoking cultures of the Old World (centered in South Asia and East Africa) and the tobacco culture of the indigenous Americas. Archaeological evidence supports the theory that African pipe-making forms and techniques crossed over to the Americas largely intact.[22] There are several instances of seventeenth-century pipes found in North America and the Caribbean that closely match correlates from distinct areas of pipe design

[19] On fumigation in Tibet, see Gillian G. Tan, "Differentiating Smoke: Smoke as *duwa* and Smoke from *bsang* on the Tibetan Plateau," *Anthropological Forum* 28 (2018): 126–36; and Daniel Berounský, "Burning the Incestuous Fox: A Tibetan Fumigation Ritual (*wa bsang*)," *Études mongoles et sibériennes, centrasiatiques et tibétaines* 50 (2019): 1–31.

[20] Ulrike Steinert, "I smell a rat! Fumigation in Mesopotamian and Hippocratic Recipes for Women's Ailments," *The Recipes Project*, March 11, 2014, accessed October 11, 2021, https://recipes.hypotheses.org/3278; René Labat, "A propos de la fumigation dans la médecine assyrienne," *Revue d'assyrologie et d'archéologie orientale* 55 (1961): 152–3. On early modern fumigation relating to pregnancy, see Wendy Perkins, *Midwifery and Medicine in Early Modern France* (Exeter: Univ. of Exeter Press, 1996), 70–1 and 85; and A. S. Weber, "Women's Early Modern Medical Almanacs in Historical Context," *English Literary Renaissance* 33 (2003): 358–402.

[21] Pliny the Elder, *The Natural History*, trans. John Bostock and H. T. Riley (London: Taylor and Francis, 1855), book 28, chap. 67.

[22] Matthew C. Emerson, "African Inspirations in a New World Art and Artifact: Decorated Tobacco Pipes from the Chesapeake," in "*I, Too, Am America*": *Archaeological Studies of African-American Life*, ed. Theresa A. Singleton (Charlottesville: Univ. Press of Virginia, 1998), 52.

in Africa, such as Mali, the Niger River, and the Kingdom of Kongo.[23] It is notable, in this context, that an early European advocate for smoking as a medical technology, and one of the earliest documented users of the word *cachimbo*, was João Curvo Semedo (1635–1719), a Lisbon apothecary and physician notable for his ties to the Africa trade. The author of numerous medical compendia, including the best-selling *Polyanthea Medicinal*, Semedo is virtually unique among early modern European medical writers in that his books extensively employ names taken directly from African languages.[24] Writing of the medical virtues of a Congolese animal horn, for instance, Semedo explains that "in the Kongo language, *abada* refers to a mythical animal similar to a unicorn."[25] In another example, he describes an antivenom that a Dutch surgeon in Portuguese Angola tested on dogs, writing, "the Africans call it *minhaminha*, because in the language of Angola *minhaminha* means 'swallows,' for it swallows up the virtues of the other medicines."[26] It seems likely that *cachimbo* made its way into Portuguese via the same circuits of the Portuguese slave trade that carried drugs—and words—like *minhaminha* and *abada* to apothecaries in Lisbon.

One of the most common pipe forms in premodern Africa was the water pipe (fig. 2), a lineage seemingly preserved in the original meaning of the word from which *cachimbo* itself was derived (a well or water reservoir). Eighteenth- and nineteenth-century Europeans tended to frame the water pipe (known as *qalyân* in Persia, *argilah* or *nargilah* in the Arab world, and as the hookah or "hubble bubble" in British India) as part of a larger complex of "Oriental" objects and practices associated with Egypt, the Levant, Persia, and India. As a result, it is often assumed that these objects had their origins in the Middle East or South Asia. However, there is evidence for a sub-Saharan African origin for the underlying technology, originally used to smoke cannabis and other combustibles.[27] (The substances smoked in water pipes were, according to an eighteenth-century account, not tobacco alone but "a paste composed of fine tobacco, betel, ambergris, and various aromatic and gently narcotic drugs.")[28] By the eighteenth century, the water pipe was thought of by Europeans as a distinctively Persian custom. In 1773, for instance, the traveler Carsten Niebuhr wrote that "the Arabs smoke a very long pipe in the manner of the Turks, or through water like the Persians . . . Those of the rich are made from glass, silver and gold and are given divers figures."[29] However, he also noted a hint of

[23] Ibid., 53. On the culture of smoking in the seventeenth-century Kingdom of Kongo and neighboring regions, see Bernard Clist, "From America to Africa: How Kongo Nobility Made Smoking Pipes Their Own," in *The Kongo Kingdom: The Origins, Dynamics and Cosmopolitan Culture of an African Polity*, ed. Koen Bostoen and Inge Brinkman (Cambridge, UK: Cambridge Univ. Press, 2018), 197–215.

[24] On Semedo's links to African medical knowledge, see Benjamin Breen, "Semedo's Sixteen Secrets: Tracing Pharmaceutical Networks in the Portuguese Tropics," in *Empires of Knowledge: Scientific Networks in the Early Modern* World, ed. Paula Findlen (London: Routledge, 2018); Breen, *The Age of Intoxication: Origins of the Modern Drug Trade* (Philadelphia: Univ. of Pennsylvania Press, 2019), 90; and Hugh Cagle, *Assembling the Tropics: Science and Medicine in Portugal's Empire, 1450–1700* (Cambridge, UK: Cambridge Univ. Press, 2018), 297–300.

[25] João Curvo Semedo, "Memorial de Varios Simplices," (circa 1710), 29. This is an undated pamphlet bound with many later editions of *Polyanthea Medicinal*. Unless otherwise noted, the translations in this article are mine.

[26] Ibid., 29.

[27] In *African Roots* (cit. n. 6), Duvall argues that "African knowledge is foundational to all water pipes."

[28] William Smellie, *Account of the Institution and Progress of the Society of the Antiquaries of Scotland* (London, 1782), 93.

[29] Carsten Niebuhr, *Description de l'Arabia d'après les observations et recherches faites dans le pays même* (Copenhagen: Nicolas Möller, 1773), 51.

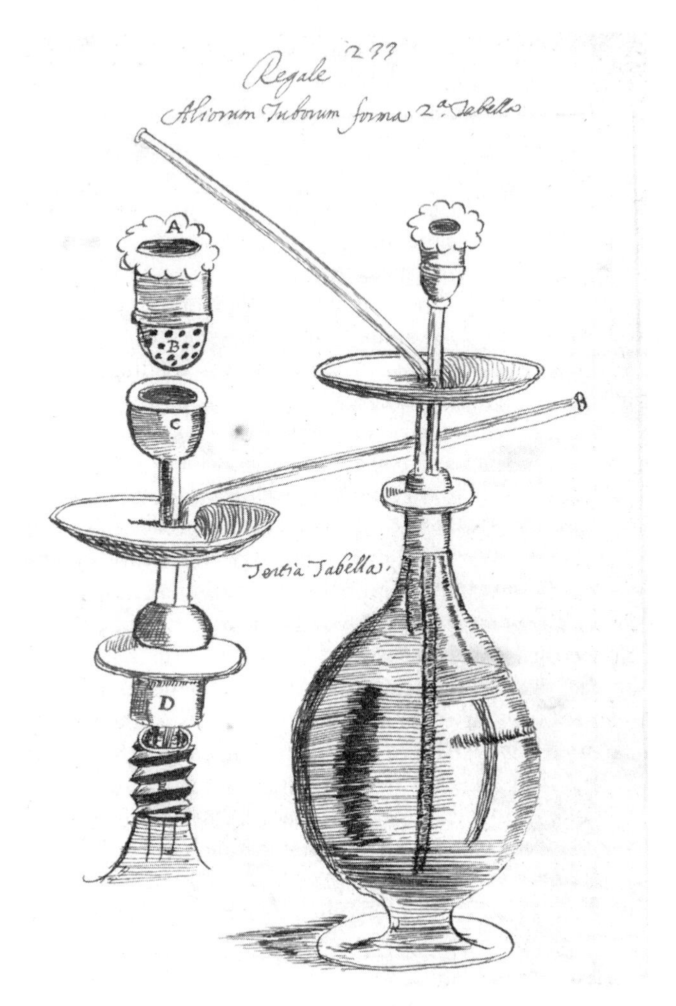

Figure 2. *A "Persian" water pipe in University of Pennsylvania MS Codex 122, "Viridarium regale" (seventeenth-century Italian), based on an engraving in Johann Neander,* Tabacologia *(Leiden, 1622). Courtesy of the Kislak Center for Rare Books and Manuscripts, University of Pennsylvania, Philadelphia, PA.*

non-Persian origin: "the common people can make one inexpensively with a coconut." One of the earliest words for this type of water pipe, *nargilah*, similarly referred to the coconut, a plant native to India and Africa but not to Persia or the Levant.

A late eighteenth-century account of "women in the southern part of Egypt" who smoked a hashish and tobacco blend from water-filled coconuts may offer an important clue for a route of transmission from East Africa to the Middle East and Persia—although the observer in question simply assumed that this practice was "a crude imitation of the *pipe à la persanne*" instead of the other way around.[30] The resulting intoxication was profound and supposedly untranslatable:

[30] Charles-Nicolas-Sigisbert Sonnini de Manoncourt, *Voyage dans la Haute et Basse Égypte: fait par ordre de l'ancien gouvernement* (Paris: F. Buisson, 1799), 103.

> In place of inebriating liquors, the Arabs and Egyptians compose diverse preparations from which is procured a sort of sweet intoxication, a state of reverie that leads to gaiety and agreeable dreams. This type of annihilation of the faculty of thought, this sleep of the soul, has nothing in common with the intoxication occasioned by wine and alcoholic spirits, and our language does not have terms to express it.[31]

Sonnini described the mixture as a smokable "pâte" made out of the "fruits" and seed capsules of the cannabis plant and mixed with honey, pepper, and nutmeg, along with tobacco. The coconut shell water pipe may itself have been a substitute for the calabash. A common water vessel throughout Africa, the calabash was also used in some of the earliest documented water pipes.[32]

PROBLEMS OF TRANSLATION: TOBACCO AND CANNABIS

When Herodotus described the purification practices of Scythians following elite burials, he wrote of a ritual involving the construction of a tent-like enclosure of "wool mats." At the center of this enclosure, the Scythians threw cannabis onto "red-hot stones, where it smolders and sends forth such fumes that no Greek vapor-bath could surpass it." According to Herodotus, "the Scythians howl in their joy at the vapor-bath."[33] The term Herodotus used here—*κάνναβις*, or *kánnabis*—was a loanword from Old Persian (*kanab*). From Greek, it made its way largely unchanged into Latin (*cannabis*) and from thence into the Romance languages and English.[34]

The very prevalence and persistence of this name was itself a profound source of confusion. The cannabis plant has long existed in two distinct phenotypes, sometimes known as *Cannabis sativa* and *Cannabis indica*. The former, often simply known as hemp in English, had long been used throughout western Asia and Europe since the Neolithic period to create cordage and textiles. (Herodotus explained that "the Thracians make garments of [*kanabis*] which closely resemble linen.") This is the more geographically widespread form of the plant, and it is the one that was most familiar to early modern European observers. The other form of cannabis—the intoxicating variant—was possibly first cultivated for medicinal and ritual use by proto-Indo-Iranian populations in Western Asia, becoming increasingly distinct from hemp over generations of artificial selection.[35] Both Old Persian (Avestan) and Sanskrit use a different root (*bang-* or *bangha*) to refer to the psychoactive variant of cannabis. Confusingly, however, in older Persian sources *bang* could apparently refer not just to *Cannabis indica* but to a larger family of psychoactive drugs including henbane (*Hyoseyamus niger*) and jimsonweed (*Datura stramonium*).[36] Both henbane and jimsonweed have ancient traditions of use in their own right, along with very different pharmacological profiles.

[31] Ibid.

[32] Van der Merwe, "Antiquity of the Smoking Habit in Africa" (cit. n. 8), 147–50.

[33] A. D. Godley, trans., *The Histories of Herodotus* (Cambridge, MA: Harvard Univ. Press, 1920), 4.74–6.

[34] The Old English *hænep*, which became modern English *hemp*, is thought to be derived from the same Indo-European *kanab*-root; Walter William Skeat, *The Concise Dictionary of English Etymology* (London: Wordsworth Editions, 1993), 200.

[35] Elizabeth Wayland Barber, *Prehistoric Textiles: The Development of Cloth in the Neolithic and Bronze Ages with Special Reference to the Aegean* (Princeton, NJ: Princeton Univ. Press, 1992), 37–8.

[36] G. Gnoli and ʿA.-A. Saʿīdī Sīrjānī, "Bang," in *Encyclopedia Iranica*, vol. 3: fasc. 7 (London, 1988), 689–91; David Stophlet Flattery and Martin Schwartz, *Haoma and Harmaline: The Botanical Identity of the Indo-Iranian Sacred Hallucinogen Soma and Its Legacy in Religion, Language, and Middle-Eastern Folklore* (Berkeley: Univ. of California Press, 1989), 124, 127.

This ancient split between the identities and functions of cannabis—as a component in a larger category of medicinal drugs or ritual sacraments on the one hand, and as a prosaic raw material on the other—caused considerable confusion among the earliest Europeans to observe the cannabis cultures of Persia and South Asia. One of the earliest European references to an intoxicating variant of cannabis comes from the Portuguese-Jewish physician Garcia da Orta. Building on knowledge gained during his medical practice in Goa, India, throughout the 1540s and 1550s, da Orta described *bangha* as a narcotic drug which, in his view, was detrimental to health and provoked "sadness and great nausea."[37] However, he also described the recreational use of the drug in candies combined with opium and sugar by Indian elites, namely the Sultan Bahadur of Gujarat, who, according to da Orta, claimed to have no need to travel to Mexico or Persia because he could do so "in his sleep" with the aid of this powerful drug.[38] Da Orta argued that *bangha* (made from the flowers of *Cannabis indica*) was distinct from the common hemp of Europe, which he called *linho alcanave*.

Jan Huyghen van Linschoten described what he called *bang* as a plant product "like Hempe, but it hath no substance whereof to make any thing." In other words, he identified the substance by describing what it *wasn't*, emphasizing that this variant of cannabis had an entirely different practical use as a drug, not as a raw material for textiles. He repeated what would become a common two-part refrain about the medicinal properties of the drug: that it was used to "provoke lust" and "a good appetite." But again, Linschoten portrayed this *bang* not as a unitary substance but as an element in compound remedies. He wrote that the Indians "put greene Arecca unto it, therewith to make a man drunke . . . sometimes they also mix it with Nutmegs and Mace . . . Others (that is to saye, the rich and welthy persons) mix it with Cloves, Camphora, Ambar, Muske, and Opium."[39] Both Linschoten and da Orta agreed that cannabis, though resembling hemp, was not quite the same thing. Neither described the smoking of the drug.[40]

In a 1702 report to the Royal Society on "East India Plants, sent from Fort St. George," the London apothecary James Petiver described several different types of "bangue," differentiating them not only according to their botanical characteristics, such as number of leaves, but to their perceived potency ("The *Hort[us] Malab[aricus]* says this sort is much stronger than the next.").[41] According to Petiver's gloss, his Indian variant of cannabis was "similar in nearly all of its uses to . . . opium."[42] The East India merchant Thomas Bowrey, from whom Petiver likely first learned of the drug, referred

[37] Garcia da Orta, *Colóquios dos simples e drogas da India* (Goa, India: 1563), 27r.

[38] Ibid., 26r.

[39] Jan Huyghen van Linschoten, *The Voyage of John Huyghen van Linschoten to the East Indies* (London: Hakluyt Society, 1885), 115. Reprint of the 1598 English translation published by John Wolfe, *Iohn Huighen van Linschoten, his discours of voyages*.

[40] The truth was a bit more complicated: although the *Cannabis indica* documented by da Orta in India typically has higher THC concentrations than *Cannabis sativa*, the two phenotypes exist in a continuum rather than as distinct species. They frequently hybridize, both with and without human intervention, and contain dozens of pharmacologically active compounds with a range of medical and recreational applications.

[41] James Petiver, "The Eighth Book of East India Plants, Sent from Fort St. George," *Philosophical Transactions* 23 (1702/3): 1453–54. Petiver was referencing the Latin materia medica compendium *Hortus Malabaricus* [Garden of Malabar]. On Petiver's connections to other early authors who described *Cannabis indica*, namely Robert Knox and Thomas Bowrey, see Anna Winterbottom, *Hybrid Knowledge in the Early East India Company World* (London: Palgrave Macmillan, 2016), 127, 144, 254n98.

[42] Petiver, "Eighth Book of East India Plants" (cit. n. 41), 1453.

to cannabis using the Portuguese form of the name, *bangue* (Bowrey was in fact called an "old Portuguese rogue" in an altercation later in his life, a term reflecting an adulthood spent in the Lusophone trading emporia of the Indian Ocean).[43] Cannabis appeared in Petiver's account as an orally consumed drug, not a smokable. Intriguingly, however, one of the earliest published references to the medicinal use of *Cannabis indica* in the seventeenth-century—Robert Hooke's praise for *bangha* that he'd purchased from Bowrey, which Hooke lauded as a potential cure for "lunaticks"—used a tobacco pipe bowl as a unit of measurement for the recommended dosage.[44]

As with the terms applied to cannabis, the Spanish word *tabaco* took time to cohere, decisively, around the concept of smokable dried leaves of the *Nicotiana tobacum* plant.[45] The Spanish colonist and writer Gonzalo Fernández de Oviedo y Valdés (1478–1557) described tobacco as "a species that is very similar to henbane." Oviedo argued that this "very bad habit . . . of taking the smoke of that which they call *tabaco*" was intoxicating like wine and was used "to take leave of the senses."[46] The earliest European botanical depiction of tobacco, published by Rembert Dodoens in Amsterdam in 1553, followed Oviedo's lead. This work depicted a tobacco plant but labeled it *Hyoscyamus luteus*, a variety of henbane native to the Mediterranean that had been known to Greco-Roman authorities.[47] Henbane, a member of the nightshade family that contains the deliriant daturine, has been an element of psychoactive unguents and potions since ancient times (Pliny warned that all forms of henbane "[have] the property of flying to the head . . . like wine").[48] This may explain why tobacco, with its very different botanical characteristics, was confused with this Old World drug.

Even after the botanical identification of tobacco had become clearer around the beginning of the seventeenth century, the social and medical function of the drug was still up for debate. King James I of England famously attacked tobacco as a kind of Trojan horse by which "heathen" customs from the Americas (and, with them, the influence of the devil) had invaded his kingdom, blaming tobacco's perceived benefits for treating "the Pox" as the source of its malicious influence. The problem of translating tobacco into the categories of early modern European culture lay at the core of its controversy. Was tobacco a medicine, to be prescribed by physicians and prepared and sold by apothecaries? Or was it something more like wine, enjoyed for recreation or sustenance? In 1635, several English tobacco merchants found

[43] Anna Winterbottom, "Producing and Using the Historical Relation of Ceylon: Robert Knox, the East India Company and the Royal Society," *British J. Hist. Sci.* 42 (2009): 536n127, citing St. Helena Archives, Consultation Books, ii, f. 175 08/06/85.

[44] Robert Hooke, "An Account of the Plant, call'd Bangue, Before the Royal Society, Dec 18. 1689," in *Philosophical Experiments and Observations of the Late Dr. Robert Hooke*, ed. William Derham (London, 1726), 201–2; Breen, *Age of Intoxication* (cit. n. 24), 101–2.

[45] The Spanish word *tabaco* existed before the Columbian voyages. It was used (not unlike the ancient Persian *bang*) as a kind of catchall term for a conceptual grouping of medicinal herbs from the early fifteenth century, derived from the Arabic *tabbaq*, attested since the ninth century.

[46] Gonzalo Fernández de Oviedo, *Coronica de las Indias: La historia general de las Indias agora nuevamente impressa corregida y emendada* (1535; repr., Salamanca: Juan de Junta, 1547), libro V, cap. II ("de los tabacos"), fol. 47r. On Oviedo, see Marcy Norton, *Sacred Gifts, Profane Pleasures: A History of Tobacco and Chocolate in the Atlantic World* (Ithaca, NY: Cornell Univ. Press, 2008), 58.

[47] Rembert Dodoens, *Trium priorum de Stirpium historia commentariorum imagines* (Antwerp, 1553), 437. For a discussion, see Melissa Morris, "Cultivating Colonies: Tobacco and the Upstart Empires, 1580–1640" (PhD diss., Columbia Univ., 2017), 75–6.

[48] Pliny the Elder, *Natural History* (cit. n. 21), book 25, chap. 17.

themselves embroiled in a legal case that put these questions to the fore. The case centered on the question of "whether tobacco is a medicine [*medicamentum*] or a food [*alimentum*]." A lawyer hired by one of the merchants argued that "[the fact] that tobacco is no victual hath appeared by the declaracion made by Divines, Civilians, and Physitians." They also appealed to the experience of "the Indians who judge tobacco not to number among their foods but to merit being referred to as a medicine," in part because "the smoke sent into the stomach . . . loosens anxieties."[49] Although the case was decided in favor of tobacco not being an "aliment," this conclusion was contradicted by the fact that tobacco *was* widely used in potions and even edible preparations in the form of electuaries made by apothecaries.

TRANSLATING EARLY MODERN TECHNOLOGIES OF SMOKING

The implements used in smoking itself posed even greater problems of translation and explanation than the classification of tobacco or cannabis (fig. 3). European authors like Oviedo, Monardes, and Dodoens could at least reach for classical analogues from Greco-Roman botany, such as henbane, and could explain the properties of "drinking tobacco" (or consuming cannabis) in comparison with alcohol or opium intoxication.[50] There were few such correlates for the practice of smoking by means of a pipe. True, the confused accounts of the earliest Spanish, Portuguese, and French chroniclers had been somewhat clarified by the end of the sixteenth century, as the practice of smoking became a lived experience in European urban centers rather than an exotic custom read about in books.[51] Yet uncertainty about what exactly to *call* the new technologies of smoke persisted. An early German compendium of materia medica from 1589 described the practice as "receiving smoke through a funnel into the mouth."[52] An Elizabethan chronicler struggled to find adequate words to describe the use of pipes; he wrote of the recent vogue for "taking in the smoke of the Indian herbe called Tobacco, by an instrument formed like a little ladell, whereby it passeth from the mouth into the head and stomach."[53] Even more confusing was the question of how to appropriately judge the medical value of this new process, which, in a very visible and performative way, translated matter itself from one elemental form (a material solid) to another (an immaterial vapor or fume).

A common theme among sixteenth-century accounts of tobacco was to describe a wide variety of preparation techniques, most of which didn't involve pipe smoking as it later became understood. The Drake Manuscript (1586) mentioned the medicinal properties of tobacco smoke, but *not* smoke inhaled into the lungs. "When the Indians are mortally wounded," the manuscript's anonymous author claimed, "one lays them on a rack and makes an oven with a tube leading to the wound of the sick man. When

[49] British Library, Add MS 33587, fol. 2r and fol. 15r.

[50] On the use of alcoholic intoxication as a kind of experiential "baseline" for European encounters with indigenous American psychoactive drugs in this period, see Breen, *Age of Intoxication* (cit. n. 24), chap. 5.

[51] "He that will refuse to take a pipe of Tobacco amonsgt his fellowes . . . is accounted peevish and no good company," wrote King James I of England in his *Counterblaste* (cit. n. 16).

[52] Johann Wittich, *Bericht von den wunderbaren bezoardischen Steinen* (Leipzig: Hans Steinmanns Erben, 1589), 108–9 ["rauch durch einen trichter in den mund empfangen"].

[53] William Harrison, "Great Chronologie," MS entry for 1573, cited in George Latimer Apperson, *The Social History of Smoking* (London, 1914), 14.

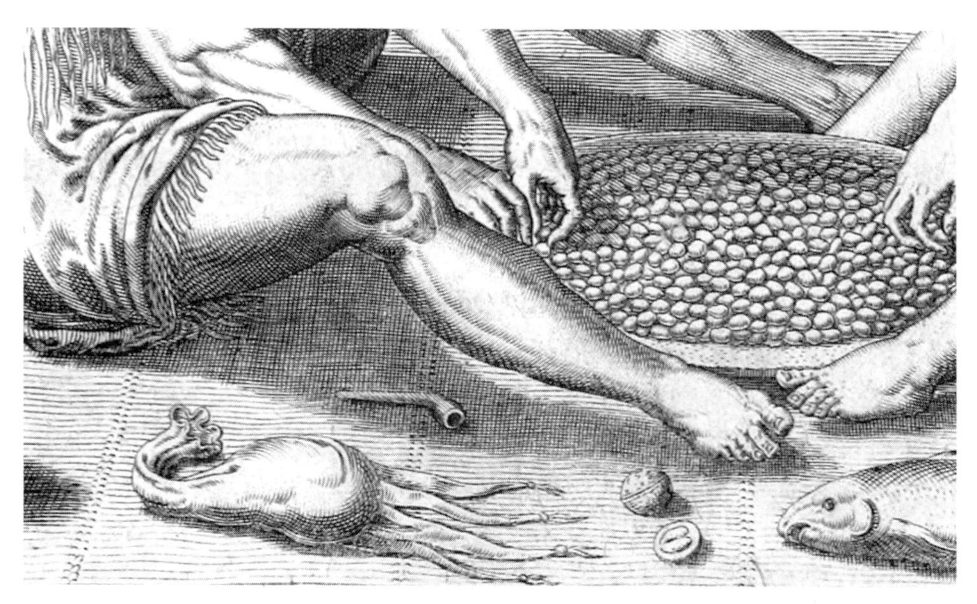

Figure 3. *Depiction of an Algonquian pipe and tobacco pouch by Theodore de Bry, in Thomas Harriot,* Admiranda narratio fida tamen, de commodis et incolarvm ritibvs Virginiæ *(Frankfurt: 1590), fol. 16. John Carter Brown Library, Providence, RI.*

the fire is lighted, they put in it a leaf of tobacco together with a resin called balsam and . . . the smoke enters the wound of the patient."[54]

Gonzalo Fernández de Oviedo, in one of the earliest and most influential accounts of the new drug to circulate among Iberian natural philosophers and colonists, had described tobacco smoking taking place via "two tubes [*cañones*] joined to become one" by which smoke was inhaled into the nostrils.[55] Thevet's famous description of tobacco (*petun*) seemed to describe a technique more akin to cigar smoking than to use of a pipe. Describing tobacco as an "herb . . . resembling our bugloss," which the native Brazilians "judge to be marvelously profitable for many things," Thevet explained:

> They carefully gather this herb, and dry it in the shade in their little cabins. The manner of using it is as follows: They wrap some quantity of this herb, being dried, in a palm leaf which is very large, and they roll it to the length of a candle, then put a flame at the base, and receive the smoke through their nose and mouth. It is very healthy, they say, to distill and consume [*distiller & consumer*] the superfluous humors of the head. . . . it is true that, if you take too much of this smoke or perfume, it goes to your head and inebriates, like the "fumes" [*fumet*] of a strong wine.[56]

Thevet likened the *fumet*, the heady vapors of a strong wine (a word deriving from Latin *fumus*, smoke), to the *fumée* of petun. The explanation that the Brazilians

[54] *Histoire Naturelle des Indes* ["The Drake Manuscript"], circa 1586, MA 3900, Morgan Library and Museum, New York, USA, fol. 92r.

[55] Oviedo, *Historia general de las Indias* (cit. n. 46), fol. 47r.

[56] Andre Thevet, *Les singularitez de la France Antarctique, autrement nommée Amerique: & de plusieurs Terres & Isles decouvertes de nostre temps* (Paris: Maurice de la Porte, 1558), 57. On early French accounts of tobacco, see Alisha Rankin, "New World Drugs and the Archive of Practice," in the present volume (*Osiris* 37).

smoke tobacco to "distill . . . the humors" was typical of sixteenth-century European attempts to translate indigenous American epistemologies of the body to humoral theory. It also allowed him to make sense of what the smoke was doing in terms that were more familiar to him: distillation of wine created powerful "fumes" that could produce drunkenness, but the "fumes" of tobacco could also "distill" the brain itself, with medicinal benefits. This conception of a "distillation" of the thoughts occupying the mind in the form of vapor became a surprisingly common trope in the visual culture of the early seventeenth century.

Elsewhere, a more familiar format of the tobacco pipe was beginning to circulate. In sixteenth-century Brazil, Portuguese colonists took note of the drug and the manner of smoking it via an implement they sometimes called a *canudo* (tube or straw). Gabriel Soares de Sousa's account of Brazil in the year 1587 noted "*Petume* is an herb which in Portugal we call *santa*." He described the practice of "drinking in the smoke of it" by

> joining together many leaves of the plant twisted around themselves, and inserting a tube [*canudo*] made of palm leaves, the fire being put to it by a brand, and just as one heats embers they put this canudo in their mouth, and they drink up the smoke into their insides until it leaves by the nostrils. All men who drink wine, drink much of this smoke and say that it improves the wine.[57]

No dedicated word for such smoking technologies existed in Portuguese until *cachimbo* began to appear in Portuguese texts, apparently in the 1640s. It is not clear whether the word was being used in these early accounts to describe a smoking implement that differed in some way from the various objects called *canudos* or *tubos* or *cañones* (reeds, tubes, long hollow objects) by early Iberian authors. What *is* clear is that the Portuguese encyclopedist Rafael Bluteau's straightforward definition of *cachimbo* in 1712 ("a long, slender straw of baked clay by which tobacco is taken") did not tell the whole story.[58]

Cachimbos were used to consume a number of different substances, not just tobacco, and were part of a larger set of smoke-related medical technologies. This was an imitation of what Europeans had observed in the Americas and Africa: in both these cultures of smoking, tobacco and cannabis (respectively) were only the most common combustibles among a variety of alternatives. For instance, the Spanish Jesuit José de Acosta praised the tobacco of New Spain as one of the New World "plants . . . of rare virtues" which rival "the drugs that come from the Orient."[59] However, the sole specific description of a tobacco-based medicine that Acosta recorded was actually a compound substance. He described an oily paste or *betún* (etymologically linked, perhaps, to the Latin *bitumen* but also an echo of Thevet's word for tobacco, *petún*) composed of tobacco ash, spiders, scorpions, millipedes, and the seeds of the hallucinogenic drug known as *ololiuhqui* (morning glory seeds).[60]

In early modern Portuguese medical writings, it sometimes seems as if pipes were used to smoke everything *but* tobacco. The prominent Lisbon physician and drug-seller

[57] Gabriel Soares de Sousa, *Tratado Descritivo do Brasil em 1587* (Rio de Janeiro: Typographia Universal de Laemmert, 1851), 206.

[58] Rafael Bluteau, *Vocabulario Portuguez, e Latino* (Coimbra: Collegio das Artes da Companhia de Jesu, 1712), 24.

[59] José de Acosta, *Historia natural y moral de las Indias* (Barcelona, 1591), 174.

[60] Ibid., 240–1.

João Curvo Semedo recommended a "most desiccating" treatment for lung complaints that featured "arsenic mixed with equal parts of styrax, terebinth, [and] *almecega* resin . . . the smoke taken by cachimbo."[61] It was not the only cachimbo-based remedy in Semedo's writings. At another point, he described an "excellent remedy for chest congestion":

> Take by cachimbo [*tomar por cachimbo*] rosemary, marjoram, ningela [*Nigella sativa?*], red roses, Indian cloves, nutmeg, spikenard, of each thing half an ounce all made into powder. The seeds of hemp infused in vinegar of strong white wine, dried and toasted over embers, and taken as smoke into the nostrils, is prodigiously effective against the flux.[62]

Gabriel Grisley, a German-born émigré to Portugal, likewise advocated for a wide range of smoke-related healing practices. The cachimbo makes a prominent appearance: Grisley recommends "the smoke of [tussilago] leaves taken with a cachimbo of tobacco" to "heal a dry cough."[63] But so too do such techniques as "inhaling the vapors" of boiled betony "with a funnel," consuming soapwort root "in the manner of tobacco," and a variant on the age-old practice of gynecological fumigation: "taking the vapor of [tormentilla] root . . . from below" to "comfort the woman's womb."[64] This wide range of "pyric" medical practices may have been considered a distinctively Portuguese practice at the time: one Portuguese-Dutch dictionary from 1714 defined *cachímbo* as "A tobacco pipe, or, a pipe, because some [other] herbs are smoked [by the Portuguese]."[65]

Nor did seventeenth-century references to the use of an implement called a cachimbo necessarily imply the familiar format of a tobacco pipe. After all, the *kixima* of the West Central Africans varied widely in its material characteristics, from the water pipe to the "earth pipe": a tunnel in soil, through which smoke passed from a burning reservoir into the mouth of the smoker. In some texts, the lines between pipe-smoking and fumigation were blurred. Jacob de Castro Sarmento, writing in 1735, described a complex cure involving the creation of an air-tight "pavilion" in which the patient sat and inhaled vaporized cinnabar. "Weak people," Sarmento added somewhat judgmentally, could be permitted "to breathe fresh air through the end of a cachimbo" stuck outside of the tent, should they feel overwhelmed.[66] Another Portuguese physician advised that people in "robust" health may want "to smoke [*cachimbar*] the powder of *pão santo, era terrestre* [*Glechoma hederacea*], rosemary, tussilago, tobacco, and orpiment" as a lung treatment.[67] The preparation echoed what a British medical pamphlet,

[61] Semedo, "Memorial" (cit. n. 25), 198.

[62] João Curvo Semedo, *Atalaya da vida contra as hostilidades da morte* (Lisbon, 1720), 144.

[63] Gabriel Grisley, *Desenganos para a medicina. Ou, botica para todo pay de familias* (Lisbon, 1656), 30r.

[64] Ibid., 43v, 10v, 130r. Grisley also recommended marjoram "mixed with powdered tobacco" as a "remedy for many ailments" (87v). In this case and in the case of the soapwort root, or *saboeira* (10v), it's unclear whether he was describing the smoking or insufflation (snorting) of these preparations.

[65] Abraham Alewyn and João Collé, *Tesóuro dos Vocábulos das dûas Línguas, Portuguéza, e Bélgica* (Amsterdam, 1714), 179 ["Een toebaks-pyp, of, een pyp, daar men eenige kruiden door rookt"]. This reference is discussed briefly in Chris Duvall, *Cannabis* (Reaktion Books, 2014), 132.

[66] Jacob de Castro Sarmento, *Materia medica physico-historico-mechanica* (London, 1735), 496–7.

[67] Francisco de Fonseca Henriques, *Medicina Lusitana, e soccorro delphico* (Amsterdam: Miguel Diaz, 1710), 457.

Abraham Miles's passive-aggressively titled *The Countryman's Friend, and No Circumventing Mountebanck*, had recommended for horses a few decades earlier. But Miles went into more detail about what "smoking" such a preparation entailed. A "pennyworth of Towsillagine, and a pennyworth of Uripigmentum [orpiment]" should be placed in a "chaffing dish of wood coals," he wrote. This burning mixture was to be covered "with a Tunnel, that the pipe of the tunnel may convey the smoake into [the horse's] nostrils."[68] In short, *cachimbo* and *pipe* in seventeenth-century accounts could mean everything from a typical tobacco pipe to various long-forgotten contraptions such as Miles's chafing dish joined to "the pipe of the tunnel."

PYRIC TECHNOLOGIES AND ELEMENTAL TRANSFORMATION

The same pyric technologies of elemental transformation that fascinated medical authorities and early modern consumers could also serve to deceive and dazzle a credulous crowd. When, in 1682, the English natural philosopher John Evelyn witnessed the mysterious glow of phosphorus derived from chemical experiments with urine, his mind immediately went to a mountebank he had seen in Bologna. This man, it seemed to Evelyn, must have been privy to a similar technology when he dazzled a crowd of onlookers with a mysteriously glowing ring. Evelyn's thoughts then ran to another form of popular display: the attendants of a Catholic mass. "I appeared in the darke like the face of the moone, or rather like some spirit, or strange apparition," Evelyn marveled after rubbing the strange substance onto his face. He worried that if this mysterious, fiery substance were to fall into the hands of Catholic priests, "what a miracle might they make it, supposing them either to rub the Consecreated Wafer with it, or washing the Priests face & hands with it, & doing the feate in some darke Church or Cloyster, proclaime it to the Neighborhood."[69]

Three years later, "dining at Mr. Pepys's," Evelyn again marveled at the "many constellations, burning most vehemently," which glowed forth in a glass of phosphorus made from "human blood and urine, elucidating the vital flame, or heat in animal bodies."[70] This fascination was not merely visual. Evelyn and other seventeenth-century Europeans encountered new technologies of flame and smoke within the context of a medical epistemology that saw flame and light as the kindred of the fiery "spirits" that animated bodies. The English writer and physician Sir Thomas Browne (1605–1682), for instance, called life a "pure flame" and spoke of the soul as "an invisible Sun within us." Indeed, Browne appears to have made this metaphorical fire the center of his cosmology. For him, the "fire and scintillation" of the soul was nothing less than the "warme gale and gentle ventilation of this Spirit [of God]." Without this divine essence, Browne believed he would be deprived of life itself, senseless even to the "heat under the Tropick" and blind to any light "though I dwelt in the body of the Sunne."[71]

[68] Abraham Miles, *The Countrymans Friend, and No Circumventing Mountebanck but a Rare Method of Chyrurgery and Physick* (London, 1662), 32.

[69] John Evelyn, *The Diary of John Evelyn*, ed. Guy de la Bédoyère (Woodbridge, UK: Boydell Press, 2004), August 4, 1681, 246.

[70] Ibid., December 13, 1685, 297. Granted, John Evelyn was more fascinated with fire and smoke than most, being the author of *Fumifugium, or, The inconveniencie of the aer and smoak of London dissipated* (London, 1661).

[71] Sir Thomas Browne, *Religio Medici* (London: Printed for Andrew Crooke, 1643), section 32.

Viewed from this perspective, the translation of smoking into European medical culture takes on a new, overtly spiritual significance. As such, pyric technologies became woven into a larger set of religious and moral debates involving the degree to which the pursuit of "novelty" (and colonial conquest) could be integrated within an existing framework of Christian virtue. Evelyn's worry about this new technology's potential misuse by actors he regarded as unethical—in this case, Catholic priests—lends credence to Hansun Hsiung's observation elsewhere in this volume that the circulation of novel techniques in the early modern world involved not only linguistic translation, but also efforts (often failed) to establish a "commensurability of virtues."[72]

The pursuit of pyric novelty often seemed to ally smoking-proponents with the figure of the mountebank in the eyes of European early commentators. John Davies, writing in 1580s England, praised tobacco as a "herbe of heavenly power . . . [from] another world" whose "sweete substantiall fume . . . is thought a gentlemanlike smell."[73] But he also connected it to the "mountebanks" who sell "oyles" in the marketplace. As chemical medicine grew in popularity, observers began to draw comparisons between the complex distillation apparatuses of alchemists and apothecaries and the contraptions involved in smoking. One of the earliest appearances of the word *cachimbo* in print (1642) poked fun at "the tobacconists of *cachimbo*, or the *cachimbistas* of tobacco" whose "smoke from their nostrils and mouths" was like "the volcanos of hell" and who were "auditioning for the part of trumpeters of Pluto." These foolish cachimbo-users, the author wrote, were victims of their desire to "dry out . . . the distillations of the head [*desecar . . . distilaciones de la cabeça*]."[74]

This language of "dry[ing] . . . distillations of the head" was part of a larger project of chemical medicine: to apply novel techniques of manipulating matter, particularly the transformation of the substance of medicinal herbs and minerals with fire and distillation, to transform the physical, mental, and spiritual characteristics of the patient. All of which brings us to the late sixteenth- and early seventeenth-century minigenre of "distillation of fools" engravings (fig. 4). The poem accompanying the print pictured in figure 4 complains of immoral city-dwellers who "by fire their brains a purge indure," which "will evaporate" the foolish contents of their minds. A flaming brazier, labelled "Ardor Divinis," and a still filled with "wholesome herbs" such as "Horestrange" and "Sage of Jerusalem," sit beside the "red hott furnace." Items like tennis rackets, drinking cups, and tobacco pipes are visualized in the smoke issuing from the foolish man's brain as the phantasmagoria of a culture in thrall to material objects. Pipes, in these prints, become part of a larger complex of pyric technologies involving the medicinal application of heat to induce elemental transformations in the body, "drying" and "distilling" the congested humors of the brain into the vaporized, and potentially heaven-bound, residue of thought itself.

In short, the early modern globalization of pipes and smoking is important not only because of its relationship to the history of tobacco. The spread of the pipe in both its African/South Asian and American variants was part of a larger grouping of early modern *technologies of elemental transformation*. Pipes, with their novel route of

[72] Hansun Hsiung, "Use Me As Your Test!," in *Osiris* 37.

[73] John Davies, *Epigrammes and elegies* (Middleborough, [1590?]), 36.

[74] Domingos Pereira Braçamonte, *Banqvete qve Apolo hizo a los embaxadores del rey de Portugal don Iuan Quarto* [João IV] (Lisbon, 1642), 112.

Figure 4. *"Doctor Panurgus" curing the folly of his patients by purgative drugs and alchemical fires. Engraving attributed to Michael Droeshout, c. 1620. Wellcome Collection, London.*

drug administration conducive to increasing the potency of psychoactive substances, held out the promise of evoking not just physical, but mental transfiguration.

A popular allegorical poem of Stuart England by Thomas Jenner circulated around the same time as these prints, and similarly played with the theme of thoughts made manifest (yet, paradoxically, also made immaterial and fleeting) in the form of smoke. "When the smoke ascends on high / Think, though beyold'st the vanity / Of worldly stuff gone with a puff." Yet Jenner's poem—like the prints themselves, which seem to be poking fun at these efforts to distill foolish thought rather than celebrating them—was not optimistic about the potential of this new technology to evoke real transformation. The thoughts of the fool may be distilled away. But a dross was left behind that could not be made immaterial. "When the Pipe grows foul within," Jenner warned, "Think on thy soul defil'd with sin, / And then the fire it doth require."[75]

Notably, every printed instance of the distillation of fools motif includes an accompanying figure depicting this physical remainder of foolishness, which falls into a

[75] The poem is discussed in Jeffrey Knapp, "Elizabethan Tobacco," in *New World Encounters*, ed. Stephen Greenblatt (Berkeley: Univ. of California Press, 1993), 307.

bedpan as excrement rather than ascending into the heavens as smoke (fig. 5). This was, perhaps, a warning that the promise of techniques which claimed to entirely transmute the physical into the immaterial smoke was a false hope.

Fears of elemental transformation and quackery may have helped tamp down the culture of experimentation that seems to have accompanied the earliest century of smoking in early modern Europe. The winnowing away of alternative modes of pipe usage—until tobacco pipes and tobacco smoking became not just normative, but the sole application of this new technology of elemental transformation—also served to conceal the influences of the African cannabis-smoking cultures with their alternative forms of the pipe. In the end, the only testament to their influence was the persistent survival of the word itself in the Lusophone world: not an analogue to the *pipa* of Galician or Castilian (evocative of the calls of birds and hence of the narrow tubes of the piper), but *cachimbo*, the African reservoir of water.

Figure 5. *Details of the purging of foolishness in two related prints:* Left: *"The doctor cures fantasy, and purges folly with drugs," Mattheus Greuter, (Paris?, 1620);* right: *the "Doctor Panurgus" print reproduced in figure 4. Wellcome Collection, London.*

PYROTECHNICS AND CIVILIZATION

James Secord has called for historians of science to "shift our focus and think about knowledge-making itself as a form of communicative action."[76] Smoking was a new technological product, and it was medicalized within a larger practice of pyric technologies. But it was also, as Secord puts it, a novel form of "communicative action": after all, the practice of smoking is hard not to notice, or smell. It is a highly performative act, and it announces itself in a distinctive, perhaps even disruptive fashion. Increasingly, perhaps, that bodily performance of novelty became divorced from expectations of proper decorum and textual knowledge-making expected of the European physician and philosopher.

Smoking technologies, I believe, also became bound up with a larger set of concerns about the New World and Africa as tropical spaces beset by malign astrological forces and fiery vapors. For the sixteenth-century Spanish missionary José de Acosta, the Americas were a world characterized by esoteric smokes and fires. He spent considerable time describing the effects of quicksilver, which he wrote had a "mortal smoke" that was transmitted "by tubes in the manner of an alembic."[77] And he remarked on the unfamiliar "smoking . . . mountains" of Mexico, which some speculated contained "the fires of hell." These were domains in which elemental transformation was, from the European perspective, unavoidable. Pipes were not, then, a discrete and clearly delimited technology, but part of a larger package of concerns about ecology, spiritual and medical authority, and the potential permeability of body and mind.

Given this, it is little wonder that smoking was a reoccurring fascination—and source of profound cultural anxiety and fear—throughout seventeenth-century Europe. When, in 1614, Francis Bacon directed the staging of a lavish celebration at King James's court, *The Masque of Flowers*, one of the lead attractions was a figure called Kawasha.[78] This was a kind of pipe-god ("the crown of his cappe a chimney . . . bases of tobacco-colour stuff cut like tobacco leaves") who was pitted against the Roman god of wine, Silenus.[79] Kawasha was, in fact, based on a real account of an Algonquian god, Kiwasa, depicted in de Bry's illustrated edition of Harriot's report of Virginia (fig. 6). In de Bry's image, "Kivvasa" is depicted simply as a seated "idol" within an enclosure. Here—perhaps inspired by the numerous instances of pre-Columbian ceremonial pipes featuring seated human figures in somewhat similar poses, with their heads as bowls—Bacon had refigured the god as a massive pipe in human form. Tobacco leaves adorned Kawasha's base, and smoke issued from his head. "[Kawasha] is come from a far country / to make our nose a chimney" went one of the lines in the masque. The masque pitted Silenus, god of wine, against Kawasha, "this great Potan [*petun*, tobacco]," who threatens to make his enemies "all snuffing, puffing, smoke, and fire." The same verse likened the smoke and fire expelled by tobacco-smokers to that of the "fell dragon," recalling another print by de Bry that depicted indigenous

[76] James A. Secord, "Knowledge in Transit," *Isis* 95 (December 2004): 654–72, on 661.

[77] Acosta, *Historia* (cit. n. 59), 342.

[78] Christine Adams, "Francis Bacon's Wedding Gift of 'A Garden of a Glorious and Strange Beauty' for the Earl and Countess of Somerset," *Garden History* 36 (2008): 36–58.

[79] Olivia Bloechl, *Native American Song at the Frontiers of Early Modern Music* (Cambridge, UK: Cambridge Univ. Press), 118–9. For the original, see *The Maske of Flowers, Presented by the Gentlemen of Graies-Inne, at the Court of White-hall* (London: Printed by N. O., 1614), n.p.

Figure 6. *The "Idol Kivvasa" from Theodor de Bry's illustrated reprint of Thomas Hariot's* Briefe and True Report of the New Found Land of Virginia *(London, 1590), based on a drawing by colonist John White. This hand-colored example is courtesy of the British Library.*

Brazilians being tormented by the flames of dragon-like demons. According to the stage directions in the printed version of the masque, Kawasha was joined in his dance by figures including a "Mountebancke" and a "Jewesse of Portugall," indicating the ambiguous status of smoking that persisted even after its rapid increase in popularity in the previous two decades.

The resulting conflict between the forces of drink and of smoke ended with a characteristic note of ambivalence. A chorus joined in song, calling smoking a "holy rite" from a new land. Yet they also connected this novel custom to the ancient practice of fumigation: ("Nothing but fumigation / doth chase away ill spirits, / Kawasha and his Nation / Found out these holy rites.")[80]

CONCLUSION

Pyric technologies played an important role in debates about the early modern drug trade, and in globalization and empire more generally. As a vehicle for the delivery of psychoactive and potentially addictive substances, pipes contributed both to a newfound recognition of psychoactivity and to new efforts of translating bodily practice, material technologies, and social formations.

[80] The quotes in this paragraph are from *Maske of Flowers* (cit. n. 79), n.p.

Although it is often assumed that the concept of smoking and of the *techne* of pipes reached Europe via the New World, this article has argued that this theory appears incomplete given the available evidence from Portuguese sources, which point to multiple points of origin for the globalization of smoking. Rather than simply seeing the pipe as part of the "Columbian Exchange" of a distinctively New World product (tobacco) that crossed over to the Old World, we should envision a multidirectional movement of pyric technologies and associated smokable drugs. There were important contributions from African and South Asian cannabis cultures, which blended with indigenous American tobacco smoking practices in the early decades of the Atlantic slave trade and, over time, became largely indistinguishable from them in the historic record.

All of these smoking technologies combined with the early modern European preoccupation with alchemical transformation of matter to create a larger context for smoking, not just as a form of recreational sociability, but as an experimental medical technology with far-reaching intellectual and material ties.

This globalization of pyric technologies was inseparable from the practices of alteration and effacement that accompanied the codification of smoking within European textual authorities. As Alisha Rankin details in the present volume, perhaps the most "iconic" early modern description of smoking—Nicolas Monardes's "effusive account of tobacco," which first appeared in print in 1569—had a remarkably tangled textual history, with medical authors in other vernacular languages such as Dutch, French, and English asserting their own competing claims about tobacco smoking alongside his own.[81] The "instability" that Rankin identifies in the textual history of Monardes's account of tobacco was compounded by the complex interplay between the elite medical discourse of licensed physicians and the vernacular worlds of sailors, slaves, and healers.[82] Shireen Hamza's careful study of vernacular Indo-Persian medical glossaries is a useful guide here, showing how close attention to vernacular terminology can illuminate embodied forms of knowledge that operated outside of elite textual traditions.[83] All too often, however, the vernacular worlds that produced words like *cachimbo* drop out of the medical and scientific record, producing a flattened and more simplistic picture over time.

This story is part of a larger—and largely untold, but not untellable—history of the circulation of African medicinal knowledge and techniques in the early modern period.[84] Figures like Robert Boyle in England and João Curvo Semedo in Portugal were highly interested in African pharmacological knowledge and regularly engaged slave traders and travelers in the region. But by and large, their interest failed to translate into the mainstream of European medicine. This was, perhaps, due to two factors. The first was the confusion of names, making it hard to trace African genealogies and differentiate them from Asian and Amerindian ones; the second, the associations between elemental transformation, quackery, and African *feiticeiros*, which made the more expansive forms of medicalized smoking on display in some seventeenth century texts

[81] Rankin, "New World Drugs" (cit. n. 56).

[82] Ibid.

[83] Shireen Hamza, "Vernacular Languages and Invisible Labor in *Ṭibb*," in *Osiris* 37.

[84] On which see Pablo F. Gómez, *The Experiential Caribbean: Creating Knowledge and Healing in the Early Modern Atlantic* (Chapel Hill: Univ. of North Carolina Press, 2017); and Gómez, "[Un]Muffled Histories" (cit. n. 14).

(the "cachimbo complex"?) fade in favor of the recreational use of tobacco via more familiar smoking implements.

How might this more capacious understanding of the early modern translations of smoking affect other questions and lines of research? For one thing, the larger historical traces and influences of the long-neglected African element in the globalization of smoking seems worthy of more investigation, particularly the opportunities it offers for integrating the history of material culture and bodily practice with textually-based histories of medicine and health. In her history of smoking in the twentieth-century Balkans, Mary Neuburger notes an apparent contradiction in the tobacco habits of Europeans: "Whereas tobacco itself arrived from the New World, the West's smoking rituals were borrowed from the East. Oriental-style smoking became an escapist symbol of wealth and excess for the West to emulate, succeeding where American 'savage' smoking rights had not . . . Smoking, like coffee and the coffeehouse, was the product of Eurasian cultural interaction and exchange."[85]

If we assume that American-style tobacco smoking is the progenitor of all European smoking cultures, this "Eastern" influence looks anomalous, perhaps a result of nineteenth-century Orientalist fantasies of hookahs and Turkish tobacco. Yet if we reconceive of smoking as a technology that became global from *multiple* distinct drug regions—the tobacco-smoking culture of the indigenous Americas and the cannabis-smoking cultures of sub-Saharan Africa and South Asia—the picture looks very different. It also prompts the question of why opium smoking emerged as a massive force of societal change and economic activity in the nineteenth century, but was relatively absent in early accounts. Why, if seventeenth-century cultures of smoking involved eclectic experimentation with smoking technologies and "compound" preparations of smokable drugs, did opium smoking apparently not become widespread until the late eighteenth and early nineteenth centuries?[86]

A final question worth exploring further relates to the role of "pyric technologies" in the history of ideas. For instance, how did smoking technologies interact with early modern theories of cognition and philosophy of mind? Margaret Cavendish argued that sense perception and the rational faculties, working in tandem, were able to "pattern out" the characteristics of the physical world, effectively creating replicas of these patterns within the mind.[87] The brain, like a lens or other instrument, was capable of magnifying these patterns and making them not only knowable intellectually but *experienceable* on a material level. Psychoactive drugs, like any other sensory patterns, thus created "copies" of themselves in the mind in this schema, transforming from the physical to the mental and back again (or from the sensitive to the rational) and thereby altering the patterns of thought itself. How did the elemental transformations and psychoactive experiences made possible by the spread of smoking technologies influence such early modern accounts of the interaction between sense perception and consciousness?

[85] Mary C. Neuburger, *Balkan Smoke: Tobacco and the Making of Modern Bulgaria* (Ithaca, NY: Cornell Univ. Press, 2012), 4–5.

[86] According to Carol Benedict, a smokable preparation of tobacco, opium, and other substances known as *madak* first emerged among Portuguese sailors in the ports of Southeast Asia in the seventeenth century, from which it spread to China. However, the trade was dwarfed by the opium commerce of the early nineteenth century, which relied on demand from the apparently novel technique of smoking unmixed opium; Benedict, *Golden-Silk Smoke* (cit. n. 4), 6, 79.

[87] Margaret Cavendish, *Observations Upon Experimental Philosophy*, ed. Eileen O'Neill (1666; repr., Cambridge Univ. Press, 2001), 175–6.

What is clear is that historians of science and medicine need to take smoking seriously as a source of experimental practice and conceptual novelty in the early modern period. Smoking was not just a method of consuming tobacco. It was a rapidly evolving and highly diverse set of material technologies and physical practices that sought to convert the material into the immaterial. Smoking implements mediate between phases of matter, but also between the realm of macroscopic plant life and the microscopic realm of the brain—and, crucially, between human beings.

Historically, pipes have been icons of both sociability and of solitary reflection. One imagines a philosopher with a pipe, but also the drunken tavern-goers of a Brueghel painting or Hogarth print, or the ritual exchange of an Algonquian calumet ceremony. The tobacco pipe, as I have argued, was and is not coterminous with these technologies of smoke in all their myriad forms. With the rise of e-cigarettes, the ancient material tradition of the pipe may well be on the wane for good. But the immaterial smoke will remain.

Translating the Inner Landscape: Anatomical *Bricolage* in Early Modern Japan

*by Daniel Trambaiolo**

ABSTRACT

The translation of European anatomical treatises is widely considered one of the pivotal processes of early modern Japanese history, introducing new ways of understanding the body and new styles of visual representation, as well as serving to launch the broader intellectual movement of "Dutch Studies" (*rangaku* 蘭学). In this article, I consider how anatomical knowledge and anatomical translations were integrated into the medical thinking of Japanese doctors whose understandings of the body continued to be informed by older East Asian traditions of medical knowledge. In contrast to the *rangaku* translators, who developed a novel anatomical lexicon in order to faithfully reproduce the meanings of European source texts, early nineteenth-century practitioners of Sino-Japanese medicine, including Kako Ranshū, Mitani Boku, and Ishizaka Sōtetsu, described the anatomical body using the language of the Chinese medical classics. At the same time, they used new understandings of the body from European anatomy to solve long-standing problems in the interpretation of those classics, identifying specific anatomical structures in the digestive system corresponding to the "triple burner" (C. *sanjiao*, J. *sanshō* 三焦) and reimagining the body's conduits for the circulation of blood and *qi* and the targets of acupuncture in terms of the veins, arteries, and nerves.

> "The image cannot be the idea, but it can play the role of a sign, or, more precisely, cohabit with the idea in a sign; and, if the idea is not yet present, reserve its future place and make its contours appear in the negative. The image is frozen, linked in a univocal way to the act of consciousness that accompanies it; but the sign, and an image that has come to act as a sign, [. . .] are already *permutable*, that is, capable of entering into successive relations with other entities."
>
> —Claude Lévi-Strauss[1]

* Department of Japanese Studies, University of Hong Kong, Centennial Campus, Pokfulam Road, Hong Kong; trambaiolo@hku.hk.

I would like to thank the organizers and participants of the workshop series "Translating Medicine in the Premodern World" and the anonymous referees for *Osiris* for their invaluable feedback on earlier versions of this essay. This work was partly supported by a Sin Wai Kin Junior Fellowship awarded by the Hong Kong Institute for Humanities and Social Sciences.

[1] Claude Lévi-Strauss, *Wild Thought: A New Translation of "La Pensée Sauvage,"* trans. Jeffrey Mehlman and John Harold Leavitt (Chicago: Univ. of Chicago Press, 2021), 24.

Osiris, volume 37, 2022. © 2022 History of Science Society. All rights reserved. Published by The University of Chicago Press for the History of Science Society. https://doi.org/10.1086/719225.

During the winter of 1809, the Japanese doctor Kako Ranshū drafted a treatise, later published as *Essentials of Anatomy and Acupuncture* (*Kaitai chin'yō* 解體鍼要, 1819), with the aim of reconciling the established practices of acupuncture with the new forms of European anatomical knowledge promulgated in Japan since the late eighteenth century (fig. 1). Kako was keenly aware of the difficulties involved in such a reconciliation. Chinese, Korean, and Japanese doctors had long considered knowledge of the body's interior to be an essential foundation for medical practice, but the new vogue for cutting up corpses and scrutinizing their contents could seem both morally questionable and a distraction from the more important task of understanding the living body. Nevertheless, Kako insisted, the new anatomy could not be dismissed as suitable only for European barbarians: "When one knows the locations of the viscera, the connecting networks of the conduit vessels, the joints of the bones, and the contents of the skins and membranes, one can grasp the origins and development of disease. [. . .] This is not without benefits for the healing arts."[2]

The translation of European anatomical treatises during the eighteenth and nineteenth centuries is one of the most familiar narratives in the history of Japanese medicine, and even in the history of Japanese modernity itself.[3] According to this narrative, Japanese doctors before the mid-eighteenth century accepted the descriptions of internal body structures in Chinese medical classics such as *The Yellow Emperor's Inner Canon* (*Huangdi neijing* 黃帝內經), as well as the illustrations of these structures in later Chinese books. During the 1750s, doctors of the "Ancient methods" (*kohō* 古方) school such as Yamawaki Tōyō (1705–62) began to challenge these accepted accounts of the body by conducting anatomical observations of their own, but the true birth of Japanese anatomy occurred two decades later, when Sugita Genpaku (1733–1817) and a group of friends made a direct comparison between the images in a European anatomical treatise and the internal organs of an executed woman, thereby establishing the superiority of European knowledge of the body.[4] Motivated by this discovery, Sugita and his friends set about the arduous task of translating a European anatomical treatise, crafting the necessary vocabulary as they went, and in the process created the new field of *rangaku* 蘭学 (lit. "Dutch studies"), so called because Dutch was the only European

Abbreviations: *SW,* Guo Aichun 郭靄春, *Huangdi neijing suwen* 黃帝內經素問校注語譯 [*The Yellow Emperor's Inner Canon: Numinous Pivot*, edited and annotated with modern translation] (Tianjin: Tianjin kexue jishu chubanshe, 1981) [cited by chapter number]; *LS,* Guo Aichun 郭靄春, *Huangdi neijing lingshu* 黃帝內經靈樞校注語譯 [*The Yellow Emperor's Inner Canon: Celestial Pivot*, edited and annotated with modern translation] (Tianjin: Tianjin kexue jishu chubanshe, 1989) [cited by chapter number]; *NJ,* Ling Yaoxing 淩耀星, ed. *Nanjing jiaozhu* 難經校注 [*The Canon of Difficulties*, edited and annotated] (Beijing: Renmin weisheng chubanshe, 1991) [cited by difficulty number].

[2] Kako Ranshū 加古蘭州, *Kaitai chin'yō* 解體鍼要 [Essentials of anatomy and acupuncture] (1819), Preface (1809), 1b-2a. In this essay, all translations from Chinese and Japanese are my own.

[3] A particularly striking example of the significance of anatomical translation in the historiography of modern Japan can be seen in James L. McClain's timeline "Japan in Revolutionary Times," which begins with the publication of Sugita Genpaku's *Kaitai shinsho* in 1774 and ends in 1903, on the eve of the Russo-Japanese war; McClain, *Japan: A Modern History* (New York: Norton , 2002), 114–8. Other typical examples include Marius Jansen, *The Making of Modern Japan* (Cambridge, MA: Harvard Univ. Press, 2000), 212–4; and Brett Walker, *A Concise History of Japan* (Cambridge, UK: Cambridge Univ. Press, 2015), 149–54.

[4] Sugita Genpaku, *Rangaku kotohajime* 蘭学事始 [The origin of Dutch learning], reprinted in *Nihon koten bungaku taikei* 日本古典文学大系, vol. 95 (Tokyo: Iwanami shoten, 1964), 469–553. For an English translation, see Sugita Genpaku, *Dawn of Western Science in Japan*, trans. Ryōzō Matsumoto (Tokyo: Hokuseido Press, 1969). On the context of this memoir, see Katagiri Kazuo 片桐一男, *Chi no kaitakusha Sugita Genpaku: "Rangaku kotohajime" to sono jidai* 知の開拓者杉田玄白:『蘭学事始』とその時代 [Sugita Genpaku, pioneer of knowledge: The origin of Dutch learning and its era] (Tokyo: Bensei shuppan, 2015).

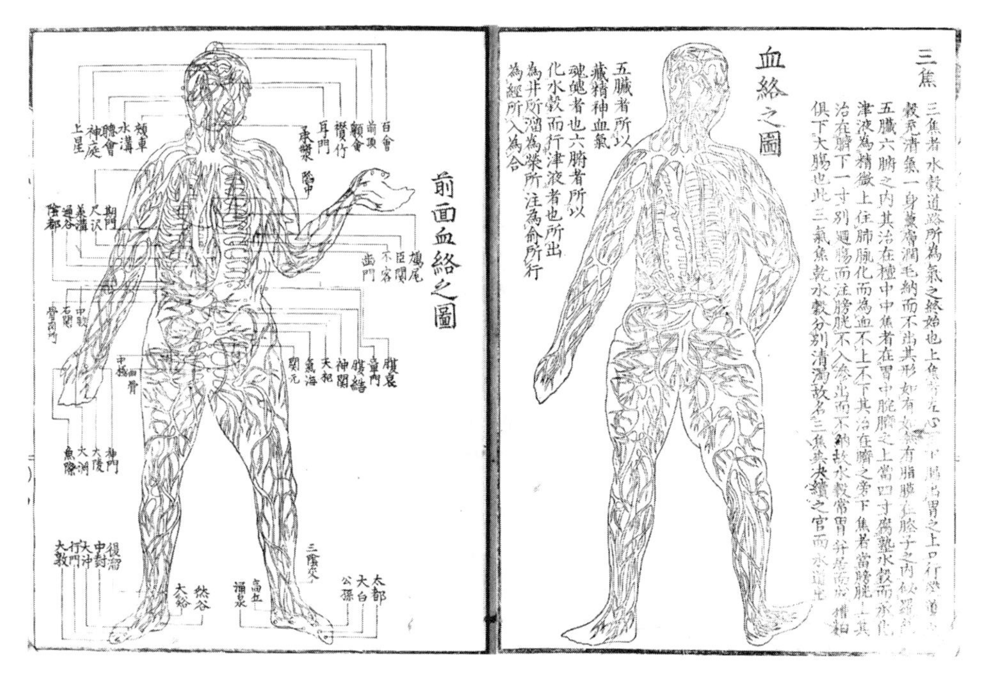

Figure 1. *Blood vessels and acupuncture points in Kako Ranshū,* Kaitai chin'yō *(1819), 3b–4a. The images are labeled "Diagrams of the blood networks." The images were based on Motoki Ryōi's flapbook (cf. fig. 2), while the locations of acupuncture points were based on Chinese diagrams of the points and conduits (cf. fig. 3). Image courtesy of the Main Library, Kyoto University, https://rmda.kulib.kyoto-u.ac.jp/item/rb00001512.*

language that Japanese doctors were able to study after the shogunate's seventeenth-century ban on contact with other Europeans).[5] Sugita's *New Book of Anatomy* (*Kaitai shinsho* 解體新書, 1774), translated from the Dutch version of a treatise by the German physician Johann Adam Kulmus, initiated a new phase of Japan's engagement with European culture, with effects that continued to shape Japanese history into the nineteenth century and beyond.

In a seminal essay on the anatomical illustrations of Sugita Genpaku's *New Book of Anatomy*, Shigehisa Kuriyama identified the central problem for the history of anatomy in eighteenth-century Japan as the question of "what it could mean to conceive of the body anatomically."[6] Kuriyama's answer to that question—that "the real novelty of Gempaku's visual world consisted not just in *what* he saw, but more fundamentally in *how* he saw, in a new perceptual style"—remains a provocative suggestion, as does his argument that Sugita and other *rangaku* scholars saw their pursuit of realistic visual mimesis as constituting a social and moral imperative as much as an epistemic one.[7] But

[5] On the development of *rangaku*, see Grant K. Goodman, *Japan and the Dutch, 1600–1853* (Richmond: Curzon Press, 2000); Timon Screech, *The Lens within the Heart: The Western Scientific Gaze and Popular Imagery in Later Edo Japan* (Honolulu: Univ. of Hawai'i Press, 2002); and Terrence Jackson, *Network of Knowledge: Western Science and the Tokugawa Information Revolution* (Honolulu: Univ. of Hawai'i Press, 2016).

[6] Shigehisa Kuriyama, "Between Mind and Eye: Japanese Anatomy in the Eighteenth Century," in *Paths to Asian Medical Knowledge*, ed. Charles Leslie and Allan Young (Berkeley: Univ. of California Press, 1992), 21–43, on 23.

[7] Ibid., 28. (Emphases in original.)

Kuriyama's analysis, whose interpretation of *rangaku* anatomy remains close to the *rangaku* scholars' own perspective, is but a starting point for exploring the history of anatomy in early modern Japan. As Timon Screech has argued, *rangaku* scholars' rhetoric about the epistemic priority of visual observation was in large part "ideological."[8] Rather than accepting this rhetoric at face value, we need to study not only these scholars' actual practices of observation, description, and depiction, but also those of their contemporaries who engaged in similar practices but came to different conclusions. At the same time, our analyses of early modern Japanese anatomical images' visual character must be matched by an analysis of their semiotic character. Anatomical images were not just pictures: they were *labeled* pictures. It was the relationships between anatomical pictures and textual descriptions, not the visual qualities of the pictures themselves, that made these pictures objects of knowledge.[9] This is why *rangaku* scholars devoted so much energy to developing and refining their translations of the texts accompanying anatomical images: translations made pictures speak.

But pictures could be made to speak in more than one way. Consider the images of the blood vessels on a double-page spread of Kako Ranshū's *Essentials of Anatomy and Acupuncture* (fig. 1). Kako drafted this book more than three decades after the publication of Sugita's *New Book of Anatomy*, at a time when *rangaku* was a mature and expanding intellectual movement. But a cursory examination of these pages reveals that Kako was using European anatomical images to illustrate a very different understanding of the human body from that of the *rangaku* anatomists. The picture on the left is titled "Frontal view of the blood networks," but its labels list not the names of blood vessels but those of important acupuncture points. The picture on the right is accompanied by two paragraphs of text: one of these quotes an overview of the organs and the acupuncture points from the *Numinous Pivot* (*Lingshu* 靈樞), an ancient Chinese medical classic; the other collates excerpts from at least three different Chinese sources on the so-called triple burner (C. *sanjiao*, J. *sanshō* 三焦), one of the six hollow organs (C. *fu*, J. *fu* 府, 腑) that played an important role in East Asian medical understandings of body function but which European books did not mention at all.[10] By surrounding the European images with text from other sources, Kako gave them an entirely new set of meanings.

At first glance, this jumbled hodgepodge of Chinese texts and European images might seem an unpromising source for investigating the history of anatomical translation. The texts, composed entirely of quotations from Chinese sources, would hardly be

[8] Timon Screech, "The Birth of the Anatomical Body," in *Births and Rebirths in Japanese Art: Essays Celebrating the Inauguration of The Sainsbury Institute for the Study of Japanese Arts and Cultures*, ed. Nicole Coolidge Rousmaniere (Leiden: Hotei, 2001), 83–140, on 104.

[9] Cf. Sachiko Kusukawa, *Picturing the Book of Nature: Image, Text, and Argument in Sixteenth-Century Human Anatomy and Medical Botany* (Chicago: Univ. of Chicago Press, 2011); Dániel Margócsy, "From Vesalius through Ivins to Latour: Imitation, Emulation, and Exactly Repeatable Pictorial Statements in the *Fabrica*," *Word & Image* 35 (2019): 315–33.

[10] The short paragraph in figure 1 quotes from *LS* 1. The longer paragraph on the triple burner interleaves excerpts from *LS* 18 and *NJ* 31. The longer paragraph also includes a small number of phrases whose textual origins I have been unable to identify; these phrases may have been added by Kako himself. On the "triple burner," see Pi Guoli 皮國立, "Tuxiang, xingzhi yu zangfu zhishi—Tang Zonghai sanjiao lun de qishi" 圖像、形質與臟腑知識——唐宗海三焦論的啟示 [Imagery, materiality, and knowledge of the viscera: Lessons from Tang Zonghai's views on the Triple Burner], *Gujin lunheng* 15 (2006): 71–98; and Sean Hsiang-lin Lei, "*Qi*-Transformation and the Steam Engine: The Incorporation of Western Anatomy and Re-Conceptualisation of the Body in Nineteenth-Century Chinese Medicine," *Asian Medicine* 7 (2012): 319–57, on 330–7.

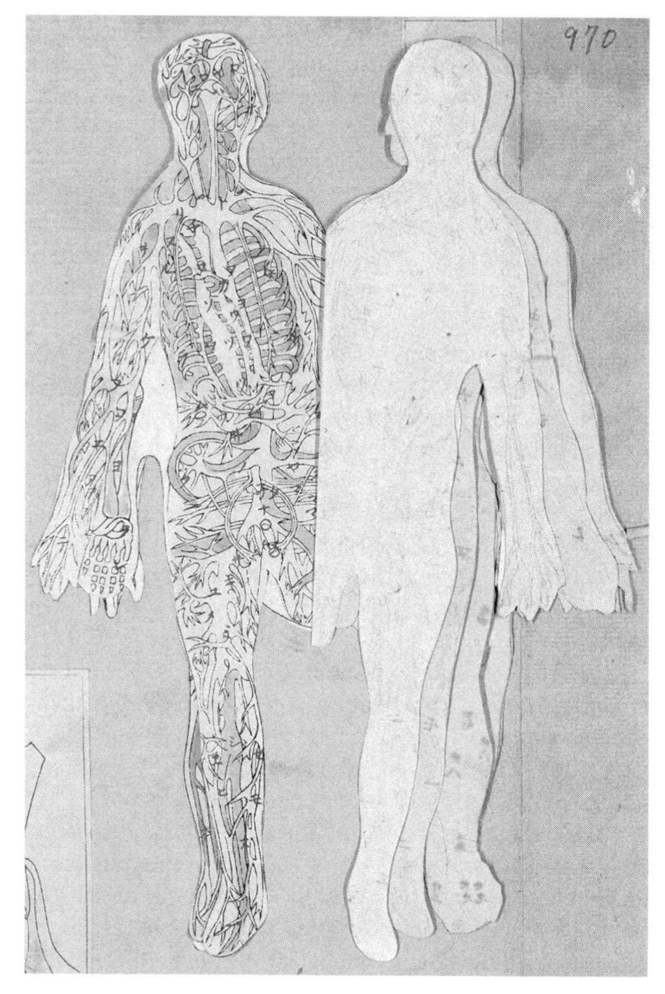

Figure 2. *Illustration of the blood vessels in Motoki Ryōi,* Oranda zenku naigai bungōzu oyobi kengō *(ca. 1680, reprinted 1772). Image courtesy of the University of Tokyo General Library, https://kotenseki.nijl.ac.jp/biblio/100273330/viewer/15.*

considered "translations" in the usual sense. Moreover, although Kako claimed to have studied anatomy since his youth and to have observed fifteen dissections, his familiarity with the ideas of the *rangaku* anatomical translators seems to have been limited: contemporary *rangaku* scholars would have considered his images of the blood vessels hopelessly inaccurate and obsolete, since they were copied not from a recent European anatomical treatise but from a flapbook produced more than a century earlier (fig. 2). Kako did not develop his argument by quoting from or engaging with *rangaku* translators' detailed explanations of anatomical concepts; instead, he implied a relationship between blood vessel anatomy and the practice of acupuncture simply by juxtaposing visual and textual elements from these disparate sources. Kako may have ended up unconvinced by the results of this attempt: he did not publish the book during his own lifetime, and we should perhaps understand it as the record of an interrupted intellectual process rather than as a definitive statement. But precisely because its constituent elements are so imperfectly integrated, its seams so clearly visible, a work like Kako's *Essentials of Anatomy and Acupuncture* offers a first glimpse of an alternative approach to understanding the history of anatomy in eighteenth- and nineteenth-century Japan.

Sugita Genpaku and Kako Ranshū had more in common than we might at first suppose. Both believed that European anatomical images offered an accurate representation of the body, and that studying these images could ultimately help medical practitioners treat their patients. For both Sugita and Kako, an important part of what it meant to "conceive of the body anatomically" was being able to identify and name the structural features of the body visible in these images and to relate these structures to ideas about function. Where Sugita and Kako diverged was in their assumptions about the legitimate range of discursive contexts into which these images could be integrated. For Sugita, the superiority of European anatomical images demanded acceptance of European anatomical concepts, and the only texts that could legitimately be associated with these images were translations from European languages. For Kako, however, it was possible to admire the precision of the images while suspending judgment on the concepts that *rangaku* scholars associated with them. These images were, to borrow the suggestive formulations of Claude Lévi-Strauss quoted above, images whose idea was "not yet present," permutable signs that remained "capable of entering into successive relations with other entities"—including even entities whose very existence was rejected by *rangaku* anatomists.

The contrast between the attitudes of Sugita and Kako as they approached the problem of constructing anatomical knowledge recalls Lévi-Strauss's dichotomy between two types of intellectual actor that he called the engineer and the *bricoleur*. Lévi-Strauss's engineer is one who "cross-examines the universe": working with concepts, he creates the tools he needs to "open a way through and situate himself *beyond* the constraints that make up a given state of civilisation."[11] The *bricoleur*, meanwhile, is one who addresses himself to "whatever is at hand": drawing on a heterogeneous but bounded set of resources, he works not with "concepts" but with "signs," and "willingly or by necessity, remains *on this side* of those constraints." In terms of this dichotomy, Sugita Genpaku and other *rangaku* scholars saw themselves as engaged in a form of conceptual "engineering": they set out to explore new worlds of knowledge, crafting new lexical and conceptual tools whenever they felt the inadequacy of the resources available within the East Asian tradition. For Kako Ranshū and other doctors who sought to integrate anatomical knowledge with East Asian medicine, the *rangaku* scholars' approach was misguided, since it severed the connection between anatomical knowledge and healing practices that provided the motivation for studying anatomy in the first place. For these doctors, any satisfactory description of European anatomical images needed to be linked through networks of signification to the ideas and practices they employed when diagnosing and treating their patients. Constructing these networks would involve not the invention of new concepts but the reinterpretation of existing ones.

There is of course a sense in which any complex semiotic object—a novel, a *noh* play, a protest movement, an *Osiris* article, a haiku—is a kind of *bricolage*. Complex communicative acts are made possible through the reorganization of existing elements

[11] Lévi-Strauss, *Wild Thought* (cit. n. 1), 23. For discussion of Lévi-Strauss's concept and its subsequent reception, see Anne Mélice, "Un concept lévi-straussien déconstruit: Le 'bricolage,'" *Les Temps Modernes* 656 (2009): 83–98; and Christopher Johnson, "*Bricoleur* and *Bricolage*: From Metaphor to Universal Concept," *Paragraph* 35 (2012): 355–72.

of meaning, and as Lévi-Strauss recognized, the difference between the *bricoleur* and the engineer is "not as absolute as we might be tempted to imagine."[12] But the impulse to break apart the elements and linkages that constitute complex semiotic objects in order to recombine them in new ways can vary dramatically for different types of semiotic objects in different cultural and historical circumstances. My claim in this article is that the impulse to shatter the semiotic networks contained in illustrated anatomical texts and reconstitute them in new ways was felt particularly strongly by a number of Japanese doctors in the late eighteenth and early nineteenth centuries. The exciting possibility of new insights into the structure of the body was mingled with discomfort at the fact that these images were coupled to accounts of somatic function that differed radically from the ideas about the body on which their own healing practices were founded. Analyzing the anatomical thinking of these doctors, most of whom have been forgotten or ignored by standard historical accounts, allows us to see important continuities running through a period of Japanese medical history that is often thought to mark a radical break from the past.

In this article, I first examine the earliest anatomical translation from Dutch into Japanese by the seventeenth-century Nagasaki translator Motoki Ryōi as an example of the initial stage of the encounter between East Asian medicine and European anatomy, when these two systems of knowledge had not yet come to be seen as fundamentally distinct. Next, using the example of Yamawaki Tōyō's *Record of the Viscera* (1759), I show that the problems of interpreting and linking anatomical terms and images were not exclusively issues of translation, but could also arise within the Sino-Japanese tradition as Japanese doctors attempted to sort through the conflicting evidence of ancient Chinese texts, later illustrations, and direct anatomical observations. I follow this with a consideration of how the *rangaku* translators constructed a sense of fundamental difference between European and East Asian understandings of the body through their translation choices, and how other Japanese doctors in the early nineteenth century sought to construct a sense of the two traditions' fundamental comparability by reinterpreting European anatomical images and concepts in Sino-Japanese medical terms. Finally, I conclude with a brief reflection on the broader implications of this history of anatomical *bricolage* in early modern Japan for understanding the relationship between East Asian and European medical accounts of the body.

THE BEGINNINGS OF ANATOMICAL TRANSLATION

The earliest Japanese translation of a European anatomical text was compiled by the Nagasaki Dutch interpreter Motoki Ryōi in the late seventeenth century (fig. 2).[13] It was a

[12] Derrida's critique of Lévi-Strauss's dichotomy on the grounds that "every discourse is *bricoleur*" and that "the engineer is a myth produced by the *bricoleur*" is thus accurate but also superfluous: once we descend from the level of philosophical abstraction to that of historical description, the myth of the engineer turns out to be analytically useful because it has shaped the behavior of historical actors. For the critique, see Jacques Derrida, "Structure, Sign, and Play in the Discourse of the Human Sciences," in *Writing and Difference*, trans. Alan Bass (London: Routledge), 278–94.

[13] Nihon Ishi Gakkai 日本医史学会, ed., *Zuroku Nihon iji bunka shiryō shūsei* 図録日本医事文化史料集成 [Illustrated compendium of cultural materials on medicine in Japan], vol. 2 (Tokyo: San'ichi shobō, 1977), 149–58; Hara Sanshin 原三信, ed., *Nihon hajimete hon'yaku shita kaibōsho* 日本で初めて翻訳した解剖書 [The first Japanese translated anatomical book] (Fukuoka: Rokudai Hara Sanshin ranpō i sanbyakunen kinen shōgakkai, 1995); Screech, "Birth of the Anatomical Body" (cit. n. 8); Kera Yoshinori 計良吉則 and Sakai Shizu 酒井シヅ, "*Oranda keiraku kinmyaku zōfu zukai* no hon'yakusho to shite no fukanzensa: yakushutsu shinakatta go no shiten kara" 『阿蘭陀経絡筋脈臓

version of Johann Remmelin's *Catoptrum microcosmographicum* (Microcosmographic mirror), a flapbook originally published in a Latin edition in 1619 and subsequently translated into a number of European languages, including a Dutch edition of 1667 that was most likely the basis for Motoki's version.[14] Unlike the two-dimensional images of standard anatomical treatises, Remmelin's images were constructed as a series of paper layers that could be peeled back one by one, gradually revealing deeper strata of the body's internal structure. It is not difficult to imagine the fascination this paper simulation of an anatomical dissection might have held for Japanese doctors during a period several decades before Japanese doctors began to conduct anatomical observations for themselves (fig. 2).

Motoki's copy of the original text was probably brought to Japan by Willem ten Rhijne, a doctor employed by the Dutch East India company and best known as the author of the earliest detailed European description of Japanese acupuncture.[15] Ten Rhijne later expressed frustration at his difficulties communicating with Motoki, who was "more experienced in medical matters than all the other interpreters," but spoke only "faltering Dutch in half-words and fragmentary expressions."[16] Nevertheless, Motoki managed to acquire a sufficient understanding of European anatomical and physiological concepts to translate the accompanying key to Remmelin's images, adding brief comments of his own based on conversations with the Dutch doctor.

We have no direct evidence concerning the intellectual processes by which Motoki made his translation. Unlike the later *rangaku* translators, Motoki left no memoir of his lexicographic travails, nor any outline of the principles he followed in his translation. We can only infer his principles from the translation itself, which consists almost entirely of labels and very little in the way of explanatory material. Motoki's translations for European anatomical concepts were periphrastic and inconsistent, with different words and phrases describing the same anatomical concepts at different locations in the key (table 1, column 1). Although he employed a small number of neologisms—most notably "brain sinew" (*nōkin* 脳筋) for the nerves and "heart conduit" (*shinkei* 心經) for the arteries—in most instances he drew freely on the existing Sino-Japanese medical lexicon. Several of his borrowings from this lexicon were terms that later came to be associated exclusively with Sino-Japanese medicine, such as the "conduits" and the "triple burner," but Motoki does not seem to have used these terms as part of a conscious attempt to develop a synthesis between European and Chinese medical concepts; rather, he used Sino-Japanese terms in this way simply because the Dutch and Chinese accounts of the body had not yet come to be seen as fundamentally distinct. As I will argue below, it was only after *rangaku* scholars introduced the idea that the European

---

腑図解』の翻訳書としての不完全さ: 訳出しなかった語の視点から [The incompleteness of *Dutch Images and Explanations of the Conduits, Networks, Sinews, Vessels, and Organs* as a translated book: A study of the untranslated terms]," *Nihon ishigaku zasshi* 58 (2012): 3–14.

[14] Kenneth F. Russell, *A Bibliography of Johann Remmelin, the Anatomist* (East St. Kilda, Victoria: J. F. Russell, 1991); Cali Buckley, "Johann Remmelin's *Catoptrum Microcosmicum* and the End of an Era," *Bodleian Library Record* 26 (2013): 18–35.

[15] John Z. Bowers and Robert W. Carrubba, "The Western World's First Detailed Treatise on Acupuncture: Willem Ten Rhijne's *De acupunctura*," *J. Hist. Med. Allied Sci.* 29 (1974): 371–99; Roberta Bivins, "Imagining Acupuncture: Images and the Early Westernisation of Asian Medical Expertise," in *Imagining Chinese Medicine*, ed. Vivienne Lo and Penelope Barrett (Leiden: Brill, 2018), 339–47; Wei Yu Wayne Tan, "Rediscovering Willem ten Rhijne's *De Acupunctura*: The Transformation of Chinese Acupuncture in Japan," in *Translation at Work: Chinese Medicine in the First Global Age*, ed. Harold J. Cook (Leiden: Brill, 2020), 108–33.

[16] Bowers and Carrubba, "Willem ten Rhijne's *De Acupunctura*" (cit. n. 15), 377.

Table 1. *Japanese translations for key European anatomical terms discussed in this essay*

	Motoki Shōdayū, *Oranda keiraku kinmyaku zōfu zukai* (MS ca. 1680, printed 1772)	Sugita Genpaku, *Kaitai shinsho* (1774)	Udagawa Genshin, *Ihan teikō* (1805)	Mitani Boku, *Kaitai hatsumō* (1813)
veins	*kei, keiraku* 經, 經絡 conduits, conduits and networks	*seimyaku* 靜脈 calm vessels	*seiketsumyaku* 靜血脈 calm blood vessels	*raku* 絡 networks
arteries	*shinkei* 心經 heart conduit	*dōmyaku* 動脈 moving vessels	*dōketsumyaku* 動血脈 moving blood vessels	*kei* 經 conduits
nerves	*suji, nōkin* 筋、腦筋 sinews, brain sinews	*shinkei* 神經 spirit conduits	*shinkei* 神經 spirit conduits	*ekidō* 液道 liquid paths
nervous fluid	腦髓汁、腦汁 *nōsuishū, nōshū* brain marrow juice, brain juice	*shinkeishū* 神經汁 spirit conduit juice	靈液 *reieki* numinous liquid	靈液 *reieki* numinous liquid
pancreas	—	*dai kiriiru* 大機里爾 large *klier* (gland)	*sui* 膵 pancreas (neologism)	*chūshō* 中焦, middle burner *chimoto-bukuro* "source bag for blood"
chyle	*kokushū, nyūshū* 穀汁、乳汁 grain juice, milky juice	*geeru* 奇縷 *gyl* (chyle)	*nyūbi* 乳糜 milky gruel	*ei* 營, nourishment *chimoto* "origin of blood"
cisterna chyli	*kokushū no fukuro* 穀汁ノ袋 bag for grain juice	*geeru kakyū* 奇縷科臼 *gyl* (chyle) pit	*nyūbi nō* 乳糜囊 milky gruel bag	*geshō* 下焦 lower burner *chimizumoto-bukuro* "source bag for blood and water"
thoracic duct	*kokushū no tōru kei* 穀汁ノ通ル經 conduit for grain juice	*geeru kan* 奇縷管 *gyl* (chyle) tube	*nyūbikan* 乳糜管 milky gruel tube	*jōshō* 上焦 upper burner, *kimoto-bukuro* "source bag for life"

anatomical body was radically distinct from the body known to Chinese medicine that a conscious syncretism began to seem possible or necessary.

The majority of the structures that Motoki labeled as "conduits" (*kei* 經) were blood vessels, but he also used the same word to label all sorts of channels for the transport of fluids, from chyle and milk to semen and urine. In some cases, he labeled blood vessels as "conduits and networks" (*keiraku* 經絡), and although he did not offer a definition of this phrase, he seems to have intended these terms, which traditionally designated respectively the major and subsidiary pathways for blood and *qi*, to refer to the major and minor blood vessels in Remmelin's images. Moreover, his text contains several clues that his "conduits" were intended as more than just a loose structural analogy.

The most suggestive of these clues are clustered together in a group of labels for the conduits of the legs. According to one of these labels, "the conduit in the middle of the calves arises from the liver; blood can be taken from this conduit."[17] This idea of a connection between the liver and a particular blood vessel in the leg was not present in Remmelin's original captions. Motoki may have acquired this idea about bloodletting from the lower leg through conversations about European bloodletting techniques with ten Rhijne; however, it is notable that the location of the blood vessel to which Motoki attached this label matches that of the liver conduit known to acupuncturists (fig. 3), suggesting that Motoki may have thought of Remmelin's blood vessels and the Chinese liver conduit as different representations of the same somatic entity.

Motoki added a similar comment in his label for a different point on one of the "conduits" of the calf: "When the lower burner is stagnant, take blood from this conduit." This remark is somewhat more difficult to interpret in terms of acupuncture: the lower burner was not traditionally considered to have a conduit of its own, and the triple burner meridian was located in the arm rather than in the leg. Before we consider in more detail what Motoki might have meant, it will be useful first to trace in more detail the history of the triple burner concept in East Asian medicine, since the nature of the triple burner was one of the major conceptual problems that later Japanese doctors hoped to solve using European anatomy.

The triple burner had long been a contentious element in classical medical accounts of the body.[18] As one of the six hollow organs, it occupied an important position in orthodox medical doctrine, but its status as a material entity was much harder to pin down. The brief descriptions in the *Basic Questions* (*Suwen* 素問) classified it as an organ involved in digestion, transmitting fluids, and storing and transforming foodstuffs, while the *Numinous Pivot* described its appearance in somewhat enigmatic terms: "The upper burner is like mist; the middle burner is like froth; the lower burner is like a drain."[19] The *Canon of Difficulties* called it "the pathway of water and grains, where *qi* ends and begins"; it specified the locations of the upper, middle, and lower burners by their proximity to the heart, stomach, and bladder, but also notoriously claimed that the triple burner "has a name, but is without form."[20] A number of later Chinese doctors vigorously criticized this doctrine that the triple burner was "without

[17] Motoki Ryōi 本木了意, *Oranda zenku naigai bungōzu oyobi kengō* 和蘭全躯内外分合図及験号 [Dutch flapbook of the interior and exterior of the body], preface by Suzuki Sōun 鈴木宗云 (1772), 6a.

[18] Pi, "Tuxiang, xingzhi yu zangfu zhishi" (cit. n. 10); Lei, "*Qi*-Transformation and the Steam Engine" (cit. n. 10).

[19] *LS* 18; see also *SW* 8, 9, 11.

[20] *NJ* 25, 31.

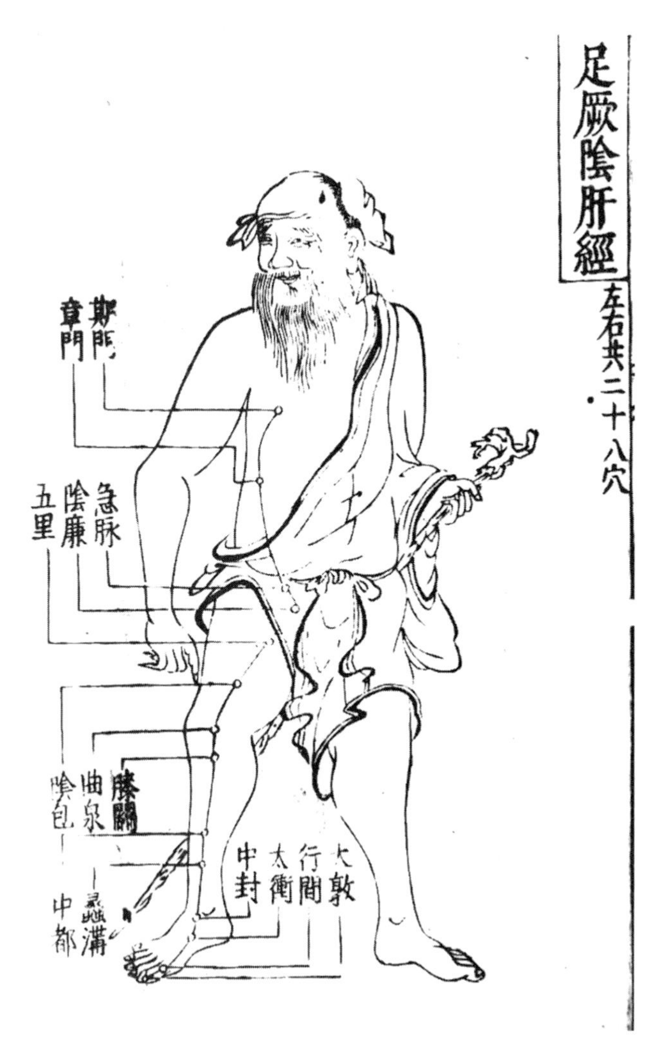

Figure 3. *Diagram showing the leg* jueyin *liver conduit. Zhang Jiebin,* Leijing tuyi (1624, *reprinted ca. 1700), 3:42b. Image courtesy of the Main Library, Kyoto University, https://rmda.kulib.kyoto-u.ac.jp/item/rb00005777.*

form," but even doctors who insisted that the triple burner possessed a definite form could not agree on what that form was. In the thirteenth century, the Southern Song doctor Chen Yan suggested a connection between the triple burner and the "fatty membrane" (C. *zhimo* 脂膜, *zhiman* 脂䏋), a structure that had been observed in dissections and was frequently depicted in images of the viscera but whose significance was rarely discussed in texts (fig. 4).[21] The late Ming dynasty doctor Zhang Jiebin, whose writings were influential in Japan during the seventeenth and early eighteenth centuries, offered a very different interpretation, describing the triple burner as a bag surrounding the other

[21] On Chinese and Japanese images of the fatty membrane, see Takashima Bun'ichi 高島文一, "Jintai naikeizu no shiman/shimaku ni tsuite" 人体内景図の脂䏋・脂膜について, in *Rekishi no naka no yamai to igaku* 歴史の中の病と医学, ed. Yamada Keiji 山田慶児 and Kuriyama Shigehisa 栗山茂久 (Kyoto: Shibunkaku, 1997), 489–502.

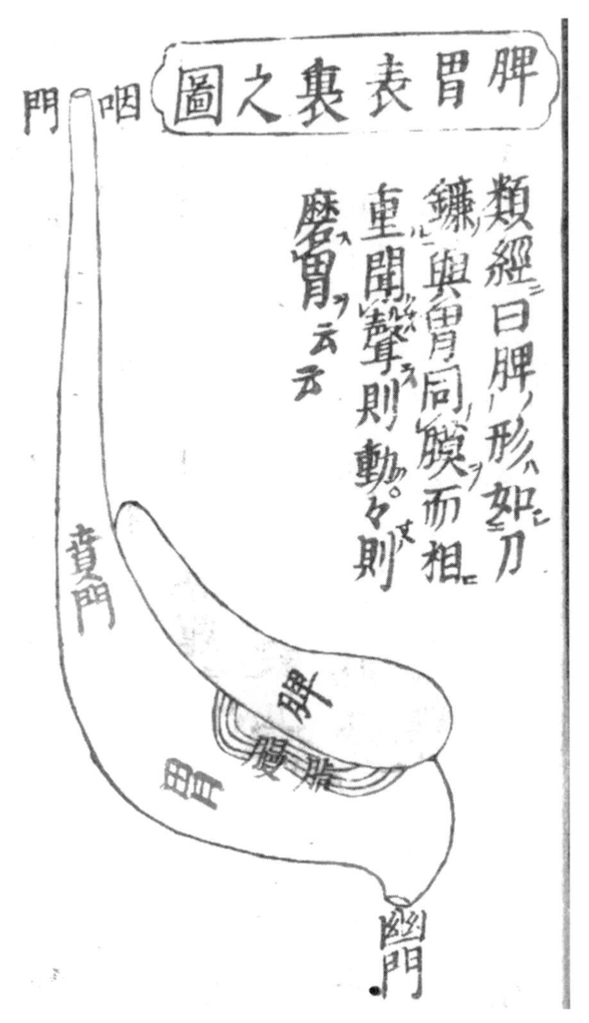

Figure 4. *Location of the "fatty membrane"* (C. zhimo/zhiman, J. shimaku/shiman) *between the stomach and spleen. Okamoto Ippō,* Zōfu keiraku shōge *(1690), 2:5a. Image courtesy of the Main Library, Kyoto University, https://rmda.kulib.kyoto-u.ac.jp/item/rb00003893.*

organs (fig. 5). Meanwhile, doctors who believed the triple burner was "without form" could still attempt to represent this visually, for example by presenting a tabulated version of the canonical description of the upper, middle, and lower burners' locations (fig. 6).[22] These debates about the physical nature of the triple burner continued into the nineteenth century and have not been fully resolved even today.[23]

Motoki's text referred to the triple burner in three places. The first, noted above, was his recommendation of bloodletting as a therapy for "stagnation in the lower burner"; the second was in a set of labels attached to Remmelin's illustrations of the skeleton

[22] Some Chinese authors conveyed the same idea by using two horizontal lines to divide the torso into three sections. See Lei, "*Qi*-transformation" (cit. n. 10), 331.

[23] Terasawa Katsutoshi 寺沢捷年, "Sanshō ni kan suru Ōtomo Kazuo gakusetsu no datōsei" 三焦に関する大友一夫学説の妥当性 [The validity of Ōtomo Kazuo's theory concerning the triple burner], *Nittō ishi* 69 (2018): 57–66.

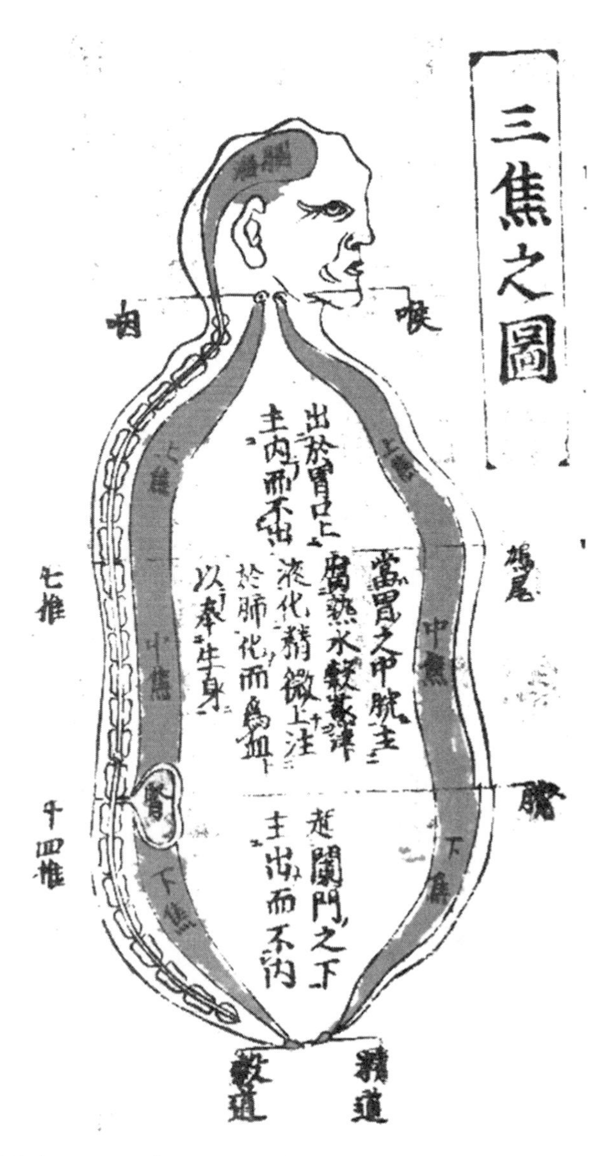

Figure 5. *The triple burner as a bag surrounding the other viscera. Okamoto Ippō,* Zōfu keiraku shōge *(1690), 4:12b. Image courtesy of the Main Library, Kyoto University, https://rmda.kulib.kyoto-u.ac.jp/item/rb00003893.*

and nerves; the third accompanied a diagram of blood vessels surrounding the digestive organs, which Motoki labeled as "conduits and networks surrounding the middle burner." Given the historical ambiguities surrounding the concept of the triple burner, these references do not allow us to definitively establish what Motoki meant by these uses of the term. Nevertheless, his attachment of the term to an image of skeleton and nerves suggests he understood the triple burner not as a concrete visible entity but rather as a partition of the interior space of the body into conceptually distinct regions. That is, he accepted the view that the triple burner was "without form" and employed it as a concept for labeling specific regions of the torso and for guiding therapeutic intervention.

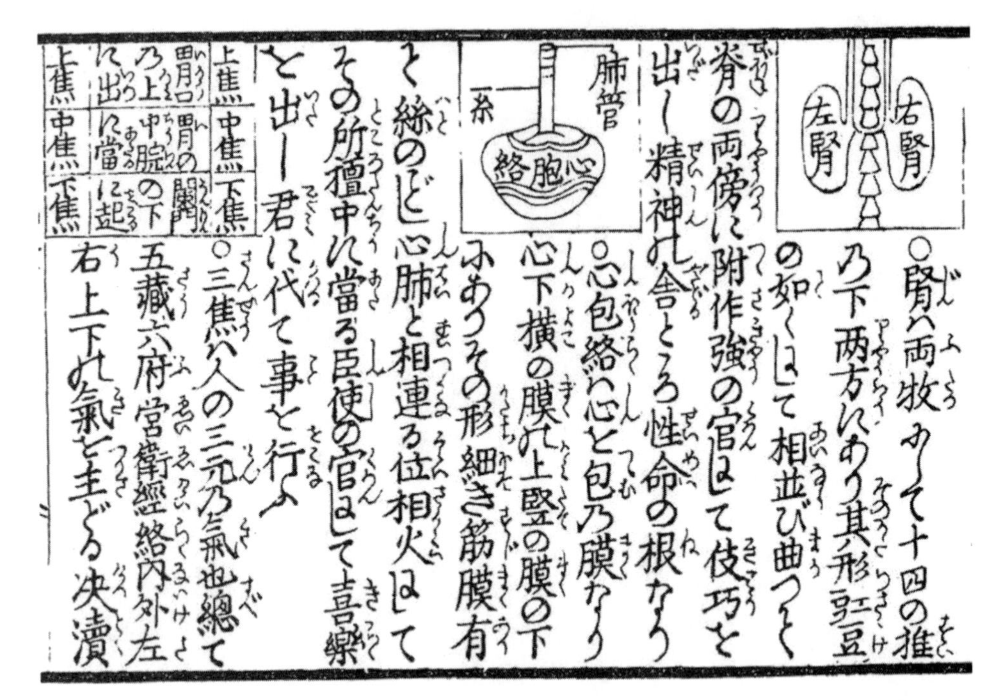

Figure 6. *The triple burner as an organ without form. In contrast to the illustrations of the kidneys and pericardium, the triple burner is represented as a schematic table in the upper left, describing the locations of the upper, middle, and lower burners in relation to the stomach and intestines. Hongō Masatoyo,* Idō nichiyō kōmoku *("Systematic outline of the Way of medicine for everyday use," 1707; repr., 1873). Image courtesy of the National Diet Library of Japan, https://dl.ndl.go.jp/info:ndljp/pid/833098.*

There is no suggestion in Motoki's text that he saw the triple burner or the conduits as ontologically problematic or incompatible with European anatomical knowledge: his labels used "conduits" and "triple burner" in much the same way that they used "spleen" or "liver." In the late seventeenth century, it was not yet accepted that European anatomy and East Asian medicine were so radically distinct that fundamental elements of the body in one tradition might be completely missing from the other. When the European and East Asian accounts of the body were found to diverge on particular issues, it was still possible to think of these divergences merely as differences of opinion, comparable to the many differences of opinion that had arisen among doctors within China, Korea, and Japan. It was only later, with the development of *rangaku*, that such divergences came to be seen as discrepancies forcing a final choice between two systems of knowledge.

YAMAWAKI TŌYŌ'S ANATOMICAL OBSERVATIONS AND THE PROBLEM OF THE INTESTINES

The earliest well-documented Japanese anatomical observations took place during the intercalary second month of 1754, when the Kyoto doctor Yamawaki Tōyō obtained permission to view the dissection of an executed criminal. Five years later, he published his observations in his *Record of the Viscera* (*Zōshi* 蔵志, 1759). Yamawaki had seen

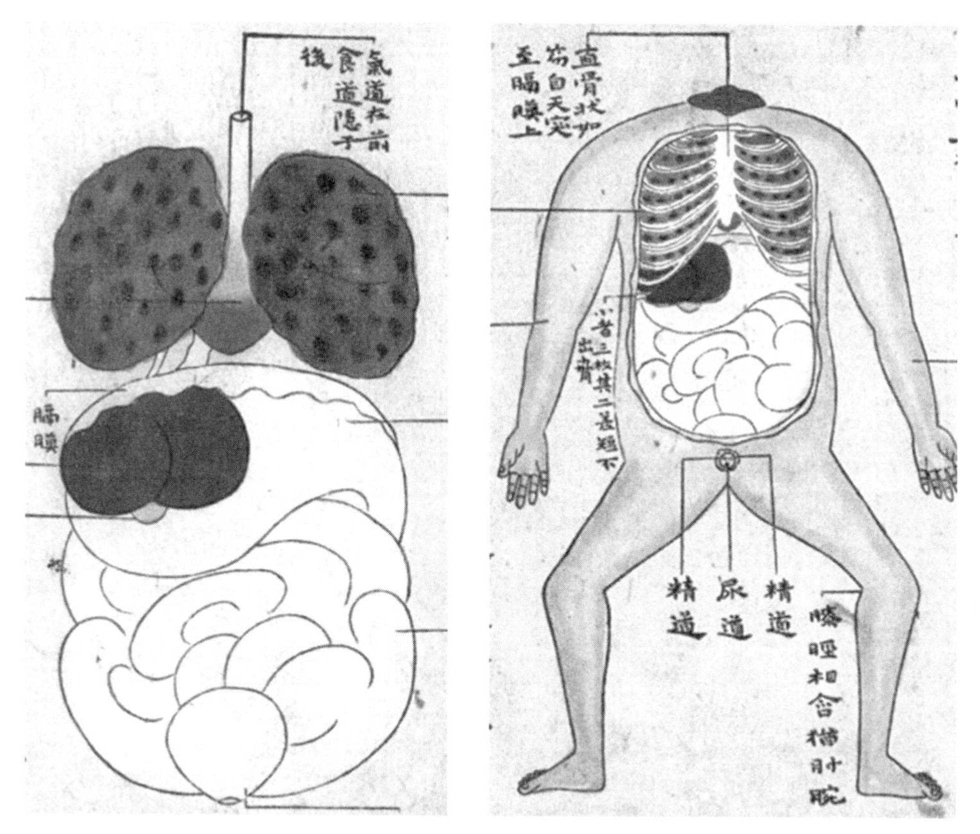

Figure 7. *Visceral anatomy based on observation of human dissection. The intestines are shown as a single mass of coils, rather than the two distinct sets of coils seen in fig. 8. Yamawaki Tōyō,* Zōshi *(1759), 1:7a–b. Image courtesy of the Main Library, Kyoto University, https://rmda.kulib.kyoto-u.ac.jp/item/rb00003884.*

anatomical images in European treatises, but his motivation to carry out these observations and publish the results grew primarily from his desire to establish the correctness of a description of the body involving "nine organs" (J. *kyūzō* 九臓) that he had arrived at through his philological research on ancient Chinese texts.[24] He believed that confirming this description, which contrasted with the more orthodox enumeration of five solid and six hollow organs (J. *gozō roppu* 五臓六腑), would also validate his own preferred style of therapeutics, using formulas from the oldest transmitted Chinese treatise on herbal therapies, Zhang Zhongjing's *Discourse on Cold Damage Disorders* (*Shanghan lun* 傷寒論, ca. 220 CE).

The illustrations in Yamawaki's treatise differed in several important ways from earlier Sino-Japanese images of the viscera (fig. 7). These were not schematic diagrams of

[24] Okamoto Takashi 岡本喬, *Kaibō kotohajime: Yamawaki Tōyō no hito to shisō* 解剖事始め：山脇東洋の人と思想 [The beginnings of anatomical dissection: Yamawaki Tōyō's life and thought] (Tokyo: Dōseisha, 1988); Maki Fukuoka, *The Premise of Fidelity: Science, Visuality, and Representing the Real in Tokugawa Japan* (Stanford: Stanford Univ. Press, 2012), 34–42; Daniel Trambaiolo, "Ancient Texts and New Medical Ideas in Eighteenth-Century Japan," in *Antiquarianism, Language, and Medical Philology: From Early Modern to Modern Sino-Japanese Medical Discourses*, ed. Benjamin Elman (Leiden: Brill, 2015), 81–104, on 90–4.

the living body's contents but explicitly mimetic depictions of a dissected corpse, showing the bloody stump of a decapitated neck and the thickness of the surface layers removed to expose the interior of the torso. Their mimetic character was further enhanced by the manual addition of color to the woodblock-printed illustrations. This attention to color was a consistent feature of the new styles of anatomical illustration in treatises by Sino-Japanese practitioners in the late eighteenth and early nineteenth centuries. The same technique of manual addition of pigments was used for Kawaguchi Shinnin's *Treatise on Dissection* (*Kaishi hen* 解屍編, 1772) and Kako Ranshū's *Essentials of Anatomy and Acupuncture* (1819); a bookseller's advertisement for the latter touted the inclusion of hand-colored pictures recording the organs' "true colors" (*shinshoku* 真色) as one of the treatise's selling points.[25] Beginning in the early nineteenth century, some anatomical treatises (including the treatise by Mitani Boku discussed below) substituted polychromatic woodblock printing techniques for manual pigmentation.

This emphasis on color contrasted with the monochromatic visual style of the printed anatomical treatises produced by *rangaku* scholars, which imitated their European sources' use of hatching and stippling to represent the three-dimensional forms of anatomical structures but made no attempt to represent their true colors. As Shigehisa Kuriyama has noted, *rangaku* rhetoric proclaimed that "in judging the usefulness of a picture, one should value resemblance."[26] But resemblance could be judged in different ways, and the authors of *rangaku* and Sino-Japanese anatomical treatises came to different conclusions about the most important criterion for accuracy in visual mimesis. *Rangaku* scholars, many of whom were practicing surgeons, preferred to emphasize three-dimensional forms. Sino-Japanese doctors, who were familiar with the many subtle ways color could help them diagnose a patient's illness, preferred to emphasize color.[27]

One of Yamawaki's major conclusions from his observations was that the body contained a single "intestine" (J. *chō* 腸) rather than the distinct small and large intestines (J. *shōchō* 小腸, *daichō* 大腸) described in Chinese medical treatises, which his son Yamawaki Tōmon dismissed as a "mirage."[28] The existence of a single intestine was an essential element in Yamawaki's argument that there were "nine organs," and it would be easy to see his claims about the lack of a distinction between the small and large intestines merely as an error caused by his lack of experience as an anatomical observer and his bias in favor of evidence that could be taken to support his own theory. But to understand the substance of Yamawaki's claim, we must consider exactly what he was rejecting when he denied the existence of the small and large intestines. For Japanese doctors during the 1750s, the terms "small intestine" and "large intestine" did not yet denote the anatomical structures that came to be known by these names a few decades later. Instead, their meaning depended in part on the practical contexts in which they were deployed. For doctors thinking about the mechanisms of visceral pathology, the small and large intestine were hollow organs paired respectively with

[25] Advertisement in the end matter of Kako Ranshū, *Sekkō yōketsu* 折肱要訣 [Essential secrets of breaking arms] (1810). The two copies of Kako's book in the Fujikawa Collection at Kyoto University Library (including the copy shown in fig. 1) do not have pigment added, but the same collection contains an undated manuscript copy with colored illustrations, titled "*Kaitai zu*" 解體圖 [Anatomical images].

[26] Kuriyama, "Between Mind and Eye" (cit. n. 6), 33.

[27] On the importance of color in visual perceptions of the body and disease, see Shigehisa Kuriyama, *The Expressiveness of the Body and the Divergence of Greek and Chinese Medicine* (New York: Zone Books, 1999), 153–94.

[28] Yamawaki Tōyō 山脇東洋, *Zōshi* 蔵志 (1759), colophon by Yamawaki Tōmon 山脇東門, 1a.

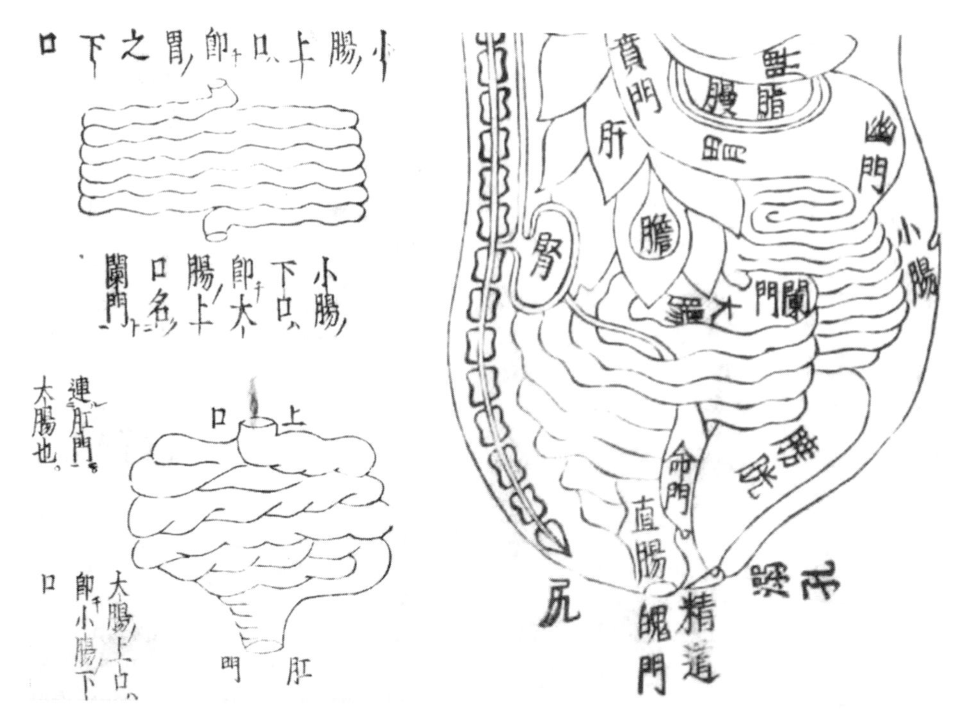

Figure 8. *The intestines as depicted in Zhang Jiebin's* Leijing tuyi (1624). (Top left) *Small intestine (3:36a).* (Bottom left) *Large intestine (3:32a).* (Right) *Detail showing the viscera of the lower abdomen (3:44a), with the small and large intestines depicted occupying distinct positions within the abdomen, the small intestine superior and ventral, and the large intestine inferior and dorsal. Image courtesy of the Main Library, Kyoto University, https://rmda.kulib.kyoto-u.ac.jp/item/rb00005777.*

the heart and with the lungs. For acupuncturists, they were targets of therapy that could be manipulated by needling specific points along their conduits in the arms. As far as their anatomical identity was concerned, Chinese illustrations typically depicted them as separate coiled structures occupying distinct regions of the abdomen (fig. 8). From the perspective of modern biomedical anatomy, Yamawaki was quite justified in denying the existence of the "small intestine" and "large intestine" depicted in Chinese illustrations; his "error" lay in his failure to recognize that these terms could be associated with the rather different pair of anatomical structures that they designate today. Assigning new anatomical meanings to the "small intestine" and "large intestine" was far from trivial: when Kawaguchi Shinnin depicted these organs as distinct structures in their anatomical treatise of 1772, he added a note admitting that the coiled forms of these organs made them difficult to distinguish and warning that the illustration was only approximate.[29]

Even doctors familiar with European anatomical images and terminology did not necessarily conclude that Yamawaki had drawn the wrong conclusions from his observations. Several years after the publication of Yamawaki's *Record of the Viscera*, the doctor Suzuki Sōun came across a reproduction of Motoki's flapbook in the possession of a merchant.[30] Motoki's flapbook had originally circulated only in manuscript, and by

[29] Kawaguchi Shinnin 河口信任, *Kaishi hen* 解屍編 [On dissecting a corpse] (1772), 17a.
[30] Suzuki Sōun, preface to Motoki, *Oranda zenku* (cit. n. 17).

the mid-eighteenth century it was largely forgotten, but Suzuki was so fascinated by his first encounter with the images that he began to hunt for additional copies in Japanese towns and cities from Nagasaki to Kyoto, seeking to find replacements for the small flaps of paper that had become detached and lost, as well as to track down a copy of the textual key providing the explanation of the images. After eventually managing to assemble a complete set of the images together with the key, he arranged for it to be reprinted in a new edition in 1772, with a colophon by Yamawaki Tōyō's nephew Shimizu Takeshi.

Despite his enthusiasm, Suzuki remained puzzled by some aspects of the flapbook and accompanying text, most notably the fact that the book's distinction between the small and large intestines apparently contradicted Yamawaki Tōyō's observations. During a visit to Nagasaki, he raised the issue with the surgeon and Dutch language interpreter Yoshio Kōgyū, who replied:

> The translation uses the old terms, fitting them to the meaning: what harm can there be in this? In the language of the foreigners, the so-called small intestine is *dunne darm*. Translating it directly, it is the place of loose stools like fermented rice. The so-called large intestine is *dikke darm*. Translating it directly, it is the place of firm stools like rice balls. These terms do not mean "large" and "small": it is one thing that fulfils two functions.[31]

Yoshio's argument rested in part on a linguistic ambiguity. Unlike the Sino-Japanese terms *shōchō* and *daichō* (whose literal meanings are close to the English "small intestine" and "large intestine"), the Dutch *dunne darm* and *dikke darm* literally meant "thin intestine" and "thick intestine"—the adjectives were equally appropriate for describing either the diameters of these organs or the consistency of their contents. Rather than using the discrepancy between the Dutch terminology and Yamawaki's claims as evidence of a fundamental conflict between European and Sino-Japanese anatomical knowledge, Yoshio proposed an interpretation of the Dutch terms that rendered them compatible with Yamawaki's claims.

This conversation between Suzuki Sōun and Yoshio Kōgyū suggests that as late as 1772, it was still unremarkable to regard European anatomy as complementary to and continuous with Chinese and Japanese knowledge of the body. Doctors who believed healing practices could be improved by investigating the body's internal structures set about their tasks of observing corpses, producing images, and naming parts, drawing eclectically on elements of European anatomical knowledge but not feeling any compulsion to study European anatomy as a systematic whole. It was the *rangaku* translators who constructed and insisted on this new understanding of European anatomy as a complete and systematic form of knowledge; once they had done so, even doctors who wished to integrate European anatomy with Sino-Japanese understandings of the body were forced to pursue their projects of anatomical *bricolage* under a new set of rules.

RETRANSLATIONS AND REINTERPRETATIONS: INTEGRATING CHINESE AND EUROPEAN ANATOMY

The *rangaku* project of anatomical translation that began with Sugita Genpaku's *New Book of Anatomy* in 1774 was guided by assumptions quite different from those of

[31] Ibid., 4a-b. The Dutch terms were transcribed in the kanbun text as kanji, with katakana indicating the pronunciations: 底武禰太留武(デンネタルン), 底都幾太留武(デツキダルン).

Motoki Ryōi a century earlier. These different assumptions had important consequences for their respective translational styles. Motoki had assumed a continuity between European and Chinese understandings of the body, allowing him to freely adapt existing Sino-Japanese words to describe the structures in Remmelin's images. *Rangaku* translators drew much stricter distinctions between two types of anatomical concepts that they encountered in European texts: those that they saw as present in both the European and East Asian traditions (e.g., heart, lungs, spleen) and those that they saw as specific to European anatomy (e.g., blood vessels, nerves, glands). For the latter group, they developed a systematic translational lexicon, avoiding the ambiguities that Motoki had introduced into his translation by borrowing concepts such as the conduits and the triple burner. It was through this novel lexicon, as much as through their novel visual style, that the *rangaku* translators sought to distinguish their new knowledge of the body from older anatomical traditions.

On a linguistic level, *rangaku* scholars' two main strategies for coining new anatomical terms were phonetic loanwords and Sino-Japanese neologisms.[32] Of these two strategies, Sino-Japanese neologisms were felt to be more elegant and comprehensible than loanwords, which tended to be used only where an appropriate neologism seemed too difficult to formulate. For example, the 1774 edition of the *New Book of Anatomy* used several loanwords, such as *geeru* 奇縷 for "chyle" (D. *gyl*) and *kiriiru* 機里爾 for "gland" (D. *klier*).[33] But most of these loanwords were dropped from later *rangaku* translations in favor of less alien-seeming alternatives (table 1, columns 2 and 3). Udagawa Genshin's *Outline of Medical Principles* (*Ihan teikō* 醫範提綱, 1805) rendered "chyle" as *nyūbi* 乳糜 (lit. "milky porridge") and "gland" as *sen* 腺 (a newly invented *kanji* composed of a "flesh" radical together with a semanto-phonetic element meaning "source" or "spring").[34] Already in the *New Book of Anatomy*, semantically transparent Sino-Japanese neologisms had been used wherever possible: *dōmyaku* 動脈 (moving vessel) for arteries, *seimyaku* 靜脈 (calm vessel) for veins, *shinkei* 神經 (spirit conduits) for nerves, *shinkeishū* (spirit conduit juice) for the "nervous fluid" believed to be the mechanism of nerve action, *shōjin* 小腎 (small kidneys) for the adrenal glands, and many others. All of these neologisms drew on elements of traditional Sino-Japanese medical vocabulary but recombined them in new ways; collectively, they constituted a lexicon that not only described the anatomical body but also carried a surplus of meaning that communicated the novelty and distinctiveness of *rangaku* anatomy as a field of knowledge.

Not all Japanese doctors were convinced by the claims of *rangaku* anatomy. The old understandings of the body based on Chinese sources were not just a set of descriptive ideas: they were embodied forms of knowledge, enacted on a daily basis through the

[32] They called the loanwords *chokuyaku* 直譯 (lit. "direct translation") and the neologisms *giyaku* 義譯 (lit. "semantic translation"). Both of these terms contrasted with simple *hon'yaku* 翻訳 ("translation"), which in its narrow sense meant the use of existing Japanese equivalents for European words. See Rebekah Clements, *A Cultural History of Translation* (Cambridge, UK: Cambridge Univ. Press, 2015), 166. On the broader context of early modern translated vocabulary, see Sugimoto Tsutomu 杉本つとむ, *Edo jidai hon'yakugo no sekai: Kindaika wo suishin shita yakugo wo kenshō suru* 江戸時代翻訳語の世界：近代化を推進した訳語を検証する [The world of translated words in the Edo period: Investigating the translated words that promoted modernization] (Tokyo: Yasaka shobō, 2015).

[33] As Martin Heijdra has noted, the *kanji* in these phonetically transcribed loanwords often had phonetic values taken from Mandarin Chinese rather than Sino-Japanese; Heijdra, "Polyglot Translators: Chinese, Dutch, and Japanese in the Introduction of Western Learning in Tokugawa Japan," in *"At the Shores of the Sky": Asian Studies for Albert Hoffstädt*, ed. Paul W. Kroll and Jonathan A. Silk (Leiden: Brill, 2020), 62–75.

[34] Mieko Macé, "Le chinois classique comme moyen d'accès à la modernité: La reception des concepts médicaux dans le Japon des XVIIIe et XIXe siècles," *Daruma* 4 (1998): 79–103, on 88.

Table 2. *Using European anatomy to reimagine Sino-Japanese medical concepts*

	Motoki Ryōi, *Oranda keiraku kinmyaku zōfu zukai* (ca. 1680, repr. 1772)	Mitani Boku, *Kaitai hatsumō* (1813)	Ishizaka Sōtetsu, *Chiyō ichigen* (1826)
upper burner (*jōshō* 上焦)	(upper region of torso)	thoracic duct	—
middle burner (*chūshō* 中焦)	(middle region of torso)	pancreas	—
lower burner (*geshō* 下焦)	(lower region of torso)	cisterna chyli	pancreas
conduits (*kei* 經)	blood vessels (and other channels)	arteries	major blood vessels
networks (*raku* 絡)	(minor) blood vessels	veins	minor blood vessels
nutrition (*ei* 營)	— —	chyle	veins
defense (*e* 衛)	— —	sweat	arteries

wide range of diagnostic and therapeutic practices that these doctors relied on to treat their patients. The images in *rangaku* anatomical treatises were compelling, especially as the practices of dissection and anatomical observation became more widespread. But the words and ideas that *rangaku* scholars attached to these images were more open to question, especially since the *rangaku* accounts seemingly failed to mention several elements of the body that were fundamental to Sino-Japanese medical practice. This discrepancy between the trust placed in European anatomical images and the trust placed in *rangaku* scholars' translated words led several Japanese doctors in the early nineteenth century to produce their own distinctive forms of anatomical *bricolage*.

In what follows, I examine the writings of Mitani Boku and Ishizaka Sōtetsu (1770–1841), two Japanese doctors who attempted in the early nineteenth century to reconcile the fundamental concepts of East Asian medicine with the new anatomical understanding of the body. These attempts at reconciliation were simultaneously acts of retranslation and acts of reinterpretation: retranslation, because they proposed existing medical terms as semantic equivalents of European anatomical terms that had been rendered as neologisms by the *rangaku* translators (table 1, column 4); and reinterpretation, because to associate these existing medical terms with newly recognized anatomical structures was to give them new meanings quite different from those they had held for earlier generations of Chinese, Korean, and Japanese practitioners (table 2).

Mitani was a Kyoto doctor who had studied materia medica under the eminent herbalist Ono Ranzan (1729–1810).[35] He had only limited personal experience of anatomical dissection: although he had observed a dissection in 1802, his knowledge of the

[35] Inoue Kiyotsune 井上清恒, "Kan-ran setchū igaku no kokoromi: Mitani Boku no *Zōfu shinsha, Kaitai hatsumō* kō" 漢蘭折衷医学の試み：三谷樸の「臓腑真写・解体発蒙」考 [An attempt at Sino-Dutch syncretic medicine: A study of Mitani Boku's *True Images of the Organs* and *Elucidation of Anatomy*], *Shōwa*

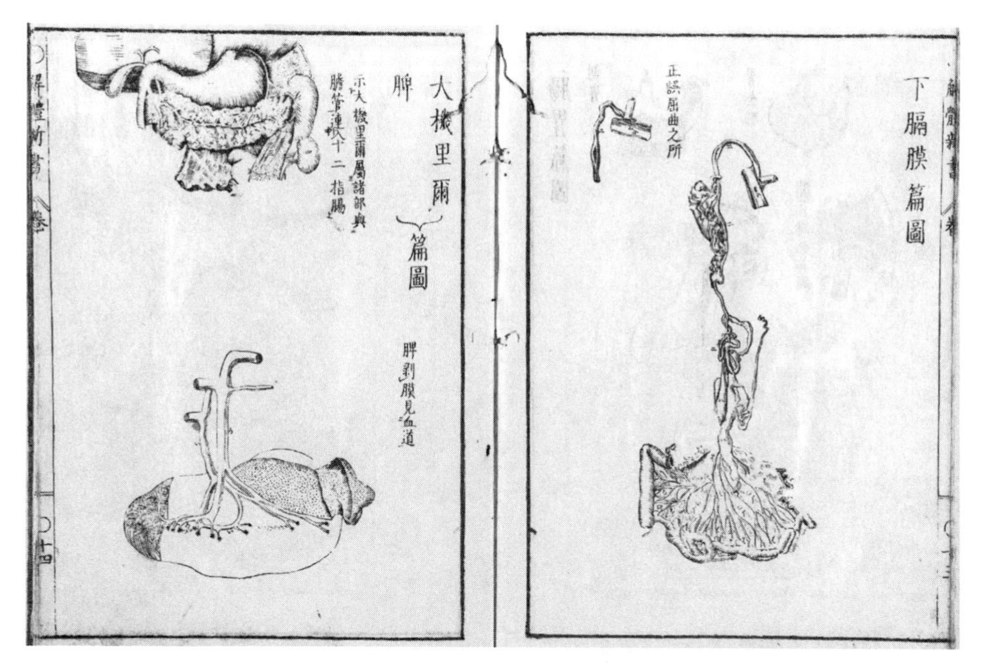

Figure 9. *Anatomy of the digestive system. Sugita Genpaku,* Kaitai shinsho *(1774), 13b-14a. Image courtesy of the Main Library, Kyoto University, https://rmda.kulib.kyoto-u.ac.jp/item/rb00001510.*

subject was drawn mainly from printed *rangaku* treatises and the hand-painted images and manuscript accounts of dissections carried out in Kyoto in 1783 and 1798.[36] A friend of Mitani's who wrote a preface for his *Elucidation of Anatomy* (*Kaitai hatsumō* 解體發蒙,1813) warned against the idea that the goal of these investigations was anatomical knowledge for its own sake: "Someone who is unaware might look at this book and call him an anatomist (*kaitaika* 解體家). . . . Materia medica and anatomy are subjects that we doctors should think about, but for Mitani they are no more than a portion of a single spot on a leopard's coat: the whole leopard is the remedies and the facts of the medical classics."[37]

Mitani's approach to anatomy drew on his background as a student of materia medica, combining an emphasis on visual observation and representation with painstaking gathering of evidence from diverse textual sources.[38] The images in his book were directly adapted from Sugita's *New Book of Anatomy*, but they employed a radically different visual style (figs. 9–13): whereas the images in Sugita's treatise embodied a communicative aesthetic of precision, those in Mitani's treatise embodied a communicative aesthetic of vividness. They were produced in color, not by the manual addition of

igakkai zasshi 32 (1972): 307–17; Sugimoto Tsutomu 杉本つとむ, *Edo no Oranda-ryū ishi* 江戸の阿蘭陀流医師 [The Dutch-style doctors of Edo] (Tokyo: Waseda daigaku shuppansha, 2004), 321–32.

[36] Mitani Boku 三谷樸, *Kaitai hatsumō* 解體發蒙 [Elucidation of anatomy] (1813), *daigen* 題言, 9b-10a.

[37] Wake Koreyuki 和気惟亨, Preface (1810) to Mitani Boku, *Kaitai hatsumō*, 3b-4a.

[38] For the relationship between materia medica (*honzōgaku* 本草學) and anatomy as forms of visual knowledge, see Fukuoka, *Premise of Fidelity* (cit. n. 24). On the broader development of *honzōgaku* as a central discipline of early modern Japanese intellectual culture, see Federico Marcon, *The Knowledge of Nature and the Nature of Knowledge in Early Modern Japan* (Chicago: Univ. of Chicago Press, 2015).

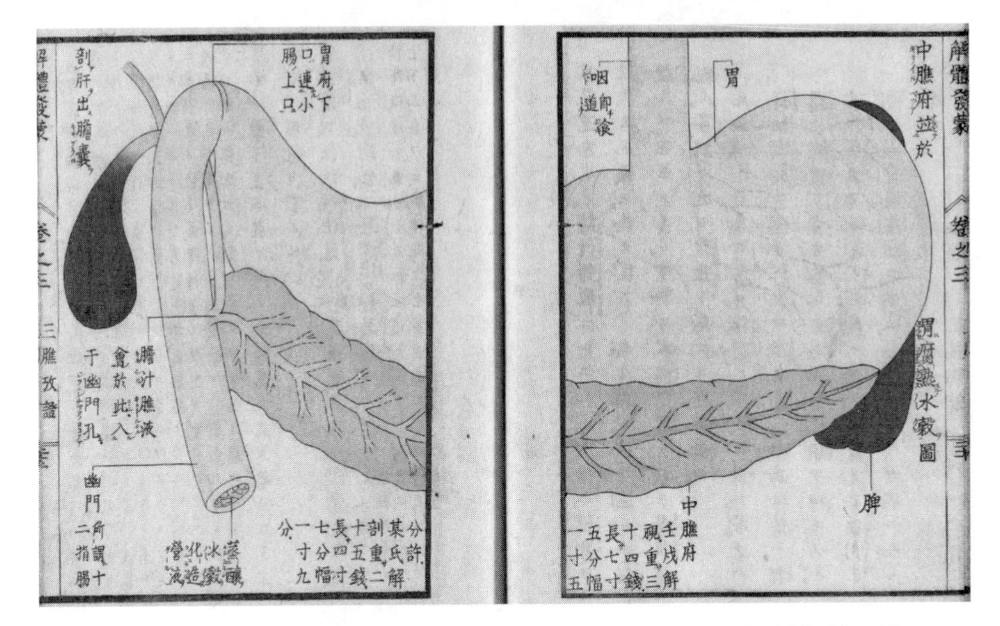

Figure 10. *Stomach, spleen, pancreas (labeled as "middle burner"), and gallbladder. (Compare with the top left image in fig. 9.) Mitani Boku,* Kaitai hatsumō *(1813), 3:21b-22a. Image courtesy of the Main Library, Kyoto University, https://rmda.kulib.kyoto-u.ac.jp/item/rb00001517.*

pigments as in eighteenth-century anatomical books, but by taking advantage of advances in the art of polychromatic woodblock printing. Unlike Sugita's *New Book of Anatomy*, which crammed multiple images of the organs onto single pages, Mitani's *Elucidation of Anatomy* devoted entire pages and even double-page spreads to the illustration of single organs, inviting viewers to appreciate them for their aesthetic value as much as for their intellectual content.

Despite being copied from European sources, Mitani's images were in some ways closer in spirit to the Chinese tradition of illustrations of the viscera. They omitted all visual allusions to the process of dissection (such as the flap cut from the outer membrane of the spleen visible in the lower left image of fig. 9) and depicted the organs floating in space rather than embedded among flesh, sinews and bones. Similarly, the static cut-away surfaces of the arm and leg in Sugita's image of the blood vessels, which derived ultimately from the European convention of depicting fragments of the body as fragmentary stone sculptures, were replaced in Mitani's version by trailing clouds, evoking the circulating *qi* of the living body (figs. 12, 13).[39]

[39] These trailing clouds followed a convention frequently seen in Chinese medical illustrations of isolated body parts; their forms were reminiscent of the *lingzhi* fungus, and they carried loosely Daoist connotations. (Motoki Ryōi had substituted this style of trailing clouds for the fig leaves covering the genitalia in Remmelin's flapbook.) On the trailing clouds, see Yi-Li Wu, "The Gendered Iconography of the *Golden Mirror, Yuzuan Yizong Jinjian* 御纂醫宗金鑑, 1742," in Lo and Barrett, *Imagining Chinese Medicine* (cit. n. 15), 111–32, on 125. On statues as a model for early modern European anatomical illustrations, see Glenn Harcourt, "Andreas Vesalius and the Anatomy of Antique Sculpture," *Representations* 17 (1987): 28–61; and Sachiko Kusukawa, "The Uses of Pictures in the Formation of Learned Knowledge: The Cases of Leonhard Fuchs and Andreas Vesalius," in *Transmitting Knowledge: Words, Images, and Instruments in Early Modern Europe*, ed. Sachiko Kusukawa and Ian Maclean (Oxford: Oxford Univ. Press, 2006), 73–96, on 85–87.

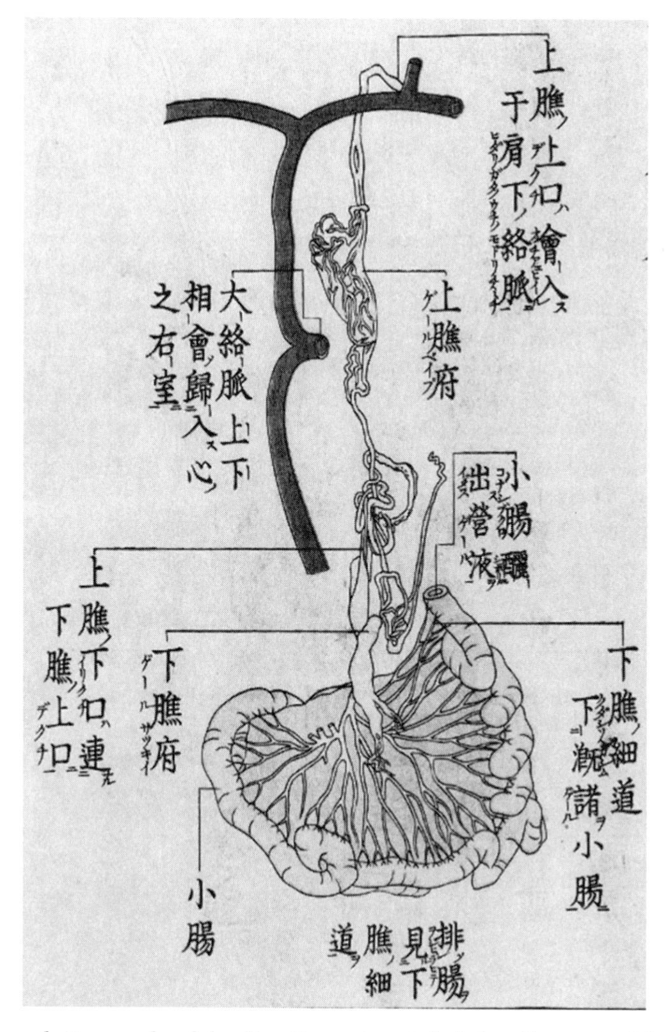

Figure 11. *Lymphatic vessels of the digestive system (labeled as "upper and lower burner"). (Compare with the right-hand image in fig. 9.) Mitani Boku,* Kaitai hatsumō *(1813), 3:12a. Image courtesy of the Main Library, Kyoto University, https://rmda.kulib.kyoto-u.ac.jp/item/rb00001517.*

Mitani's detailed discussions of the triple burner, the conduits, and networks present a dramatic contrast with Motoki's casual use of the same terms. Mitani identified the upper, middle, and lower burners with the thoracic duct, pancreas, and cisterna chyli, supporting these identifications through detailed comparison of the descriptions of the organs in the *New Book of Anatomy* with passages from the Chinese medical classics: for example, the *Numinous Pivot*'s description of the middle burner as "like froth" corresponded to Sugita's description of the foamy consistency of the pancreatic interior, while its claim that "the middle burner emits *qi* like dew" corresponded to the pancreas's secretion of digestive juices.[40] For Mitani, European anatomical knowledge

[40] Mitani, *Kaitai hatsumō* (cit. n. 36), 3:15a–b.

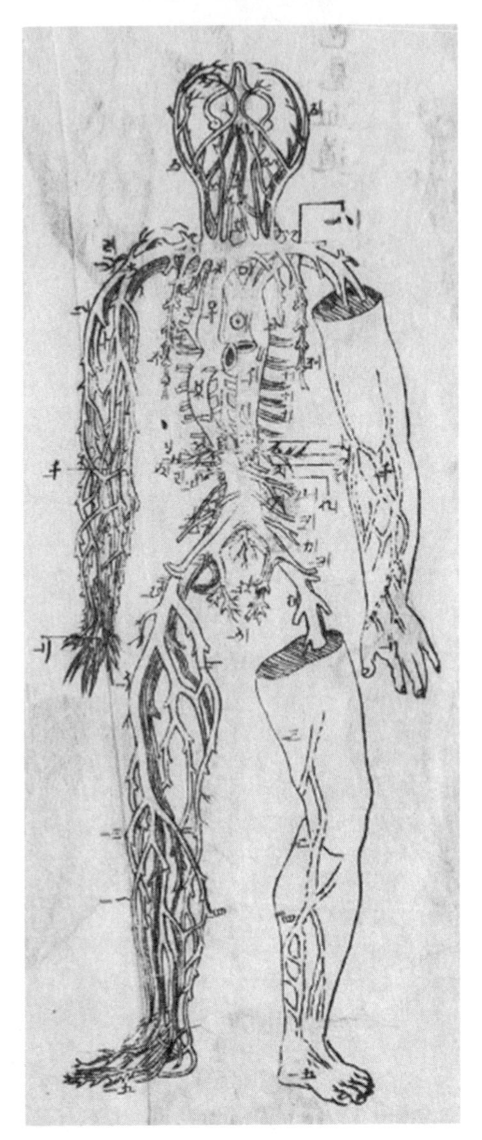

Figure 12. *Arteries and veins. Sugita Genpaku,* Kaitai shinsho *(1774). Image courtesy of the Main Library, Kyoto University, https://rmda.kulib.kyoto-u.ac.jp/item/rb00001510.*

thus provided a definitive resolution to the old debate about whether the triple burner possessed a form.[41]

Mitani's assimilation of the conduits and networks and the European blood vessels implied a still more radical reinterpretation of these orthodox medical concepts. Mitani's image of the blood vessels was a close reproduction of Sugita's (figs. 12, 13), but his captions articulated the image's meanings very differently: whereas Sugita had

[41] Later Chinese and Japanese doctors proposed alternative anatomical identities for the triple burner, including the peritoneum and the intestinal mesentery. See Lei, "*Qi*-transformation" (cit. n. 10); Terasawa, "Sanshō" (cit. n. 23).

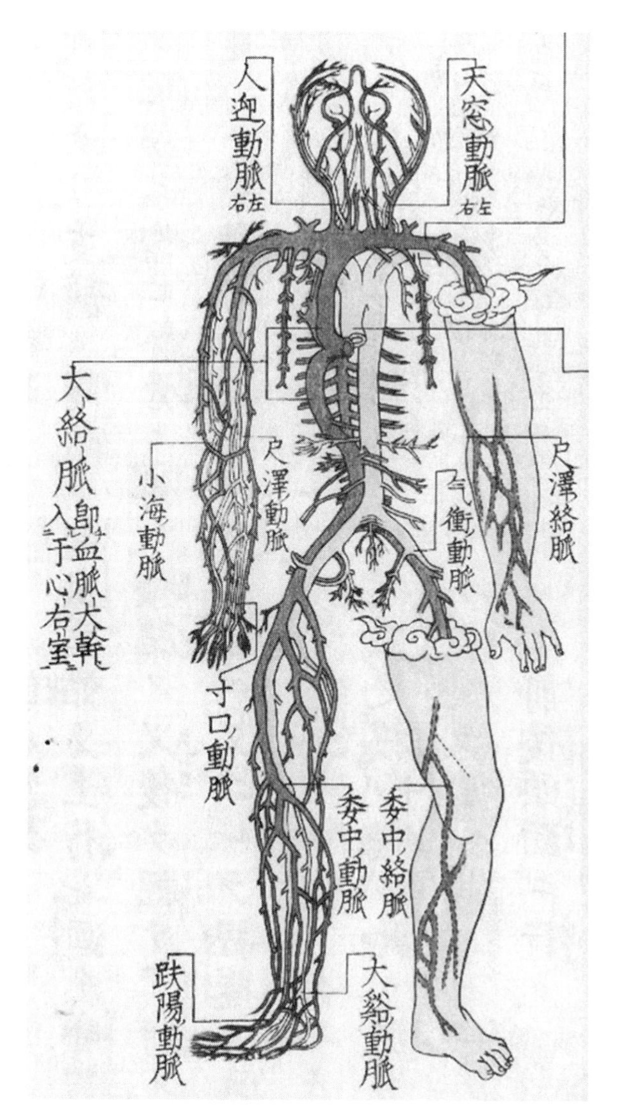

Figure 13. *Arteries and veins. Mitani Boku,* Kaitai hatsumō *(1813), 4:2a. Image courtesy of the Main Library, Kyoto University, https://rmda.kulib.kyoto-u.ac.jp/item/rb00001517.*

drawn attention to the connections between the circulatory system and the body's internal organs, Mitani's captions indicated the location of traditional acupoints, especially those where a pulse could be felt or where a vein was visible on the surface of the arm or leg.[42]

Mitani emphasized these aspects of the image because they were directly connected to his innovative identification of the arteries with the conduits (*kei* 經) and the veins with the networks (*raku* 絡). Just as he had done for the triple burner, Mitani provided extensive citations from the *Inner Canon* to support this identification. Mitani interpreted the *Numinous Pivot*'s statements that "the conduit vessels usually cannot be seen"

[42] Sugita Genpaku, *Kaitai shinsho* 解體新書 (1774), 3:8b–14a.

and that "the vessels that can be seen are all network vessels" as a reference to the deeper anatomical location of the arteries compared to the superficial veins visible on the arms and legs.[43] He further noted that the *Numinous Pivot*'s term for bloodletting, "piercing the networks" (C. *ciluo*, J. *shiraku* 刺絡), was consistent with the European practice of drawing blood by cutting veins, a practice that had attracted growing interest among Japanese doctors since the mid-eighteenth century.[44] New anatomical knowledge from translated European texts, Mitani argued, made it possible to recover the true meanings of the medical classics that had been long forgotten in China: "Few of the famous Chinese scholars of middle antiquity were able to discern these things clearly. Eventually the true and the false became muddled and were treated as the same thing. This is why the tradition of the *Inner Canon* was lost."[45]

The complex ambiguities of the medical classics meant that any attempt to reconcile them with European anatomical knowledge was always underdetermined: the same semiotic set of visual and textual elements could be reorganized and related to practical experience in multiple distinct ways. Just over a decade after the publication of Mitani's treatise, the Edo acupuncturist Ishizaka Sōtetsu pursued similar strategies to Mitani, seeking to show how European images could be reconfigured as illustrations of Sino-Japanese medical knowledge, but coming to a very different set of conclusions.[46] As an acupuncturist, Ishizaka was intrigued by the possibilities anatomy offered for reconceptualizing the body's pathways for the flow of blood and *qi*. Where Mitani's discussion of the blood vessels had focused on the canonical distinction between the conduits and networks, Ishizaka prioritized a different pair of concepts from Chinese medicine: the principle of nutrition (C. *ying*, J. *ei* 榮), which Ishizaka identified with the arteries, and the principle of defense (C. *wei*, J. *e* 衛), which he identified with the veins. In addition to these two, Ishizaka emphasized the significance of the nerves, which he thought of as pathways for the transmission of "ancestral *qi*" (*sōki* 宗気)—a phrase that Ishizaka appropriated from the medical classics and developed into his own innovative concept, the origin of essence (C. *jing*, J. *sei* 精), spirit (C. *shen*, J. *shin* 神), and ultimately all other bodily substances and activities.

European illustrations of the arteries, veins, and nerves as intricate branching networks provided Ishizaka with a new vision of pathways connecting different parts of the body quite distinct from the classical acupuncture conduits. These images suggested a conceptual foundation for an innovative style of acupuncture that disregarded

[43] Mitani, *Kaitai hatsumō* (cit. n. 36), 4:6b.

[44] Ibid., 4:7b–10a. On the practice of bloodletting in late eighteenth-century Japan and its relationship to Chinese, European, and local practices, see Trambaiolo, "Epidemics and Epistemology in Early Modern Japan: Japanese Responses to Chinese Writings on Warm Epidemics and Sand-Rashes," in Cook, *Translation at Work* (cit. n. 15), 157–75, on 167–9.

[45] Mitani, *Kaitai hatsumō* (cit. n. 36), 4:6a.

[46] On Ishizaka Sōtetsu, see Mieko Macé, "The Medicine of Ishizaka Sôtetsu (1770–1841) as Cultural Pattern of the Edo Period: Based on the Example of *Ei e chûkei zu* (1825)," *Studia Humana et Naturalia* 28 (1994): 73–90; Mathias Vigouroux and Machi Senjurō 町泉寿郎, "19 seiki Yōroppa no shinkyū no juyō ni okeru Shiiboruto to Ishizaka Sōtetsu no kōken ni tsuite: Shiiboruto kyūzō no shinkyū kankei shiryō no hikaku chōsa wo chūshin ni" 19世紀ヨーロッパの鍼灸の受容におけるシーボルトと石坂宗哲の貢献について：シーボルト旧蔵の鍼灸関係資料の比較調査を中心に [The contribution of Ishizaka Sōtetsu and Siebold to the reception of acupuncture and moxibustion in nineteenth-century Europe: A comparative investigation of materials on acupuncture and moxibustion in the former collection of Siebold], *Nihon Ishigaku zasshi* 57 (2011): 325–42; and Vigouroux, "The Surgeon's Acupuncturist: Philipp Franz von Siebold's Encounter with Ishizaka Sōtetsu and Nineteenth-Century Japanese Acupuncture," *Revue d'histoire des sciences* 70 (2017): 79–108.

the conduits and instead emphasized localized needling. In a summary of his ideas that Ishizaka published in 1826, he described the principles of this style of acupuncture as follows: "When you pierce the place where the disease is located with a gold, silver, or iron needle, it is like when inflammation occurs around a wooden or bamboo splinter: the essence, spirit, nutrition, and defense gather forcefully beneath the needle. [. . .] When you withdraw the needle, the assembled essence, spirit, nutrition, and defense scatter the disease pathogen, like wind blowing away clouds."[47] But as Machi Senjūrō and Mathias Vigouroux have shown, the published version of this text was the result of a complex and tentative process of conceptual and terminological experimentation.[48] Where the printed version refers to "essence, spirit, nutrition, and defense" (*seishin eiei* 精神榮衛), earlier manuscript versions had referred to "life *qi* and blood vessels" (*shōki ketsumyaku* 生氣血脈), or "essence, spirit, and blood vessels" (*seishin ketsumyaku* 精神血脈). Ishizaka had not changed his understanding of how acupuncture operated at the structural level: although the revised version was phrased in Ishizaka's somewhat peculiar terminology, "nutrition and defense" was simply his new term for the "blood vessels." But this change in terminology reflected a shift in emphasis, as well as a different way of thinking about how to make anatomical concepts relevant to practice. "Veins" and "arteries" were purely structural concepts derived from European anatomy, but by relabeling these structures as "nutrition" and "defense," Ishizaka linked them to a broader universe of East Asian medical discourse, hinting at the ways these structures might be connected to health, disease, and therapeutic possibilities, while at the same time suggesting new ways of interpreting the medical classics themselves.

CONCLUSION

Exploring the history of anatomical translation in early modern Japan forces us to confront difficult questions about how to understand the relationship between East Asian and European medical understandings of the body. It is common today to assume that the knowledge claims of East Asian and Western medicine are in some sense incommensurable.[49] Most obviously, the crucial term *qi* is usually left untranslated in English, since it is believed to designate a concept that cannot be adequately expressed in biomedical terms.[50] But the two traditions are also held to diverge in their basic inventories of the body's components. The European medical body has adrenal glands, a pancreas, arteries, veins, and nerves; the East Asian medical body has a triple burner, acupuncture points, conduits, and networks. Even for components seemingly recognized by both traditions, such as the liver, kidneys, or spleen, divergent claims about the functions and relationships of these components raise doubts about whether the corresponding

[47] Ishizaka Sōtetsu 石坂宗哲, *Chiyō ichigen* 知要一言 (1826), 3b–4a.

[48] Vigouroux and Machi, "19 seiki Yōroppa no shinkyū no juyō" (cit. n. 46), 335.

[49] The literature on this topic is extensive: representative works include Manfred Porkert, *The Theoretical Foundations of Chinese Medicine: Systems of Correspondence* (Cambridge, MA: MIT Press, 1974); Margaret M. Lock, *East Asian Medicine in Urban Japan* (Berkeley: Univ. of California Press, 1980); Nathan Sivin, *Traditional Medicine in Contemporary China* (Ann Arbor: Center for Chinese Studies, Univ. of Michigan, 1987); Judith Farquhar, *Knowing Practice: The Clinical Encounter of Chinese Medicine* (Boulder, CO: Westview Press, 1994); and Kuriyama, *Expressiveness of the Body* (cit. n. 27).

[50] On the (un)translatability of *qi*, see Qiong Zhang, "Demystifying *Qi*: The Politics of Cultural Translation and Interpretation in the Early Jesuit Mission to China," in *Tokens of Exchange*, ed. Lydia H. Liu (Durham, NC: Duke Univ. Press, 1999), 74–106.

terms in each of the two traditions can truly be interpreted as referring to the same somatic entities.[51]

But as a growing body of scholarship demonstrates, such skepticism about the possibility of translation between European and East Asian descriptions of the body is a comparatively recent historical phenomenon. It is a product of the rhetorical strategies employed by Chinese, Korean, and Japanese actors during the nineteenth and twentieth centuries, when competing claims about the similarity or incommensurability of East Asian and European medicine were advanced by different groups of doctors trying to assert authority over particular domains of knowledge and practice.[52] As Volker Scheid has argued, the skepticism in contemporary Chinese medicine about engaging too closely with anatomical understandings of the body derives from this period when "the contest with Western biomedicine compelled them to choose between structure and function," and we should not project it backward into our assumptions about earlier historical periods.[53]

During the seventeenth and eighteenth centuries, both East Asians and Europeans were much more likely to assume that translating fundamental concepts about the body was not only possible but desirable.[54] Europeans visiting Japan learned about techniques such as acupuncture and moxibustion; Japanese learned from Europeans about anatomy and surgery. Participants on both side of this exchange were aware that their understanding of the other side's ideas was imperfect, but this awareness did not tend towards skepticism about the possibility of knowledge about the body applying equally across different traditions. Motoki Ryōi's early anatomical translation provides a revealing example of how novel European anatomical images could be described while drawing extensively on the lexical and conceptual resources at hand: although unsystematic and occasionally inconsistent in its labels, it offered new knowledge while hinting at possible ways that the new and the old knowledge might be linked. A century later, Sugita Genpaku's *New Book of Anatomy* closed off such possibilities, its precise neologisms carefully constructed to avoid confusion between European and East Asian concepts that Sugita and his colleagues wished to keep distinct. Later doctors such as Mitani Boku and Ishizaka Sōtetsu, who sought to reconcile the new anatomy with East Asian medical concepts, could achieve this only by rejecting the *rangaku* translators' labels and reasserting the continuity of different forms of body knowledge.

[51] Farquhar, *Knowing Practice* (cit. n. 49), 92–4. Anglophone scholarship on East Asian medicine commonly capitalizes the names of the solid and hollow organs of East Asian medicine (Liver, Kidneys, Spleen, etc.) to emphasize that they do not denote the same concepts as the corresponding anatomical terms.

[52] Kim Taylor, *Chinese Medicine in Early Communist China, 1945–63: A Medicine of Revolution* (London: RoutledgeCurzon, 2005); Sean Lei, *Neither Donkey nor Horse: Medicine in the Struggle over China's Modernity* (Chicago: Univ. of Chicago Press, 2014); Bridie Andrews, *The Making of Modern Chinese Medicine, 1850–1960* (Vancouver: Univ. of British Columbia Press, 2014).

[53] Volker Scheid, "Promoting Free Flow in the Networks: Reimagining the Body in Early Modern Suzhou," *Hist. Sci.* 56 (2018): 131–67, on 133.

[54] Roberta Bivins, *Acupuncture, Expertise, and Cross-Cultural Medicine* (Houndsmill, Basingstoke: Palgrave, 2000); Hugh Shapiro, "How Different Are Western and Chinese Medicine? The Case of Nerves," in *Medicine across Cultures: History and Practice of Medicine in Non-Western Cultures*, ed. Helaine Selin (Dordrecht: Kluwer Academic Publishers, 2003), 351–72; Marta Hanson and Gianna Pomata, "Travels of a Pulse Treatise: The Latin and French Translations of the *Tuzhu maijue bianzhen* 圖註脈訣辨真 (1650s–1730s)," in Cook, *Translation at Work* (cit. n. 15), 23–57; Motoichi Terada, "The Montpellier Version of Sphygmology: Classical Chinese Medicine and Vitalism," in Cook, *Translation at Work* (cit. n. 15), 176–205.

Both temporally and conceptually, early modern Japanese doctors' efforts at anatomical *bricolage* occupy a position intermediate between the earliest encounters of European and East Asian medicine and the fraught political confrontations of the twentieth century. They illustrate a range of possibilities for anatomical thinking about the body in an East Asian medical tradition that was informed, but not yet dominated, by its encounter with Western medicine. At the same time, they also show how the meanings of anatomical and medical terms were never completely stable but remained open to repeated and sometimes radical reinterpretation. These acts of reinterpretation were as fundamental to the development of Japanese medical ideas in the early modern period as the act of translation itself: they mobilized the textual and visual resources of the medical archive, the experiences of medical practice, and the evidence of anatomical observations to create new possibilities for understanding and interacting with the body.

Casting Blood Circulations:
Translatability and Braiding Sciences in Colonial Bengal

*by Projit Bihari Mukharji**

ABSTRACT

In this article I examine three Bengali translations of texts about blood circulation from the last decade and a half of the nineteenth century. Through a close examination of these three translations, I demonstrate the importance of what Antonio Gramsci called "translatability" and what I have elsewhere called "braided sciences." Taken together, these two heuristic devices demonstrate a relationship between social hierarchies and translation, on the one hand, and a selective alignment of various strands of epistemic, technological, and narrative traditions on the other. One of the principal objectives of the article is to recenter the importance of caste and social hierarchy in understanding colonial knowledge production.

INTRODUCTION

Any translation involves crossing a boundary. Usually, this boundary is understood to be a linguistic one: the crossing from one language into another. More rarely, it might be understood as generic, such as from poetry to prose, or as a media boundary, for instance, from image to text. But increasingly, the boundary is also understood as cultural or epistemic. Focusing on the act of boundary-crossing, however, risks reifying the boundary itself, making it seem more natural than it is. Antonio Gramsci's notion of *translatability* seeks to take a step back and interrogate the apparent preconstitution of these boundaries, not through a sociological preface to the act of translation, but rather within the translated work itself.

This approach is particularly illuminating when dealing with late nineteenth-century medical translations in colonial Bengal. As the heartland of the British Empire in South Asia, Bengal, and, more specifically, Calcutta, the capital city of the British Raj, became a melting pot of people, ideas, cultures, languages, and knowledges. In this "colonial contact zone" numerous translations, explicitly so designated, occurred. Many of these works, even though reprinted extensively and repeatedly, were written by obscure authors whose biographies are lost to us. The details of what books they were translating

* Department of History and Sociology of Science, University of Pennsylvania, 327 Claudia Cohen Hall, 249 S. 36th St., Philadelphia, PA 19104-6304, USA; mukharji@sas.upenn.edu.

Osiris, volume 37, 2022. © 2022 History of Science Society. All rights reserved. Published by The University of Chicago Press for the History of Science Society. https://doi.org/10.1086/719226.

from are also often not entirely clear. Yet, these works are both self-confessedly translations and, despite often being in the same language, enormously different from one another.

By reading these works through the analytic prism of *translatability*, we can access the operations that constituted the boundaries which the specific translation then crossed. For Gramsci, translatability was eventually engendered in the "reality of human relationships of knowledge as an element of political 'hegemony.'"[1] So it is these "human relationships of knowledge," as they stand within a larger regime of political hegemony, that will eventually shape the nature of the translation.

By thus linking works of medical translation back to questions of political hegemony, I want to forestall the tendency to reify the divide between the colonizer and the colonized as the preeminent fault line in colonized societies, particularly in South Asia. Instead, I want to draw attention to the persistent, and indeed reinvigorated, power of caste within the regime of colonial political hegemony. I argue that by overemphasizing the divide between the colonizer and the colonized, medical historians of South Asia have largely neglected the ways in which caste refracted medical epistemologies.

Translatability, moreover, demonstrates the pitfalls of talk about "hybridity." Designating a variety of colonial knowledges as "hybrids" reifies and naturalizes two orthogonal cultural, political, and epistemic wholes ranged across the colonial caesura from which such hybrids apparently derive. On the one hand, this obfuscates the plurality and friction within each of these allegedly monolithic parental entities. On the other hand, it flattens the specificities of particular forms of epistemic interrelating across the colonial divide. A collaboration between upper-caste textual canons and reductionist and technocratic elements within the imperial milieu thereby becomes indistinguishable from an alliance between a subversive, socially marginal colonized tradition with nonhegemonic strands of European knowledge. This is both historically misleading and politically barren.

Instead, I have proposed the notion of "braided sciences." Braiding allows us to pursue two important goals simultaneously. First, it forestalls any totalizations that homogenize cultures and epistemologies or naturalize the colonial caesura. Second, it allows us to appreciate the intricate epistemic choices, political stakes, and, above all, the creativity that underpins translations of knowledge in colonial situations.[2]

One of those axes of power that the rubric of braiding allows us to illuminate and which hybridity obscures is that of caste. Yet it does so in a way that does not in turn freeze the relationship between caste and knowledge in ahistorical structural myths. Instead, it looks at the relationship itself as something that was repeatedly reconstituted through the uneven and selective movements (i.e., *translatio*) of knowledge.

This is where the Gramscian notion of *translatability* is complementary. If braiding is a concept metaphor that captures the epistemic process of the selective combination of different strands of knowledge, translatability shines a spotlight on the underlying "human relationships of knowledge" that render particular braids possible, meaningful, and redolent, while discouraging or even foreclosing other kinds of braids.

[1] Antonio Gramsci, "Science, Logic and Translatability," in *Further Selections from the Prison Notebooks*, trans. Derek Boothman (London: ElecBook, 1999), 450.

[2] Projit Bihari Mukharji, *Doctoring Traditions: Ayurveda, Small Technologies, and Braided Sciences* (Chicago: Univ. of Chicago Press, 2016), 25–7.

Taken together, braiding and translatability are a tool kit that helps to exorcise the ghost of knowledge "circulation," and the "hydraulic sensibilities" that attend to it, in many discussions of scientific translation.[3] Instead, they help us chart a course through the uneven, shifting, and differentially connected planetary space over which distinct and particular knowledges moved.

In this article, for the sake of precision and manageability, I will focus on a single physiological concept—blood circulation—in translation. I will compare the translations of this single concept in three Bengali texts that were all published within a decade of each other in the 1880s and 1890s. These texts are a book on human physiology, an Ayurvedic book written in verse, and, finally, an enigmatic little book of less than a hundred pages proposing a "new materio-spiritual" understanding of the human body. I will divide the discussion into three distinct sections devoted to each of these renditions of blood circulation. These sections will describe both the specifically braided renditions of blood circulation and the strands that make up the braids. A conclusion will draw the discussions together and locate the texts within the epistemic pluriverse of colonial Bengal.

ASHUTOSH MITRA'S VERNACULARIZED PHYSIOLOGY

The first text we will examine is a short treatise entitled *Nara Shareer Bidhan* (The system of the human body) by Ashutosh Mitra.[4] The text was published in 1882 and, like many of its contemporaries, had a parallel English title, *Human Physiology*. In the preface to the book, the author mentions, "Everyday, we are eating, getting tired, falling asleep, constantly inhaling and exhaling . . . it is both entertaining (*amodjanok*) and educative (*sikshaprada*) to study in detail (*parjalochana*) the machines by which and how these wonderous actions (*bismayakar karjabali*) are being completed." Continuing, he states that in other countries all scholars (*pandit*), even if they are not physicians, consider their education incomplete unless they have studied "biology" (*jeebtattwa*). Therefore, to remain completely unacquainted with the mechanisms of the body was, for Mitra, a national disgrace (*lojjar bishoy*). He wrote his book with a view to filling this gap in education and helping scholars get acquainted with the science of human physiology.[5]

Mitra was still a young man in 1882, but he had already emerged as one of the most promising Bengali doctors. Born in 1858 into an upper-caste family of physicians, he had entered the Calcutta Medical College, Asia's oldest college devoted to the teaching of western medicine, at the age of eighteen in 1876. At college he excelled, and Dr. George King, one of the famous professors to have taught him, remarked that Mitra was "one of the best students he knew."[6] Armed with such praise, Mitra travelled to Great Britain in 1883, the year after he had published *Nara Shareer Bidhan*, to pursue further medical studies. After spending some time attached briefly to various London hospitals and being elected Fellow of the Obstetrical Society of London, he eventually obtained dual

[3] Warwick Anderson, "Making Global Health History: The Postcolonial Worldliness of Biomedicine," *Soc. Hist. Med.* 27 (2014): 372–84.

[4] Ashutosh Mitra, *Nara Shareer Bidhan* [The system of the human body] (Calcutta: Radhagobinda Kar, 1882). Unless otherwise noted, the translations in this article are mine.

[5] Ibid., 1.

[6] Quoted in "Rai Bahadur A. Mitra, L.R.C.P.&S. Ed., F.O.S. (Lond.) Chief Medical Officer, Kashmir," *The Medical Reporter* 5 (1895): 66.

medical qualifications in medicine and surgery from the Royal Colleges in Edinburgh. When he returned to Calcutta in 1884 as one of fewer than ten Bengali physicians educated in Britain, he rapidly established a lucrative practice and was also appointed Health Officer at the Calcutta Public Health Society. The following year, in 1885, he was appointed the Chief Medical Officer of the then-independent Princely State of Kashmir.[7]

From 1885 to the end of his life in October 1914, Mitra remained in Kashmir.[8] During this time he established an efficient medical system and hospital. He headed the newly formed Srinagar municipality, where he introduced sanitary reforms, and he contributed regularly to the Indian and international medical press on issues relating to Kashmiri health. He also established a school in Srinagar and eventually became a high-ranking minister in the government of the Maharaja of Kashmir.[9] One biographer said, "Dr. Mitra's life and work testify to what an educated Indian gentleman can do in making himself not only useful, but necessary in popularizing the European system of medicine amongst people who are slow to appreciate it."[10]

Also remarkable was the identity of the publisher of *Nara Shareer Bidhan* [hereafter *Bidhan*]. It was published privately by Radha Gobinda Kar, who, like Mitra, was born into a family of physicians. Kar's father, Durgadas, had been one of the Bengali pioneers of western medicine. Indeed, he was one of the first Bengalis to begin publishing medical textbooks in Bengali according to the new, western medical tradition.[11] His son carried on in the same vein and authored a number of Bengali works on western medicine.[12] Radha Gobinda would eventually go on to establish his own private medical school, which is today one of the largest state-run hospitals in Kolkata (as Calcutta is now called) and is named after him. Though Kar's contemporary celebrity surpasses Mitra's, their biographies were remarkably similar. Kar, too, upon completion of his first medical degree from the Calcutta Medical College, had traveled to Britain and obtained a second degree from Edinburgh.[13] In fact, Kar and Mitra had overlapped at both the Calcutta Medical College and in Edinburgh, and one press report, from April 1883, suggests that they had traveled there together.[14] Kar had returned in 1886 and immediately set about establishing his medical school in Calcutta. By the time of his death in 1918, he had already established a successful and growing teaching hospital where the instruction was done in Bengali.[15]

The Kars, as I have shown elsewhere, were prominent participants in the late nineteenth-century process through which western medicine was vernacularized in Bengal.[16] The vernacularization unfolded at the intersections of medical education in Bengali, a burgeoning medical market, and a thriving print culture. Working at the intersection of

[7] Ibid.

[8] "Kashmere Minister's Death," *Times of India*, October 30, 1914.

[9] "Rai Bahadur A. Mitra" (cit. n. 6), 67; "Dr. A. Mitra, M.B.: A Sketch of His Life and Career by Pandit Anand Koul," *Indian Medical Gazette* 62 (1927): 228.

[10] "Rai Bahadur A. Mitra" (cit. n. 6), 66.

[11] Projit Bihari Mukharji, *Nationalizing the Body: The Medical Market, Print and Daktari Medicine* (London: Anthem Press, 2009), 90–1.

[12] Ibid., 77.

[13] Dilip Kumar Chakrabarti et al., "R. G. Kar Medical College, Kolkata—A Premier Institute of India," *Indian Journal of Surgery* 73 (2011): 390–3.

[14] "Indians in England," *Madras Mail*, April 28, 1883.

[15] Chakrabarti et al., "R. G. Kar Medical College" (cit. n. 13).

[16] Mukharji, *Nationalizing the Body* (cit. n. 11).

print culture and the medical market, they crafted both a vernacular identity for the system of western medicine, *daktari*, and a new social identity for physicians working within that system, *daktars*. The success of this process of vernacularization relied on its inclusiveness. Daktars were not a homogenous group: they included men, and occasionally women, with widely varied qualifications, social standing, and wealth. Books such as *Bidhan* were the product of both this process and the infrastructure that sustained it.

Such books allowed autodidacts and students from the gradually growing numbers of private medical colleges to acquaint themselves with the core principles of western medicine. They also enabled the interested lay public to acquire a basic understanding of the new ways of conceptualizing the body and its workings, thus allowing for a gradual displacement of various older therapeutic idioms. While vernacularization was certainly not limited to linguistic translations—indeed, physicians such as Mitra and Kar also wrote in English—linguistic translations undeniably played an important role in the process.

The importance of translations from English meant that men such as Mitra and Kar, who came from privileged, upper-caste families and thus had greater access to English-language scholarship, dominated the *daktari* profession. Whereas many men of lower rank and limited proficiency in English certainly practiced *daktari* medicine, the luminaries of the profession were all men who were bilingual (if not polylingual). It was they who produced the basic infrastructure for the profession, in the form of popular textbooks or lectures in the handful of vernacular medical schools established by the colonial state.

Multiple biographers and commentators writing about Mitra therefore explicitly underlined how his life was exemplary of the necessary role that "educated Indian gentlemen" had to play in order to popularize the "European system of medicine amongst people who are slow to appreciate it."[17] Likewise, one early review of *Bidhan* stated, ambitiously, that "every human being ought to know the mechanism of his own body" and that producing "vernacular science primers" such as the *Bidhan* was an act of supreme "patriotism and enterprise of the author and the publisher."[18] Men like Mitra both produced the actual resources for *daktars* and provided symbolic resources that allowed them to imagine a vernacularized western medical practice and practitioner.

The very first chapter in the *Bidhan* was devoted to blood (*rakta*). Immediately after stating how there was blood everywhere in the body, since any cut anywhere always brings forth blood, Mitra asserted the importance of a microscopic view of blood: "Looking at any object through a microscope (*anubikshan-jantra*) allows us to get a magnified view of it. Therefore, its true structure (*prakrita gathan*), which is impossible to see by our eyes, becomes clearly visible through the microscope."[19]

Gyan Prakash, a historian of science and one of the founders of the Subaltern Studies Collective, has identified the importance of what he calls a "second sight" in elite Indian translations of western science. "Translation entails the undoing of boundaries and borders entailed in the authorization of discourse," he writes, before adding that "it locates the formation of a modern Indian elite as a counter-hegemonic force in those productive in-between strategies and spaces that come into existence when 'our optic

[17] "Critical Notices: Human Physiology by Ashutosh Mitra," *Calcutta Review* 76 (n.d.): xxx; "Rai Bahadur A. Mitra" (cit. n. 6).

[18] "Critical Notices" (cit. n. 17).

[19] Mitra, *Nara Shareer Bidhan* (cit. n. 4), 24.

powers' are relocated as 'second sight.'"[20] Prakash points out that the colonized elite used translations to position itself as a counterhegemonic force to the imperial authority. But to do so effectively, elites had to also assert that their understanding of scientific truths was still superior to that of the general subaltern classes in colonized society. Otherwise, there would be no basis for a superior and hence hegemonic authority. What I want to add to this astute observation is the technological mediation of this vision. Mitra's assertion that the "true structure" of blood can only be understood through a microscopically enabled vision, while clearly disqualifying any lay, naked-eyed view of blood, also attaches his discursive authority to the use of the microscope.[21]

Mitra supplemented his microscopically mediated view of blood by adding woodcut illustrations of what blood cells "truly" looked like (see fig. 1).[22] The images, naturally, don't show what blood looks like to the naked eye. And because the woodcuts are simple and basic, neither do they look like anything one would see through a microscope. They were idealizations meant not for the actual training of a microscopic eye but rather as icons of the new techno-modern second sight.[23]

Having thus established at length a microscopic view of blood, including differentiations of the various physiological constituents of the blood such as the red blood cells (*lohit konika*), white blood cells (*shwet konika*), serum (*raktarash*), and so forth, Mitra proceeded to his second chapter, *rakta-sanchalan* (blood circulation). He inaugurated this chapter on a historical note: "Though the disciplines of anatomy and physiology (*dehatattwa o shareer-bidhan shastra*) have long been studied, until three hundred years ago physiologists (*shareertattwabidera*) had been ignorant of blood circulation."[24] Notwithstanding the enormous importance of the subject, Mitra lamented, it had been "covered in darkness" until "at long last the famous Englishman, Harvey, through great toil discovered blood circulation in 1618."[25]

Having waxed eloquent on Harvey's discovery, Mitra then made a surprising volte-face. In the very next paragraph, he asserted, "But centuries before the birth of Harvey this was not unknown to the Hindu physiologists (Hindu *shareertattwabid*). Like all the other disciplines that were known to the Hindu scholars, physiology too in course of time became covered in darkness through a lack of practice. But is it [not] a matter of remarkable intellectual achievement and progress to have known what surprised scholars in the seventeenth century so many, many hundreds of years earlier?"[26]

This excursion into the glories of ancient Hindu science, however, was but a hiatus, and Mitra did not tarry long on it. Instead, he proceeded to describe the pumping action of the heart and the differences between arterial and venous blood. Commenting on the

[20] Gyan Prakash, *Another Reason: Science and the Imagination of Modern India* (New Delhi: Oxford Univ. Press, 2000), 51.

[21] This is not at all surprising given that Edinburgh was the leading center in Britain for the incorporation of microscopic anatomy into regular medical education in precisely the period that Mitra studied there. See L. S. Jacyna, "'A Host of Experienced Microscopists': The Establishment of Histology in Nineteenth-Century Edinburgh," *Bull. Hist. Med.* 75 (2001): 225–53.

[22] Mitra, *Nara Shareer Bidhan* (cit. n. 4), 25.

[23] See also Mukharji, *Doctoring Traditions* (cit. n. 2), 177–8.

[24] Mitra, *Nara Shareer Bidhan* (cit. n. 4), 29.

[25] Ibid., 29. Though Harvey's discovery is usually dated to the publication of his book in 1628, it is likely that he had actually made the discovery in 1618–19. See *Encyclopedia Britannica Online*, Academic ed., s.v., "William Harvey," accessed December 17, 2021, http://academic-eb-com.proxy.library.upenn.edu/levels/collegiate/article/William-Harvey/106277#article-contributors.

[26] Mitra, *Nara Shareer Bidhan* (cit. n. 4), 30.

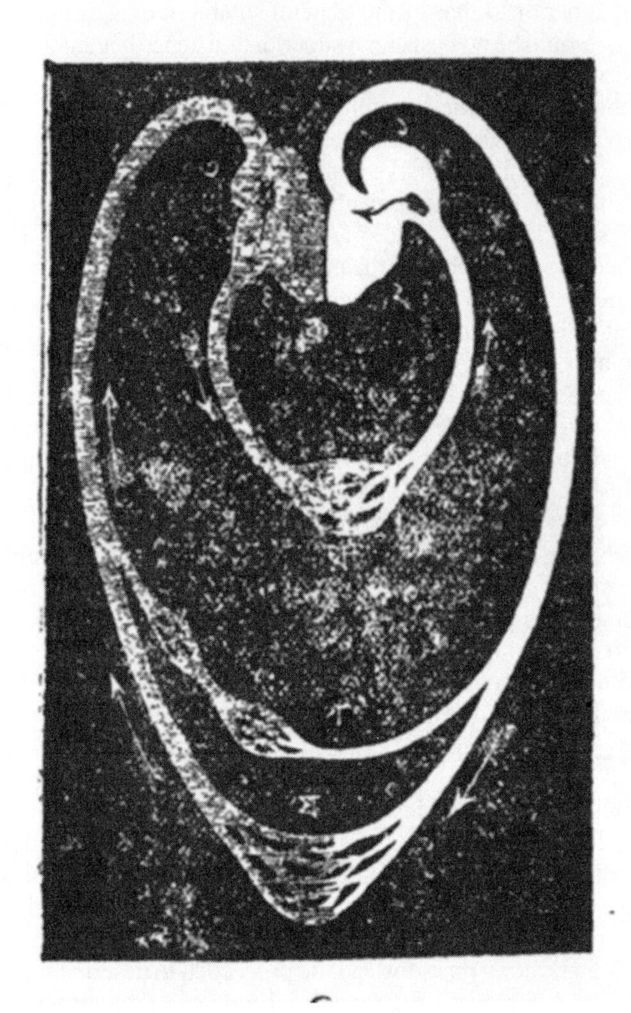

Figure 1. *Image of blood circulation in Ashutosh Mitra,* Nara Shareer Bidhan *(The system of the human body) (Calcutta: Radhagobinda Kar, 1882), 33. Author's personal collection.*

structure of circulation on the whole, he mentioned, "The earth is spherical (*golakar*). Hence, sailors traveling in one direction, after having circumambulated the world, return to the same place. Similarly, blood travels from one place and after traveling around the body returns to the same place. That is why movement (*sanchalan*) of blood is known as blood circulation (*rakter brittakare bhraman*)."[27]

Furthermore, returning to the differences between arterial and venous blood and the valve mechanisms to keep them strictly separated, Mitra invoked the microscope once

[27] Ibid.

more, asserting that the whole circulatory system might be observed in action by observing the tail of a tadpole under a microscope: "The skin on the tadpole's tail is extremely clear (*swaccha*), so we can clearly see through it with the microscope. Had our own skins been as clear, we would have been able to see our blood flowing as well."[28]

Mitra also used the elaborate discussion of the heart's pumping mechanism to clarify that the pulse (*nadi*) depended upon the beating of the heart. The pulse increased or decreased depending on the blood flow through the heart. He also gave the numerical values of normal pulse rates per minute for infants, toddlers, youth, the elderly, and the very old. Such discussions were particularly important since pulse diagnosis (*nadipariksha*) was one of the foremost diagnostic techniques in the majority of the precolonial medical traditions. Moreover, the pulse was often ascribed a quasi-mystical status. Indeed, there was a very robust tradition in precolonial pulse texts on "death pulses" that claimed to be able to foretell people's deaths years into the future. Others claimed that virtuoso physicians could hear the footfalls of death or destiny in the pulse.[29] By implicating the pulse in a purely mechanical system of blood flow and enumeration, Mitra was clearly repudiating these widely held and powerful understandings. It is also worth recognizing that Mitra was certainly doing this deliberately, since earlier in the book he had himself cited a sixteenth-century Sanskrit text, *Bhavaprakasa*, that discussed Ayurvedic pulse diagnosis in detail.[30]

Blood, for Mitra, therefore, was a microscopically analyzable physiological entity circulating around the body through a network of arteries and veins and pumped by the heart. The description is historically derived from William Harvey's seventeenth-century descriptions, but we are told that this had already been known to ancient Hindu physiologists. Notwithstanding this invocation of classical knowledge later lost, actually extant ideas about the pulse were silently dismissed. It is of course notable that the classical invocation was explicit and conspicuous, if somewhat aberrant to the narrative, whereas the dismissal was entirely silent. Despite the prominence of *nadipariksha* in local medical and lay ideas about health, Mitra did not mention it while discussing the pulse.

As the reviewer in the *Calcutta Review* had observed, the project was clearly framed in national terms. Mitra's book was intended to overcome a national disgrace, and it constantly reminded the nation of its past glories. Needless to say, this nation was entirely conceptualized as an explicitly "Hindu" one. This framing itself drew on a tradition of "medical Orientalism" that identified a classical Sanskrit corpus of medical writing as an ancient "Hindu" and "Aryan" one.[31] There was little room here for the non-Hindu, and it is doubtful what degree of pride the lower castes might feel in this avowedly Sanskritic and upper-caste "Golden Age."[32]

BINODBIHARI RAY'S MODERNIZED AYURVEDA

The second text I will turn to is a serialized essay titled *Shareer Kriya Bigyan* (Science of physiology) in a medical journal called *Chikitsak* (Physician) published in 1890. The author of the essay, Binodbihari Ray, was also the editor of the journal and wrote all the

[28] Ibid., 32.
[29] Mukharji, *Doctoring Traditions* (cit. n. 2), 77–116.
[30] Mitra, *Nara Shareer Bidhan* (cit. n. 4), 15; Sarva Dev Upadhyay, *Nadi-Vijnana: Ancient Pulse Science* (Varanasi: Chaukhamba Sanskrit Pratishthan, 1986), 54–8.
[31] Mukharji, *Doctoring Traditions* (cit. n. 2), 55–9.
[32] On this point, see Projit Bihari Mukharji, "Historicizing 'Indian Systems of Knowledge': Ayurveda, Exotic Foods, and Contemporary Antihistorical Holisms," *Osiris* 35 (2020): 228–48.

content himself. The text can therefore also be broadly considered a serialized book rather than what we now think of as a periodical. Ray, throughout his life, wrote and published an enormous amount, but he is much less well known than Mitra and his colleagues. Being based in a small town in the Rajshahi district of present-day Bangladesh, he did not have the social and cultural prominence that Mitra enjoyed.

From the little we know about Ray, he was born in 1862 and at some point earned a VLMS diploma. The VLMS, or Vernacular Licentiate of Medicine and Surgery, had been introduced by the colonial state in 1851 as a shorter, purely vernacular route to *daktari* medicine. It was introduced to meet the growing demand for education in western-style medicine from candidates who could not devote the five years necessary for the MB or LMS qualifications and did not have the linguistic competence necessary to pursue their education entirely in English. Two different vernacular linguistic routes to the VLMS were initially available through Hindustani and Bengali. If a candidate successfully completed the three-year course, he became eligible for appointment at the rank of a Hospital Assistant at one of the government hospitals, or at one of the growing numbers of mostly rural dispensaries that were partly state-funded. Though originally available only at the Campbell Medical School in Calcutta, by 1874 the demand for the new qualification led to the option being introduced at the Dhaka Medical School. Unfortunately, it is not clear where exactly Ray obtained his diploma or, indeed, when he got it.[33]

The vernacularity institutionalized in the VLMS degree also marked it as being socially and intellectually subordinate to the MB and MD degrees that men like Mitra obtained. This meant that while VLMS degree holders formed the bulk of the *daktari* profession, they were always already subordinated to those who had been educated in English. This institutionalized linguistic hierarchy engendered a new and uniquely colonial-modern rank of social differentiation. The hierarchization is a testament therefore to both how colonialism created new social ranks and how language competence itself became a marker of medical and social authority.

Ray continued to publish throughout his life, and his last known work appeared in 1941. His later works were all wide-ranging cosmological meditations that had little to do with medicine per se. Yet in both the journal *Chikitsak* and in a later work in verse, published in 1908, he was precocious in developing an explicit agenda for modernizing Ayurvedic medicine, a reform that would be much more successful in the twentieth century.[34] Explaining his motivation for publishing *Chikitsak*, he wrote in the first issue, "The medical discipline of our own country (*swadeshiya chikitsashastra*), from which all other medical disciplines have obtained their origins, that Ayurvedic medical discipline has fallen to a state of great poverty—leave alone foreigners, even our own countrymen hate it and insult it [by calling it] unscientific (*abaigyanik*)."[35] Having laid out this national context and invoked the classic "medical Orientalist" tropes of Ayurveda's global primacy and contemporary decline, Ray then proceeded to outline what was to be done about it. He admitted that many scholars, and especially many Kobirajes

[33] Projit Bihari Mukharji, "From Physiograms to Cosmograms: Daktar Binodbihari Ray Kabiraj and the Metaphorics of the Nineteenth-Century Ayurvedic Body," in *Progress and Pathology: Medicine and Culture in the Nineteenth Century*, ed. Sally Shuttleworth, Melissa Dickson, and Emilie Taylor-Brown (Manchester: Manchester Univ. Press, 2020), 220.

[34] Mukharji, *Doctoring Traditions* (cit. n. 2).

[35] Binodbihari Ray, "Abataranika [Introduction]," *Chikitsak* 1 (1890): 2. I have tried to remain as close as possible to Ray's idiosyncratic, obscure, and often circuitous idiom.

(Ayurvedic physicians), had come forward for the uplift of Ayurvedic medicine. But "external improvements" (*bajhyik unnati*) alone would not suffice. A more thoroughgoing "internal improvement" (*abhyantarik unnati*) of Ayurvedic medicine was necessary, and this, moreover, could not be accomplished by Kobirajes alone. The help of "*daktars* of this country" (*ettodeshiya daktargan*), Ray argued, was crucial, although he lamented that many *daktars* began to despise (*ghrina*) Ayurveda from the day they first stepped into the Medical College. This, Ray insisted, was a grave error. After all, "if there is anything at all in medicine which we might truly call our own, it is Ayurveda."[36] It was with the intention of doing his duty as a "*daktar* of this country," therefore, that Ray undertook the publication, presenting it as a tool for the revitalization of Ayurveda. Indeed, in his publications, Ray chose to identify himself as "Daktar Binodbihari Ray Kobiraj."

Ray was far from alone in such a move, and many similar figures emerged in different parts of British India. K. N. Panikkar, Kavita Sivaramakrishnan, and Rachel Berger have variously mapped the rise of Ayurvedic revivalism in Kerala, Punjab, and the United Provinces, respectively. In each of these contexts, we notice varying degrees of interactions with *daktari* medicine.[37] Other historians have described structurally similar initiatives to revitalize some of the other scholarly medical traditions of South Asia, such as Unani Tibb and Siddha.[38] Given that the earliest and most prominent institutions for western-style medical education were located in Bengal, this trend toward revitalization was, predictably, more intense there. Several *daktars*, such as Binodlal Sen, Udaychand Dutt, Surendranath Goswami, J. B. Ray, and Gananath Sen, all, like Ray, turned their attentions to the revitalization of Ayurvedic practices and theories.[39]

Notwithstanding such links between Ayurveda and *daktari*, it is also important to recall the ways in which Ray's project differed from that of Mitra. Where Mitra had seamlessly aligned myths about ancient Hindu physiological knowledge with praise for William Harvey, Ray mocked the fact that even other Bengalis were turning away from Ayurveda for *daktari* medicine. Mitra, no doubt, was precisely the class of man that Ray was scornfully referring to: those who had turned their backs on Ayurveda and taken up *daktari* medicine. In Ray's view the truly patriotic thing would be for *daktars* to endeavor instead to revitalize Ayurvedic medicine. It was thus that the new colonial modern hierarchies of knowledge inspired revanchist angst in those it incorporated at the bottom of its medical totem pole.

Where Mitra had sought to reconcile competing historical narratives to produce a singular and universal medical present, Ray provincialized *daktari* medical universalism and reasserted a plurality of therapeutic options. Similarly, where Mitra had commenced

[36] Ibid., 2–3.

[37] K. N. Panikkar, "Indigenous Medicine and Cultural Hegemony: A Study of the Revitalization Movement in Keralam," *Studies in History* 8 (1992): 283–308; Kavita Sivaramakrishnan, *Old Potions, New Bottles: Recasting Indigenous Medicine in Colonial Punjab (1850–1945)* (Hyderabad: Orient Longman, 2006); Rachel Berger, *Ayurveda Made Modern: Political Histories of Indigenous Medicine in North India, 1900–1955* (Basingstoke: Palgrave Macmillan, 2013).

[38] Guy N. A. Attewell, *Refiguring Unani Tibb: Plural Healing in Late Colonial India* (New Delhi: Orient Longman, 2007); Seema Alavi, *Islam and Healing: Loss and Recovery of an Indo-Muslim Medical Tradition, 1600–1900* (Basingstoke: Palgrave Macmillan, 2010); Richard S Weiss, *Recipes for Immortality: Medicine, Religion, and Community in South India* (Oxford: Oxford Univ. Press, 2009).

[39] Brahmananda Gupta, "Indigenous Medicine in Nineteenth- and Twentieth-Century Bengal," in *Asian Medical Systems: A Comparative Study*, ed. Charles M. Leslie (Berkeley: Univ. of California Press, 1977), 368–82; Mukharji, *Doctoring Traditions* (cit. n. 2).

his description of blood by peering into a microscope—an iconic "western" scientific instrument—Ray began his with his fingers in his ears.

Structured as a dialogue between the author and a rustic lay interlocutor named Banchharam, Ray's discussion explained how the sound one heard upon plugging both one's ears with one's fingers was in fact the sound of blood circulating. He explained that just as mighty rivers, such as the Padma and Gandak, make a distinct sound as they rush on, so too does the blood rushing in our veins. By blocking out the external sounds with our fingers, this is the sound we can hear.[40] The material embodiment of western scientific advance, the microscope, thus became redundant for an elaboration of blood circulation. This too was a kind of "second sight," but not one that needed the microscope as a prop.

Instead, Ray's discussion leaned upon a litany of classical Sanskrit quotations. The narrator and Banchharam debated by citing rival verses from the *Susruta-samhita* and the *Bhavaprakasha*, both centuries-old classics considered canonical in erudite Ayurvedic circles. The citations were often very general and did not speak directly to the point Ray was seeking to establish. He cited two verses, for instance, in support of his argument that the sound in our ears attests to the circulation of blood, but all that these verses actually said was that blood was mobile (*chalansheel*).[41] Patently, the reasons for such invocation resided in his attempts to align his discussion with the textual canon rather than actually derive his arguments from them.

In explaining the mechanism of circulation, once again we find Ray following the same general strategies of relying on empirical demonstrations that did not require technological mediation and interpolating a smattering of redundant classical references. He explains to Banchharam that the best way to understand the mechanism of circulation is by vivisecting a mouse. As long as the mouse is alive, he tells his auditor, we can see a small, intensely red organ beating in its chest. This is the heart. He then points out that this organ resembles a *kurma* (tortoise), which is mentioned in most *nadipariksha* manuals as being the center of the system of bodily channels or *nadis*.[42]

Ayurvedic pulse diagnosis, or *nadipariksha*, however, conceptualized the body and the act of pulse taking very differently. One of the key differences was that the system of *nadis* conceptualized in *nadipariksha* was centered on the navel, rather than the chest. Moreover, depending on the gender of the person whose pulse was being felt, the entire navel-centric system was inverted.[43] The whole geometry of the bodily space within which *nadis* were placed was therefore quite distinct from the chest- and heart-centric anatomical space that Ray was explaining. Indeed, Banchharam points out that the "tortoise" resides at the navel. Some authors, he says, hold that the tortoise migrates upward into the chest, but that happens only at the point of death. Ray's narrator, however, insists that based on the evidence from the vivisected mouse and by feeling the chest of any person whose ribs are prominent, we can dismiss the idea of the tortoise migrating and instead accept that the tortoise mentioned in Sanskrit sources was in fact the "primary machine for circulating blood."[44]

[40] Binodbihari Ray, "Shareer Kriya Bigyan [Physiological science], Part VI," *Chikitsak* 1 (1890): 120.
[41] Ibid.
[42] Binodbihari Ray, "Shareer Kriya Bigyan [Physiological science], Part VII," *Chikitsak* 1 (1890): 161–2.
[43] Mukharji, *Doctoring Traditions* (cit. n. 2), 97.
[44] Ray, "Shareer Kriya Bigyan [Physiological science], Part VII" (cit. n. 42), 162.

Ray explained the circulatory mechanism further and its similarities to the tortoise in the *nadipariksha* texts through detailed discussions of a series of diagrams. Unfortunately, either the images were never in fact printed or they have simply not survived in the sole extant copies of the first three issues of Ray's journal. It is still highly significant that, like Mitra, he too depended on illustrations. Precolonial Ayurvedic medicine, despite its prolific and diverse textual tradition, did not seem to have developed any significant visual tradition in the way several other neighboring medical and nonmedical traditions of thinking about the body did. As Dominik Wujastyk, one of the foremost scholars of classical Ayurvedic medicine, confesses, "In fact, all Ayurvedic manuscripts I have studied have been empty of accompanying illustrative materials. There are small sketches of chemical apparatuses in some alchemical manuscripts . . . But no single manuscript I have seen contains even so much as an anatomical sketch, a line drawing for surgical guidance, or any other visual representation of the medical body."[45] He wonders suggestively whether the Ayurvedic approach might have been akin to the Buddhist discourse on the body where illustrations are avoided precisely because "process is privileged over substance."[46] The earliest illustration of the medical body with organs that Wujastyk was able to find was in an eighteenth-century manuscript likely from Nepal. In that case, however, some interaction or influence of European, Perso-Arabic, or, indeed, as Wujastyk himself suggests, Tibetan traditions of illustration might well have inspired the image. The point remains, though, that illustrations were highly uncommon in the Ayurvedic tradition, especially detailed illustrations of bodily organs. Ray's practice was hence innovative, even independently of what the images actually depicted.

In later installments of the essay, Ray further developed the chemical description of blood, outlining the differences between arterial and venous blood and describing the various components of the blood, among other things. He also observed that the main reason for the circulation of blood was the existence of an innate power to contract (*sankochan*) in some of the blood vessels.[47]

Unfortunately, some of the installments of Ray's serialized essay have not survived in libraries. This makes it difficult to reconstruct the entirety of his descriptions. But this is where some of his later writings are useful. His versified Ayurvedic textbook, *Podye Ayurved Siksha* (Ayurvedic education in verse), for instance, includes a remarkable observation about the blood. In a section describing the three Ayurvedic parahumors, *bayu, pitta*, and *kaph*, Ray writes:

> So long as the railway engine remains cold,
> It remains stationary like a corpse.
> When it heats upon the addition of water and fire,
> See the machine comes alive then.
> Lacking fire, it remains cold as a corpse,
> Its liveliness will disappear.
> Likewise, so long as the human body is warm,
> It is alive, Shanti, only so long.
> I have kept a substance in the blood,
> That is what stores heat at all times.

[45] Dominik Wujastyk, "Interpreting the Image of the Human Body in Premodern India," *International Journal of Hindu Studies* 13 (2009): 189–228, on 207.
[46] Ibid., 206.
[47] Binodbihari Ray, "Shareer Kriya Bigyan [Physiological science], Part IX," *Chikitsak* 1 (1890): 184–7.

> Flowing in the blood, the heat spreads everywhere,
> Thus, the body remains uniformly hot.
> This, Shanti, is called *pitta* in Ayurveda.[48]

I have discussed Ray's use of the metaphor of the railway engine elsewhere,[49] but here I want to draw attention to the alternative materiality that he attributes to blood by locating within it an entity called *pitta*, which is one of the Ayurvedic parahumors.[50] Clearly, in this Ray went beyond the kind of second sight that Mitra had invoked and asserted the presence in the blood of something that modern instruments such as the microscope could not detect. This is where Mitra's seamless conflation of a past Ayurvedic Golden Age with a monoculturally *daktari* present and Ray's pursuit of a revitalized Ayurveda that would trounce *daktari* clearly parted ways. Ray insisted that not only were the *daktari* truths prefigured in an Ayurvedic past but moreover that Ayurveda exceeded the *daktari* understanding of human physiology. In this Ray is articulating a position that Banu Subramaniam has called an "archaic modernity," a modernity that is entirely compatible but also in excess of the rationalities of modern science.[51]

Interestingly and conspicuously, Ray also included a photograph of himself as the frontispiece of his 1908 book. The publication of such photographs was not remarkable in itself at the time: what was significant was Ray's choice to have himself photographed bare-chested so that his sacred thread was clearly visible across his breast. He also sat at an unusual angle so as to make the sacred tuft of hair behind his head visible. Both of these were conspicuous visual markers of upper-caste, and usually, though not necessarily, Brahmin status. The use of photographs to make complex claims to caste, social, and religious authority in medical texts had emerged from the third quarter of the nineteenth century,[52] but Ray's use of this mechanism is particularly redolent with the kinds of intellectual claims to archaic modernity that he was fashioning.

BROJONATH SHAHA'S HYLOZOIC SCIENCE

Our third and final text is the most singular. It identified neither with *daktari* nor with Ayurvedic medicine. This curious little book is *Dehatmik Tattva: A Treatise on Materio-Spiritualism Based on Science and Religion*, and it was published from Calcutta in 1891—the year after Ray's and nine years after Mitra's. The author was one Dr. Brojonath Shaha, though he identified himself merely as "Dr. Shaha" in the book.

Unfortunately, notwithstanding his own prolific writing and publication record, little is known about Shaha as a person. He came from a reasonably well-off but lower caste

[48] Binodbihari Ray, *Podye Ayurbbed Siksha* (Calcutta: Leela Printing Works, 1908), 8. The poem is organized as a dialogue between the god Shiva (also called Mahadeb) and a man named Shantiram, or Shanti for short.

[49] Mukharji, "From Physiograms to Cosmograms" (cit. n. 33).

[50] Three *doshas* (faults) play a central role in much of the Ayurvedic tradition: *Vayu, Pitta*, and *Kaph*, usually translated as Wind, Bile, and Phlegm. In the nineteenth century "medical Orientalists" described them as "humors." Designating them thus, however, is misleading, since they are in many ways not at all akin to the classical Greek medical humors. Moreover, later modernizers of Ayurveda have variously translated them as "hormones," and even lately as "genomes." These latter translations have explicitly criticized the earlier translations as humors. I have therefore chosen to call them *parahumors*. See Mukharji, *Doctoring Traditions* (cit. n. 2).

[51] Banu Subramaniam, "Archaic Modernities: Science, Secularism, and Religion in Modern India," *Social Text* 18 (2000): 67–86; Projit Bihari Mukharji, "Hylozoic Anticolonialism: Archaic Modernity, Internationalism, and Electromagnetism in British Bengal, 1909–1940," *Osiris* 34 (2019): 101–20.

[52] Mukharji, *Doctoring Traditions* (cit. n. 2), 66–9.

family and had obtained a Licentiate of Medicine and Surgery (LMS) degree, possibly from the Calcutta Medical College, before entering the Indian Medical Service (IMS).[53] While there were several South Asians, and especially Bengalis, in the lower echelons of the IMS by the last quarter of the nineteenth century, most of these men belonged to the upper castes.[54] Shaha was unusual in that respect. He hailed from a mercantile caste, *Gandha Banik*, which was ranked somewhere in the middle of the lower of the four major caste or *Varna* groupings, namely, the Sudras. Brahmins were permitted to accept a drink of water from them but not eat food touched by them. Most of the *Gandha Baniks* ran small shops selling spices, sugar, *ghee*, medicines, and so on. Some of them also retailed opium, *charas*, and cannabis. There were few among them who were rich, but many were comfortably off petty traders, despite their lowly ritual status.[55]

We first hear of Shaha in the pages of the *Indian Medical Gazette* in 1880, when he was already a member of the IMS, and we know that he retired from service in 1904. During his service in the IMS he rose to a fairly high position and served as the Civil Medical Officer of Rangamati, Chittagong, in present-day Bangladesh. At the time, Rangamati was a frontier region very far away from the center of political and cultural authority in Calcutta, which also happened to be Shaha's hometown. It was therefore not likely to have been a very desirable posting.

Notwithstanding its remoteness and probable undesirability, Shaha made the most of his posting. He taught himself the local language and produced the very first grammar of the local Lushai (Mizo) language.[56] Linguists working on the region to this day refer to Shaha's grammar. His publications in the *Indian Medical Gazette* suggest that Shaha had mastered several other languages as well, such as German and possibly French. Clearly, he had a voracious intellectual appetite. Besides works on medicine and the Mizo language, he also published on what he dubbed the "stylography" of the English language, wherein he sought to find a mathematical formula that could be used by those with limited knowledge of English to discern the meanings of sentences.[57] The British colonial government eventually honored him with the title of Rai Bahadur: one of the honors given out by the colonial government, it was comparable, though of slightly lower rank, to the knighthood still given to British subjects. Mitra too had obtained this honor.

Shaha's book was structured as a discourse delivered by an erudite, eclectic, and highly unorthodox Brahmin called Darshanraj Chakrabarti to his devoted, intellectually curious, and low-caste servant-cum-disciple, Bholanath. Shaha explicitly stated that Bholanath belonged to the Kaibarta caste. This was a very significant component of Shaha's text.

The Kaibartas were themselves divided into four subgroups, two of which were regarded as ritually entirely "unclean," and two that were "nearly clean." Financially, they were better off than some of the other lower castes, but their situation had worsened by

[53] Some broad details about Shaha's family background and his caste are given in a memoir written by Shaha's nephew, Taraknath Sadhu, *Smriti-katha* [Reminiscences] (Calcutta: Allen Printing Works, 1340 Bengali Era [1933]).

[54] Mukharji, *Nationalizing the Body* (cit. n. 11).

[55] Jogendra Nath Bhattacharya, *Hindu Castes and Sects: An Exposition of the Origin of the Hindu Caste System and the Bearing of the Sects towards Each Other and towards Other Religious Systems* (Calcutta: Thacker, Spink, 1896), 201–2.

[56] Brojo Nath Shaha, *A Grammar of the Lushai Language, to Which Are Appended a Few Illustrations of the Zau or Lushai Popular Songs and Translations from Aesop's Fables* (Calcutta: Bengal Secretariat Press, 1884).

[57] Brojonath Shaha, *The Stylography of the English Language* (Calcutta: Patrick Press, 1897).

the end of the nineteenth century. Many of the poorer members of the caste worked as domestic servants, just as Bholanath did in the text. Irrespective of their financial condition, however, they were barred from any form of formal education because of their alleged ritual "uncleanness."[58]

The transmission of knowledge in orthodox Hindu society was regulated by the principle of *Adhikari-bheda* (Doctrine of the distinction of rights), or the Sanskritic dogma that different castes were entitled to different kinds of knowledge. In effect, this meant that the lower castes were ritually and socially barred from acquiring most higher philosophical learning. This was monopolized instead by the Brahmins and a small handful of higher castes. As the social historian Sumit Sarkar explains, "*Adhikari-bheda* had emerged as a formal doctrine in the seventeenth-eighteenth century as a high-Brahmanical way of accommodating difference in philosophy, belief and ritual."[59] Though one aspect of this doctrine was opening up what Sarkar calls "living spaces" in which subaltern groups had substantial autonomy of practice and livelihood by demarcating specialized social and occupational niches, it also served to bolster the overall hierarchy of the caste system. "In official, high-caste doctrine," Sarkar continues, "*adhikari-bheda* often becomes synonymous with, not fluidity or openness, but neat compartmentalization, the drawing-up of more definite boundaries, and the arrangement of the various philosophies, rituals, beliefs and *sampradayas* in a fixed hierarchy culminating in high-Brahman practices and Advaita Vedantist [nondualist] philosophy."[60]

The nineteenth century, with its new social and material technologies ranging from new associations to the printing press, witnessed a new intensification of the attempts to break as well as reinforce *Adhikari-bheda*.[61] Shaha's choosing to explicitly specify the caste of Bholanath strongly suggests that he was deliberately intervening in these struggles around *Adhikari-bheda*. Indeed, as we shall see, his entire framework might be read as a challenge to the *Advaita* (nondualism) that *Adhikari-bheda* privileged.

Shaha described a ritual performance that seemed remarkably similar to certain Tantric and Hatha-yogic rituals whereby Chakrabarti was able to extract the various "life forces" from within an ascetic named Sadananda, who, in the process, is sacrificed in the pursuit of knowledge. The extracted life forces then successively reveal their role in the constitution and functioning of the body. The whole process of successive extractions and revelations starts with Chakrabarti commanding Sadananda's soul thus: "I command thee, Great Life (Life, Spirit, Soul), to exit Sadananda's body and seat yourself on the seat laid out for you." The Great Life (*mahapran)* then speaks from within Sadananda: "I, Great Life, am not distinct from the other Lives. You may call out any of my three female companions [*sahachari*]. Any one of them existing will enable my exit. Do you not know that I have no power to forsake my companions?" Continuing further, Great Life clarifies that her three companions were, "*Mastishka-pran* (Brain-life)," "*Raktasanchalan-pran* (Circulation of Blood-life)," and "*Swaspraswas-pran* (Respiration-life)."[62]

Blood circulation, far from being a process defined by the chemical makeup of blood or the pumping mechanism of the heart, in Shaha's rendition immediately becomes a powerful goddess materialized within the body. Once the three goddesses emerge from

[58] Bhattacharya, *Hindu Castes and Sects* (cit. n. 55), 279–81.
[59] Sumit Sarkar, *Writing Social History* (New Delhi: Oxford Univ. Press, 2009), 325.
[60] Ibid., 326.
[61] Ibid., 325–7.
[62] Dr. Shaha, *Dehatmik Tattva: A Discourse on Materio-Spiritualism Based on Science and Religion* (Calcutta: Kalikata Printing Works, 1891), 23–4.

Sadananda's body, Bholanath can hear but not see them. At this point *Raktasanchalan-pran* is described as emitting a sound akin to that made by a fire-engine, though it is also glossed in English parenthetically as "heart sound."[63] Later, when Chakrabarti gives Bholanath a special pair of mystical glasses, he is able to see the goddess as well. At that point the goddess clarifies that her name is in fact *RaSa*, made from the first letters of the Bengali words *rakta* (blood) and *sanchalan* (mobility/circulation). Together, however, the word *rasa* means both juice and one of the key bodily substances (*dhatu*) in Ayurveda. Her physical appearance is said to be insubstantial and shadow-like.[64]

Unlike the first goddess, *Brain-life*, who is the mother of several children, *RaSa* does not have any children. Upon Bholanath's enquiring about this, *RaSa* states that she was in fact the midwife/wet-nurse (*dhai*). She says she has been feeding *Brain-life* with milk ever since Sadananda was in his mother's womb. Before proceeding further with *RaSa's* self-description, it is important to note that the figure of the *dhai* is itself an important and, once again, most often a lower-caste figure. As several scholars have pointed out, these rural midwives gradually came to be demonized in both medical and middle-class reformist circles as the harbingers of maternal and infant mortality and the symbols of retrograde practices.[65]

Continuing with the description, *RaSa* mentions that the milk she fed *Brain-life* was obtained from the single cow she owned, called *Paripakshakti* (lit. "digestive power"), glossed in English as "digestion, a lower vital power." The milk was collected and distributed in a vessel called "Assimilation."[66] Though it is not clearly stated, Shaha's description seems to suggest that "blood" was in fact the milk of the cow named *Paripakshakti.*

Shaha never clarified whether his descriptions were intended as an elaborate allegory or something more akin to a "real" description. Yet, three things are amply clear. First, he conceptualized blood circulation as an essentially vital process regulated by vital powers that were agentic and willful rather than mere mute forces. Second, he strenuously avoided the kind of metaphors about machines, such as Ray's railway engines and Mitra's sailing ships, that mechanized the bodily processes. Finally, notwithstanding his avowed "materio-spiritualism," he worked within a framework where dualism, namely the distinction between gods and humans, rather than the erudite nondualism *Adhikari-bheda* authorized, dominated.

Most striking, though, was Shaha's utter avoidance of any invocation of a bygone Hindu Golden Age. He also strenuously pursued a narrative strategy that undermined the binaries of East and West, or national and foreign. The only boundaries his translations explicitly sought to cross were those of caste and orthodoxy.

Shaha's imagination of the body was also remarkably different from those of Mitra and Ray. The difference lay not only in the details but in its very hyperphysical form. Gopinath Kobiraj, an early twentieth-century scholar of Tantrism, had identified the

[63] Ibid., 26.

[64] Ibid., 30.

[65] Geraldine Forbes, "Managing Midwifery in India," in *Contesting Colonial Hegemony: State and Society in Africa and India*, ed. Dagmar Engels and Shula Marks (London: British Academic Press, 1994), 152–72; Anshu Malhotra, "Of Dais and Midwives: 'Middle-Class' Interventions in the Management of Women's Reproductive Health—A Study from Colonial Punjab," *Indian Journal of Gender Studies* 10 (2003): 229–59; Seán Lang, "Drop the Demon Dai: Maternal Mortality and the State in Colonial Madras, 1840–1875," *Soc. Hist. Med.* 18 (2005): 357–78; Ambalika Guha, *Colonial Modernities: Midwifery in Bengal, c. 1860–1947* (Abingdon, UK: Routledge, 2018); Sarah Pinto, "Development without Institutions: Ersatz Medicine and the Politics of Everyday Life in Rural North India," *Cult. Anthropol.* 19 (2004): 337–64.

[66] Shaha, *Dehatmik Tattva* (cit. n. 62), 32.

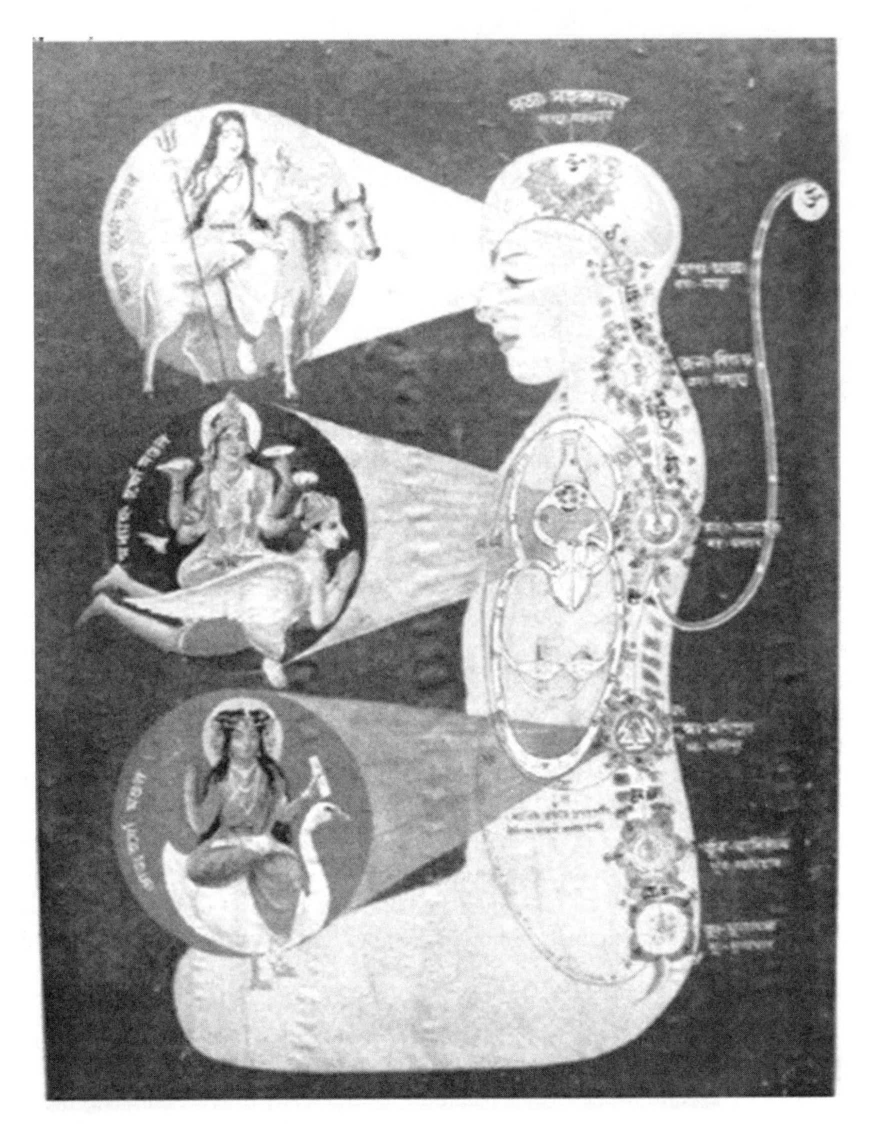

Figure 2. *Undated chart showing the hyperphysical body. Image conceptualized by Yogaprakash Brahmachari and published by Swami Yogeshwarananda Saraswati as part of the series* Sachitra Sadhan Vigyan *(Illustrated science of austere endeavor) (Calcutta: Navajeevan Press, n.d.). Author's personal collection.*

"hyperphysical body" as a modular type of body available in varying forms in Tantric and Tantra-influenced texts. These bodies were conceptualized as being more-than-physical in their constitution. They also undermined the strict demarcation and hierarchy between the human and the divine. The body itself became a sort of *mandala* within which various gods, goddesses, and other divine beings operated (see fig. 2).[67] These divine creatures were all agentic entities. While one could attribute to them patterns of behavior, they were not the blind physiological forces that were unerringly subject to the

[67] Mukharji, *Doctoring Traditions* (cit. n. 2), 267–71.

"natural laws." Such an imagination clearly and conspicuously ran counter to both Mitra's vernacularized *daktari* body and Ray's modernized Ayurvedic body.

CONCLUSION: BRAIDED SCIENCES

The foregoing discussions make it clear that even though Mitra, Ray, and Shaha produced their translations of blood circulation within a few years of each other, they are vastly different. How should we understand these differences, and what do they tell us?

What I have demonstrated is that these differences are connected to social hierarchies of caste and status. These hierarchies themselves were being reinvented during the colonial period. Anglophony, institutionalized in medical education itself, came to form a new ranking system that differentiated the upper castes further into the Anglophone upper castes, like Mitra, and the vernacular upper castes, like Ray. But an English education failed to raise the lower caste Shaha to parity with the Anglophone upper castes. What was happening, then, was that caste hierarchies were being subtly reworked, though not at all subverted, through linguistic and scientific education.

Blood circulation had to be translated across these shifting but significant hierarchies and the persistence of caste. These hierarchies provided the map of human power relationships and effectively shaped the *translatability* of the blood circulation. In other words, what could be translated and how was only conceivable with reference to these underlying relationships of power. These relationships of power were engendered in the twin axes of linguistic hierarchies established by the colonial state in its medical apparatus and the doctrine of *Adhikari-bheda* that informed the ritualized social distribution of knowledge in Bengali Hindu society. The intersections between these two axes shaped how blood circulation was approached and what strands of South Asian traditions it was braided with.

Mitra, who stood at the pinnacle of both axes of power, sought to braid together claims of an ancient Hindu Golden Age, derived incidentally through medical Orientalist writings, with a singular and hegemonic medical present. This present, for Mitra, had little space for non-*daktari* therapeutics and was mediated and authorized by instruments like the microscope. By contrast, Ray braided together claims of a Golden Age with a fractured medical present. In Ray's view therapeutic traditions were distinctly national, and the preeminent source of authority was classical textual learning, rather than microscopes. Distinct from both these braids was the one that Shaha braided.

Shaha drew not upon the hegemonic accounts of blood circulation available in the textbooks taught at medical school but rather on a more eclectic version inflected by spiritualized chemistry—which might, potentially, have resonated with efforts in certain theosophical circles in British India at the time.[68] Unlike his two peers, he referred neither to the Golden Age of Hindu Science nor indeed to the Advaita Vedantic nondualistic philosophy that underwrote most upper-caste versions of Hinduism. Instead, he drew explicitly on Hatha-yogic and Tantric imagery and repudiated *Adhikari-bheda.*

The braids were not woven merely by drawing selectively on different epistemic strands. They also operationalized different narrative genres. Mitra, who was interested in positing a singular medical modernity, wrote in straight, descriptive prose, but both

[68] Egil Asprem, *The Problem of Disenchantment: Scientific Naturalism and Esoteric Discourse, 1900–1939* (Leiden: Brill, 2014), 444–80.

Ray and Shaha in their own ways sought to fracture this singular and monolithic medical modernity by deploying a dialogic narrative mode.

Notwithstanding this commonality, it is also worth underlining the differences in the way Ray and Shaha structured their dialogues. In Ray's case, a socially superior narrator pontificates to a rustic peasant who in turn espouses "traditional" views but then eventually changes his mind. In Shaha's dialogue, the instructor Darshanraj, though himself a Brahmin, is cast out by his caste-fellows and explicitly refutes the diktats of caste. His interlocutor, Bholanath, is certainly socially his inferior, yet contrary to upper-caste stereotyping of subaltern figures, he is neither a slave to tradition nor incapable of acquiring higher forms of knowledge. In fact, we are told that Bholanath himself sought out Darshanraj because of his thirst for learning. He is unlettered but wise and follows Darshanraj not from want of livelihood but in a quest for higher learning. Darshanraj recognizes in Bholanath a worthy disciple whom he can make his intellectual heir.

As Charu Singh points out, "Dialogical genres—narratives driven by conversational exchange—have long been used as a mode of religious and philosophical exposition in Sanskrit and Hindi, to stake didactic, moral or political arguments." These dialogic genres were enthusiastically mobilized in South Asian vernacular science writings. As Singh perceptively argues, they were part of carefully thought-out literary experiments that "tactically" deployed culturally appropriate literary genres and narrative frames, usually in pursuit of an "ideological fantasy" that attempted to create space for the cohabitation of the authority of caste and experimental science.[69] Shaha's dialogue departs from, and indeed largely subverts, this ideological fantasy. But in order to do this, it still relies on the dialogic genre, which he chooses above the straight prose description of Mitra.

The braided sciences were therefore created not just by selectively pulling on distinct conceptual traditions but also through the choice of specific narrative and literary genres. The resultant iteration of blood circulation in each case was a rich tapestry that was shaped by the underlying matrices of human relations of knowledge and power. The coexistence of these various braided knowledges in turn demonstrates the plurality of colonial medical modernity.

In this regard, it is worth noting that all three of our authors were men who had been exposed to the colonial state's medical education apparatus. They were not practitioners of "subaltern therapeutics," existing in the shadows of the state apparatus, whom I have described elsewhere.[70] These men were all, in one way or another, connected quite intimately with the colonial state. That they still diverged so far in their medical epistemologies demonstrates that the medical plurality in the colonial era, and possibly later too, was not entirely a matter of inclusion or exclusion from the state apparatus. Rather, it followed from the simple fact that both knowledge and knowers existed within highly straitened and hierarchized caste societies. Consequently, on the one hand, older, precolonial knowledge traditions were unevenly accessible to different knowers, and on the other hand, new knowledges of the colonial era were always differentially translatable depending on who knew what.

[69] Charu Singh, "The Shastri and the Air-Pump: Experimental Fiction and Fictions of Experiment for Hindi Readers, 1915–1919," *Hist. Sci.* (2021), https://doi.org/10.1177/0073275320987421.

[70] Projit Bihari Mukharji, "Going Beyond Elite Medical Traditions: The Case of Chandshi," *Asian Medicine* 2 (2006): 277–91; Mukharji, "Chandshir Chikitsha: A Nomadology of Subaltern Medicine," in *Medical Marginality in South Asia: Situating Subaltern Therapeutics*, ed. David Hardiman and Projit Bihari Mukharji (Abingdon, UK: Routledge, 2013), 85–108; Mukharji, "Subaltern Surgeries: Colonial Law and the Regulation of Traditional Medicines in the British Raj and Beyond," *Osiris* 36 (2021): 89–112.

TRANSLATOR-PRACTITIONERS, EXPERTISE, AND AUTHORITY

Female Authority in Translation:
Medieval Catalan Texts on Women's Health

by Montserrat Cabré*

ABSTRACT

This article addresses the entangled histories of translation and gendered medical authority in medieval Western Europe, exploring the vernacularization of medicine from the perspective of Catalan literature. Instead of focusing on authorship or on the authenticity of the medieval attributions, it explores how women were recognized as a source of medical knowledge and how female personal names were employed as a means of conveying notions of authority on women's health. Latin medicine created its own celebrity around the acclaimed healer Trota of Salerno, although her original name was almost written out of the historical record in favor of Trotula and the label *Trotula* that flourished after her name. I study a wealth of traces showing that late medieval Catalan medicine retained a notion of female authority on women's health through the use of her name and that both Trota and *Trotula* came to authorize a significant part of medieval women's medicine in Catalan.

While checking a Catalan translation of a healthcare treatise with its Latin source, an anonymous mid-fifteenth-century editor noticed a difference between the two renderings. It was a small, two-letter deviation, but it was perceived to be important enough to merit inserting a comment in the text, right after the puzzling word:

> And as it happened, Trotula was called in—*the Latin says Trota*—as a master of the operation that was about to be performed on a girl suffering from this ailment, and she was astonished.[1]

* Facultad de Medicina, Cardenal Herrera Oria, s/n, Universidad de Cantabria, 39011 Santander, Spain; montserrat.cabre@unican.es.

It is my pleasure to express my deepest gratitude to the editors of this volume, Tara Alberts, Sietske Fransen, and Elaine Leong, for their intellectual stimulus and unwavering trust throughout the process of writing this article. They and the rest of the contributors to this volume provided many rich discussions and profitable suggestions. Comments by the *Osiris* reviewers and general editors were also very useful and helped me sharpen my arguments. Support provided by Caroline Wilson was crucial at key moments. As ever, Fernando Salmón was close by my side all along. My indebtedness to the work of Monica H. Green is expressed throughout this article; nevertheless, I wish to spell it out here as well as a further tribute to her foundational Trotula scholarship. This article is part of project PID2019-107671GB-I00 funded by the Spanish MCIN/AEI/10.13039/501100011033.

[1] "E esdevén-sa que és apellada Tròtula—*lo llatí diu Trota*— axí com a maestra de la hobra la qual devia hom tayllar per aquesta infirmitat a huna donçella, e maravellà-sse'n," Biblioteca Riccardiana, Florence, MS 2827, fols. 169v–70r (emphasis added). Unless otherwise noted, translations in this article are mine. For the edition and English translation of this passage as it appeared in the original *De curis mulierum*, see Monica H. Green, "Reconstructing the 'Oeuvre' of Trota of Salerno," in *La scuola medica salernitana: Gli autori e i testi*, ed. Danielle Jacquart and Agostino Paravicini Bagliani

Osiris, volume 37, 2022. © 2022 History of Science Society. All rights reserved. Published by The University of Chicago Press for the History of Science Society. https://doi.org/10.1086/719227.

The text continues describing how the woman healer took the girl to her own house to examine her in private. There, she determined that the patient had previously been misdiagnosed and so gave her a treatment for windiness in the womb that eventually proved successful. The editor's intertextual observations are a testimony to the keen scrutiny that they applied in validating a translation, particularly including their concern to accurately identify the name of a healer who by then had long been the most well-known female figure of medical authority in medieval Western Europe. From the late twelfth century onward, the name of Trota, the reputed Salernitan healer and medical author, was barely visible in the literate cultures of physicians, surgeons, barbers and apothecaries; instead, a derivative form of her name, Trotula—which literally means "little Trota"—had taken over the identity of the historical woman.

This seemingly incidental anecdote has the power to bring to the fore important issues regarding translation and the gendered construction of authority in Western medieval medicine. On the one hand, it shows the impact of the individual acts of translators, compilers, editors, and copyists of manuscripts in the processes of knowledge transmission and, therefore, in the textual configurations of authority over it. On the other, the vignette evinces a unique quality of female medical authority in the Middle Ages: the extent to which female characters were used as a means to convey expertise in matters of women's health. Trota of Salerno illustrates this distinctive trait like no other medieval woman. Even though her attested medical competence covered a broad variety of medical conditions concerning male and female health, she was nevertheless widely acknowledged not as a medical expert but as an expert on women's health.

In this article I explore whether the translation of healthcare texts from one linguistic tradition into another may have been a vehicle for articulating and disarticulating gendered notions of authority over knowledge. As an epistemic technology, translation involved resorting to a diverse set of techniques that ranged from the modification of the master text to changing or deleting authors' names or using titles to make distinctive imprints in the newly refashioned texts. These techniques framed the texts in certain ways, and included the use of labels relating their contents to authoritative figures gendered male or female.

Historiographical discussions of gender and medical writing in the late antique and medieval European worlds have often centered on crucial questions regarding the authorship of the texts or the authenticity of their authorial attributions.[2] Here, instead, I embark on a different route. I want to take what current scholarship understands as right and wrong attributions alike as traces that share an equivalent value for revealing notions of authority in the circulation of knowledge. Therefore I foreground the

---

(Florence: SISMEL edizioni del Galluzzo, 2007), 183–233, on 211–3; the standardized version is in Monica H. Green, ed., *The Trotula: A Medieval Compendium of Women's Medicine* (Philadelphia: Univ. of Pennsylvania Press, 2001), 126–7. I am indebted to Monica H. Green for generously sharing her full unpublished edition of the Latin original version with me.

[2] John Benton, "Trotula, Women's Problems and the Professionalization of Medicine in the Middle Ages," *Bull. Hist. Med.* 59 (1985): 30–53; Monica H. Green, "The Development of the *Trotula*," *Revue d'Histoire des Textes* 26 (1996): 119–203; Laurence Moulinier, "Hildegarde ou Pseudo-Hildegarde? Réflexions sur l'authenticité du traité *Cause et cure*," in *"Im Angesicht Gottes suche der Mensch sich selbst": Hildegard von Bingen (1098–1179)*, ed. Rainer Berndt (Berlin: Akademie Verlag, 2001), 115–46; Rebecca Flemming, "Women, Writing and Medicine in the Classical World," *Classical Quarterly* 57 (2007): 257–79; Green, *Making Women's Medicine Masculine: The Rise of Male Authority in Pre-Modern Gynaecology* (Oxford: Oxford Univ. Press, 2008).

history of translation as a witness to the transference, suppression, or reconfiguration of a notion of female authority in matters of healthcare.

My project is undertaken from the point of view of Catalan literature, a rich tradition with a significant presence in the Mediterranean region. Research into Catalan-language sources has added significantly to the corpus of known extant texts on medieval women's medicine and can offer fresh perspectives on an issue that has been the subject of comprehensive scholarship. Working with an impressive amount of source material, Monica H. Green has shown that the late medieval vernacularization of medicine in Western Europe, while dramatically increasing the production of medical texts and widening their audiences, did not bring an improvement in women's access to literate medicine. Nor did it help to empower them as medical practitioners, particularly in regard to gynecological knowledge. Rather, it testified to the increase of male control over a medical system that was starting to be organized through formal educational institutions and to be regulated by law.[3] My contribution does not deny these general trends, which narrowed women's opportunities as practitioners and eventually excluded them from regular medical practice. However, it emphasizes another current that runs in parallel with them. While centering my attention on the ways gendered authority over knowledge was envisioned in the new body of medical Catalan texts, I have been able to trace persistent marks of female authority over knowledge on women's health that reconfigured as well as retained the earlier traditions.

In this article I undertake a journey through the entangled histories of translation and gendered medical authority. I begin by tracing the different ways medical texts granted authority to women in the medieval West. Rather than pursuing questions pertaining to the authenticity of the attributions or the authorship of texts, which were quite extraneous to the medieval endeavor, I focus on the recognitions of women as original sources of healthcare knowledge and how they cut across processes of textual transmission. Second, I address how, while embracing female figures of medical authority from different and earlier linguistic traditions, Latin medicine created its own celebrity around the acclaimed healer Trota of Salerno. I then follow her presence over the late medieval Catalan corpus of medical texts, where I recover translated imprints of her authority. On occasion, these marks transcribed the original form of her name, but usually, they took its most famous version, that of Trotula. Retaining meaning from the Latin traditions, in Catalan this latter designation even came to identify a whole new genre of texts on women's health.

GENDER AND THE ACKNOWLEDGMENT OF MEDICAL AUTHORITY

Medieval medical treatises in Western Europe routinely recognized authority in matters of healthcare by associating the knowledge that they were communicating with a particular person who held it, as in the opening vignette. The notion of authorship was subsequent and subordinate to the idea of authority: a person bore authority not because s/he was the author of a text but because s/he was recognized as the origin of knowledge or expertise deemed to deserve textualization so that it could be further preserved.[4] Women

[3] Green, *Making Women's Medicine Masculine* (cit. n. 2).

[4] The classic studies are Alastair Minnis, *Medieval Theory of Authorship: Scholastic Literary Attitudes in the Later Middle Ages*, 2nd ed. (London: Aldershot, 1984; Philadelphia: Univ. of Pennsylvania Press, 1988), and Pamela Long, *Openness, Secrecy, Authorship: Technical Arts and the Culture of Knowledge from Antiquity to the Renaissance* (Baltimore: Johns Hopkins Univ. Press, 2001).

as well as men were accorded authority in matters of healthcare; nevertheless, in textual practices medical authority was fundamentally gendered in a number of ways. Both before and after the scholastic method was used as a form of elaboration and validation of medical knowledge, men played the leading role as medical authorities. Certain key names, such as Hippocrates and Galen, occur again and again across various healthcare treatises, where they appear as putative experts in all different spheres of healthcare, from fevers to gynecology. Conversely, women appear much less often in the texts. On the one hand, individual women identified by their personal names tended to be associated primarily with authoritative knowledge in matters of women's health, even if their expertise was broader. On the other, women were sometimes designated simply by their gender, or in conjunction with a peculiar feature, but not by their distinctive names. Often these anonymous women were reported to hold expertise on an ample range of health issues, from treatments for pain to ocular problems, and their authoritative presence in the sources, however meager, was not reduced to the specific recognition of their knowledge on female health.

Before printing favored the standardization of language and the stabilization of texts and of their authorial attributions, processes of transmission made all texts very labile and susceptible to alterations; unlike our Catalan editor, scribes and translators often left their interventions unrecorded. Translation exposed texts to modification, and ascriptions of authorship—the association of texts with an author name—proved to be particularly vulnerable to change. Dealing with the translation process from Greek to Latin of the Hippocratic gynecological tradition in the early Middle Ages, Laurence Totelin has traced the uses and transformation of recipes, showing a stable pharmacological practice through the centuries in combination with an inconsistent authorial attribution, which, in the corpus she analyzes, was always connected to male personal names.[5] Yet as we will see in the Trota/Trotula case, the Latin Middle Ages also embraced the ascription to female authorities of treatises on women's health. In association with this medical theme, female names were modified, disappeared, or migrated from text to text, only to resurface in another set of texts.

Another example of this process is seen in the case of a late antique Byzantine practical compendium ascribed to Metrodora. This was translated from Greek into Latin in the late eleventh century, when a first corpus of texts on women's health in Latin was created in southern Italy—which has recently been identified as the Cassinese corpus.[6] Two different versions of this text, known as the *De passionibus mulierum*, circulated in Western Europe, and excerpts from both were absorbed into other compendia on women's health. However, the authorial ascription to Metrodora is not retained in any of the extant copies of the Latin renderings of the original Greek text. Nevertheless, in the same corpus that supressed Metrodora's authority, the ancient figure of Cleopatra appears in connection with sections of the texts that dealt with women's health. Thus, during that process of translation, a woman's name was lost, but the notion of female authority in issues of women's medicine was retained through its embodiment

[5] Laurence Totelin, "Old Recipes, New Practice? The Latin Adaptations of the Hippocratic Gynaecological Treatises," *Soc. Hist. Med.* 24 (2011): 74–91.

[6] Monica H. Green, "Recovering 'Ancient' Gynaecology: The Humanist Rediscovery of the Eleventh-Century Gynaecological Corpus," in *Transmission of Knowledge in the Late Middle Ages and the Renaissance*, ed. Outi Merisalo, Miika Kuha, and Susanna Niiranen (Turnhout: Brepols, 2019), 45–54.

in another personal name. The later transmission of the treatises that circulated under Cleopatra's authorial ascription and their use of her name remains to be studied; however, her royal status and Galen's mention of her could have favored its retention.[7] Moreover, there are instances suggesting that some sort of transference of Cleopatra's name also occurred in other languages, attesting to the widespread use of her as a figure of medical authority.[8]

Along with female names coming from the Greek traditions, unnamed women experts made their imprint on Latin medical literature through the translation of Arabic handbooks. In these texts, women were clearly identified as the source of medical knowledge, normally with specific treatments being attributed to them.[9] For instance, in the section on cures for migraines, Gerard of Cremona's translation of Ibn Sīnā's *Canon* includes the ascription to an unnamed woman of a particularly valued cure: "[C]ertain physicians report a remedy for temporary migraines that they learned from a woman and whose effects are well tested."[10] Another example from the same text similarly attributes the creation of some highly esteemed eyedrops to a woman: "Collyrium that is called Fakis. It was composed by a woman for the most intense pain of a queen."[11] These moments of recognition and acknowledgment of anonymous individual women as the originators of knowledge surface here and there in healthcare texts, although since the women are deprived of personal names, they become almost unnoticeable in medical literature. Certainly, they were more invisible than the handful of female personal names acknowledged as the source of certain remedies in Galenic pharmacology.[12] Their continued presence, however, should be seen as a call to historians for further consideration.

Nevertheless, through the translation of texts these acknowledgments did travel from one culture to another and, also, from authoritative text to authoritative text within the same culture. For instance, the treatment for migraine by Ibn Sīnā mentioned above

[7] Flemming, "Women, Writing and Medicine" (cit. n. 2), 278–80.

[8] For vestiges of the recognition of Cleopatra in the Arabic and Castilian traditions, see Montserrat Cabré, "Beautiful Bodies," in *A Cultural History of the Human Body in the Medieval Age*, ed. Linda Kalof (Oxford: Berg, 2010; repr., London: Bloomsbury, 2014), 121–39, 244–8, on 134–5. Mentions of her in the Old French and Anglo-Norman traditions suggest that Cleopatra's authority may have been acknowledged more broadly in Western Europe. Thirteenth-century Old French and Anglo-Norman versions of the *Liber de sinthomatibus mulierum*, one of the three components of the *Trotula* compendium, incorporated Cleopatra as one of the alleged authorities. See Green, *Making Women's Medicine Masculine* (cit. n. 2), 169–70n11.

[9] For two named Middle Eastern women identified as experts on certain medical treatments in religious texts and in physicians' biographies, see Ahmed Ragab, "Epistemic Authority of Women in the Medieval Middle East," *Journal of Women of the Middle East and the Islamic World* 8 (2010): 181–216, on 209–12.

[10] "Quidam preterea medici narraverunt medicaminem emigranee temporalis expertum iuvativum acceptum a muliere. Et illud quidem est ut decoquantur radices cucumeris asinini et absinthium in oleo et aqua donec dissolvantur et embrocetur pars dolorosa ex aqua et oleo calidis et fiat emplastrum ex fece et mulier quidem illa quotiens administrabat hoc sanabat emigranea sive esset cum febre sive sine febre. Et ex emplastris quidem non est sicut emplastrum sinapis"; Avicenna, *Liber canonis tocius medicine* (Venice, 1527; repr., Brussels: Collectaneis Medicinae Historia, 1971), lib. 3, fen 1, tract. 2, cap. 38, fol. 143va. I am indebted to Fernando Salmón for this reference.

[11] "Collirium quod dicitur Fakis. Et composuit ipsam mulier regine faciens ad dolores vehementes cuius hec est permixtio. Climie aureos XV; ceruse ablute aureos XL; amili et dragaganti et acatie et opii, omnium ana aureos II; gummi aureos XII. Conficiatur cum aqua pluviali. Si autem advenit hora in quo faci ex eo collirium proisce super albumen ovi recentis et utem eo"; Avicenna, *Liber canonis* (cit. n. 10), lib. 5, suma 1, tract. 2, fol. 415rb.

[12] Flemming, "Women, Writing and Medicine" (cit. n. 2), 263–7.

appears in Simon of Genoa's *Clavis sanationis*, a late thirteenth-century Latin-Greek-Arabic medical dictionary.[13] Simon's *Clavis* is particularly important as an illustration of how authors and compilers of medical texts were at times dependent on the knowledge of particular women. When Simon introduces his work he recognizes by name all the authoritative male textual sources he relied on to compose his dictionary: Hippocrates, Serapion, Pliny, Muscio, Oribasius, Rasis, Isidore of Seville, and many more. But he then goes on to report that an unnamed "old woman from Crete" went across the mountains with him to identify all the different plants by their Greek names so that he could match with the natural world the information provided by Dioscorides's *De materia medica*, a revered pharmacological work.[14]

On most occasions, these acknowledgments of the value of women's knowledge offer little or no clue to any personal trait of the women involved, as in the examples above. But texts often do provide proof of efficacy or an indication of why the woman's knowledge was particularly valuable. It is indeed very rare that these women are singled out by their personal or family names; they are usually described through other features: the place they were born, their age, where they lived, and often their religious identities. For instance, one anonymous twelfth-century Latin author claims that "[he] saw a certain Saracen woman from Sicily curing infinite numbers of people [of mouth odor] with this medicine alone," and the recipe follows.[15]

Sometimes, female local knowledge was called upon. The Salernitan women—the *mulieres salernitanae*—for example, were widely credited in learned medical literature with having particular medical expertise.[16] Similarly, the "Ladies of Apulia," the women of the region in southern Italy, were credited with expertise in the development of cosmetic treatments, as acknowledged by the anonymous author of a mid-thirteenth-century Anglo-Norman text.[17] Many texts portray women as being the origin of valuable knowledge through simple references to their customary practices: "what Saracen and Jewish women do." Frequently, these beauty treatments are presented as "treatments of," but compilers in their writings also offer them as "treatments for," often making it difficult

[13] "Sicidia et sicidis et sici et sicui aliquando reperitur pro cucumere asinino. Paulus ca. de sterilitate sicio ager idest elacterio et cetera. Item idem ca. de emigranea in emplastro habito a muliere pro ea egritudine radicem agrii sicci .i. cucumeris asinini," *Simonis Ianuensis opusculum cui nomen clavis sanationis*; see "Sicidia," Simon Online, accessed July 20, 2020, https://simonofgenoa.org. "Agrusichi. Agrusichi, radix ponitur in quodam emplastro ad emigraneam a Paulo quod dicit se habuisse a quadam vetula: quod etiam scribit A. in eodem casu et est cucumer agrestis," *Simonis Ianuensis opusculum cui nomen clavis sanationis*; see "Agrusichi," Simon Online, accessed July 20, 2020, https://simonofgenoa.org.

[14] "Nec his solum contentus sed ad diversas mundi partes per sedulos viros indagare ab advenis sciscitari non piguit usque adeo quod per montes arduos nemorosas convalles campos ripasque sepe lustrando aliquando comitem me feci cuisdam anicule cretensis ad modum sciole non modo in dignoscendis herbis et nominibus grecis exponendis, verum etiam in ipsis herbarum virtutibus secundum Dya. sententiam explicandis. Omnia tentavi quam tum ingenii mei paupertas sineret ut opus efficeretur excultum," *Simonis Ianuensis opusculum cui nomen clavis sanationis*; see Preface 4, Simon Online, accessed July 20, 2020, https://simonofgenoa.org.

[15] Green, *The Trotula* (cit. n. 1), 46.

[16] Green, *Making Women's Medicine Masculine* (cit. n. 2), 64–9; Green, *The Trotula* (cit. n. 1), 48–51; Green, "Reconstructing the 'Oeuvre'" (cit. n. 1).

[17] Pierre Ruelle, *L'Ornement des dames (Ornatus Mulierum): Texte anglo-normand du XIIIème siècle* (Brussels: Presses Universitaires de Bruxelles, 1967), 58–60. For other attributions to women in this text, see Montserrat Cabré, "Autoras sin nombre, autoridad femenina (siglo XIII)," in *Las sabias mujeres II (siglos III–XVI): Homenaje a Lola Luna*, ed. María del Mar Graña Cid (Madrid: Al Mudayna, 1995), 59–72.

to interpret whose original knowledge it was. These delicate, in-passing allusions make such acknowledgments almost invisible in medical texts. Nonetheless, these references circulated widely and also crossed linguistic boundaries, as we have seen. Carmen Caballero-Navas has shown that Hebrew texts contain references to these types of unnamed women's medical practices.[18] If we gather them together and consider them significant rather than incidental, they point to a collective attribution of authority to women in certain fields of expertise, most notably cosmetics.[19]

Whether originally written in Arabic, Latin, Hebrew, or in any of the Western mother tongues, many healthcare handbooks in use in medieval Europe contained sections devoted to cosmetics and beauty care, and independent treatises devoted to the issue were also widely known. Christian Latin sources unambiguously distinguished Muslim women's expertise in the art of beauty treatments. However, their authority as a female group was eventually lost in favor of other women.

The Salernitan text *De ornatu mulierum* (Women's cosmetics), a twelfth-century treatise written in Latin by an anonymous male author, presents several of the recipes as being the practices of Muslim women. The earlier versions associate the treatments with Saracen women, but later renderings relate them to the women of Salerno, the *mulieres salernitanae* whose collective expertise I referred to above.[20] The change was neither an exception nor a simple adaptation of a recipe for new audiences, but rather a moment in a complex process of erasure of Muslim women's authority in Western healthcare literature. Obliteration was, however, gradual, subtle, and never complete during the Middle Ages. Even two centuries later, a Castilian household handbook attributed beauty treatments to both Moorish women and to the presumably Christian "ladies" (*señoras*) who were interested in knowing about products to beautify the face. The expertise of Moorish women was called on in relation to recipes containing lead and mercury, dangerous ingredients whose use had worried physicians and surgeons for centuries, particularly for their potentially noxious effects on gums and teeth. The compiler of the text warns about them, advises that they be used carefully, and suggests safe practices.[21]

These attributions of authority to unnamed women were transferred from text to text in healthcare literature and traveled over linguistic borders. Nevertheless, anonymous women were more vulnerable to invisibility and oblivion than the figures of authority who, identified by their female names, feature more prominently in Western medical literature.

A NEW FIGURE OF FEMALE AUTHORITY IN LATIN MEDICINE

Personal names were conspicuous and easily discernible even if they were not the only way women's expertise in healthcare was recognized, as we have just seen. When the vernaculars made their debut as a means of conveying medical literature, Trota—more often the diminutive form, Trotula—had already emerged as a female figure of authority on women's healthcare in Latin medical writings. Learned readers on healthcare issues were familiar with her figure even if sometimes the different spellings of the name

[18] Carmen Caballero-Navas, "Virtuous and Wise: Apprehending Female Medical Practice from Hebrew Texts on Women's Healthcare," *Soc. Hist. Med.* 32 (2019): 691–711.
[19] Cabré, "Beautiful Bodies" (cit. n. 8), 134–6.
[20] Green, *The Trotula* (cit. n. 1), 46–8, 169, 246n6.
[21] *Vergel de señores*, Biblioteca Nacional de España, Madrid, MS 8565, lib. 3, cap. 9, fols. 134r, 139v–40v.

could have confused them. Such was the case of the careful Catalan editor who made a record of the discrepancy detected between two different renderings, as the opening quote indicates. The reviewer of that translation was more attentive than certain scribes who used the two designations without noticing the divergence, apparently taking the variations of the name as interchangeable.[22] It was not unusual in the Middle Ages to use diminutive forms of an author's name to identify their works; the surgery of Roger Frugardi (fl. ca. 1170), for instance, was often called *Rogerina*.[23]

Trotula was the most widespread form of the name of Trota, the famous twelfth-century Salernitan healer, and the author of the *Practica secundum Trotam* (Practical medicine according to Trota), a text on therapeutics that covered a wide range of afflictions, including, but not limited to, women's particular illnesses.[24] Monica H. Green has painstainkingly reconstructed the empirical milieu where she worked as a vibrant medical culture that operated fundamentally in the realm of the oral. Indeed, what is known about her practice may have been written by her pupils and not directly by her, as some of the texts that acknowledge her treatments use the third person and highlight her role as a teacher, as a *magistra*.[25]

Like Trota, other women healers practiced in the city of Salerno, also in dialogue and cooperation with male practitioners; however, hers was the only female personal name that was passed on in medical literature.[26] Trota's reputation spread widely and quickly, and particularly strong testimonies to her acclaim came from Normandy and Anglo-Norman England.[27] However, her medical authority was not based on her broad general medical knowledge, attested to in the *Practica* as well as in another Salernitan medical text, *De egritudinum curatione* (On the treatment of illnesses), which includes some of her remedies for ungendered conditions.[28] Rather, it was constructed

[22] The double mentions are found in manuscripts identified and described by Monica H. Green, "A Handlist of the Latin and Vernacular Manuscripts of the So-Called Trotula Texts. Part I: The Latin Manuscripts," *Scriptorium* 50 (1996): 137–75, numbers 25, 41, 42, 44, 64, 67, 70, 73, 79, 80, 120, and 121 of her list.

[23] Piero Cantalupo, "L'inedito opusculo di pratica terapeutica della medichessa salernitana Trota. La *Practica secundum Trotam*: Testo, traduzione, appendici e glossario," *Bolletino storico di Salerno e Principato Citra* 13 (1995): 1–103, on 9–10.

[24] The work was first identified by Benton, "Trotula" (cit. n. 2). In addition to the copy discovered by Benton in Madrid, Biblioteca de la Universidad Complutense, MS 119, fols. 140r–144r, Monica Green has located another fragmentary copy in Oxford, Bodleian Library, MS Rawlinson C 506, fols. 146v–47v, and the mention of a *Practica domine Trote ad provocanda menstrua* in a fourteenth-century book catalogue at the Christ Church Priory of Canterbury; see Green, "Reconstructing the 'Oeuvre'" (cit. n. 1), on 187–8. The Madrid text, an early thirteenth-century manuscript that contains both forms of the name, was transcribed by Cantalupo, "L'inedito opusculo" (cit. n. 23).

[25] Green, "Development of the *Trotula*" (cit. n. 2), 136–7; for the known Trota's testimonies, see Green, "Reconstructing the 'Oeuvre'" (cit. n. 1), 211–26.

[26] Green, *Making Women's Medicine Masculine* (cit. n. 2), 58–69.

[27] Monica H. Green, "Salerno on the Thames: The Genesis of Anglo-Norman Medical Literature," in *Language and Culture in Medieval Britain*, ed. Jocelyn Wogan-Browne et al. (Woodbridge, UK: York Medieval, 2009), 220–31; Green, "Rethinking the Manuscript Basis of Salvatore De Renzi's 'Collectio Salernitana': The Corpus of Medical Writings in the 'Long' Twelfth Century," in *La 'Collectio Salernitana' di Salvatore De Renzi*, ed. Danielle Jacquart and Agostino Paravicini Bagliani (Firenze: SISMEL, edizioni del Galluzzo, 2008), 15–60, on 55.

[28] For the concordances between the *Practica secundum Trotam*, the *De curis mulierum*, and the *De egritudinum curatione*, see Green, "Reconstructing the 'Oeuvre'" (cit. n. 1), 219–21. On its correlations with other Salernitan texts, see Alberto Alonso Guardo, "La 'Practica secundum Trotam' y el poema médico de la 'Collectio Salernitana' (IV, 1–176)," in *En Doiro antr'o Porto e Gaia: Estudos de Literatura Medieval Ibérica*, ed. José Carlos Ribeiro Miranda (Porto: Estratégias Criativas, 2017), 151–65.

solely on the fame achieved by her cures for women: female personal names bore authority only on women's issues even if the historical healers carrying those names were acclaimed general practitioners.

The attribution to Trota of original treatments for women and her fame as an expert healer was passed on through the extensive transmission of the anonymous text *De curis mulierum* (On treatments for women). As the example of the Catalan editor in the opening account of this article attests, the text contained the explicit association of her name with a successful cure for windiness in the womb. Other sections of the *De curis* also derive from Trota's teachings, although her name remained silenced. This unique passage containing her name was solely responsible for expanding her reputation greatly.

Although *De curis mulierum*, with its personal and specific mention of Trota, had a limited independent circulation, it soon started to be linked with two other Salernitan tracts on women's health, both by anonymous male authors: the *De sinthomatibus mulierum* (Book on the conditions of women) and the *De ornatu mulierum* (On women's cosmetics). The process of unifying the three texts took place early on, and already by the end of the twelfth century they were circulating together as a compilation. By the mid-thirteenth century, the ensemble had taken on a stable format: this standard version enjoyed an extensive dissemination in manuscript form and served as the basis of the text that finally reached the printing press with a completely new prologue.[29]

All the while the process of medical translation into the European tongues was starting to take place, the personal name of Trota was being propelled from the short anecdotal passage in the *De curis* to authorizing a comprehensive grouping of texts. The majority of the more than 140 medieval Latin copies already identified of the compendium use the name Trotula instead of Trota to name the famous healer. But, most importantly, this was not just a change in naming a reputed woman practitioner and author but rather a whole reconfiguration of her authority.

Under the name of Trotula, a female figure emerges who accorded authority to the whole compendium on women's health while reciprocally widening her fame and recognition as an expert. Nevertheless, in this process, Trota's name almost disappeared in favor of Trotula. Some instances of the compendium contain the two names; in fewer cases, only the original Trota is retained, testifying without doubt to the rendering of any of the oldest versions of the text.[30] Although unaware of the complex underlying story, the Catalan editor we met at the beginning was using one of these ancient versions to check the integrity of the translation that was being revised.

The new figure of medical authority—Trotula—was being created upon the knowledge of a historically recognized woman healer—Trota—who had direct access to the female body. This access was unique to women healers but out of reach for Salernitan male physicians.[31] The twelfth-century sources reveal a practitioner who treated a wide range of health problems, from migraines to alopecia, from digestive upsets to eye diseases. Despite this, Trotula—and Trota when her original name was retained—was

[29] Green, "Development of the *Trotula*" (cit. n. 2); Green, *The Trotula* (cit. n. 1), 60; I follow the English translation of the titles from this edition.

[30] These are numbers 11, 34, 36, 62, 72, 74, 76, and 115 in Green, "Handlist Part I: Latin" (cit. n. 22).

[31] Green, "Reconstructing the 'Oeuvre'" (cit. n. 1), 191–2; Green, *Making Women's Medicine Masculine* (cit. n. 2).

envisioned solely as an expert in the care of the health and beauty of women. Prominently visible in the compendium that came to be distinguished as *Trotula*, her figure was present in learned medical libraries around Western Europe in a number of ways. Nevertheless, as is often the case with all sorts of medieval texts, it is not always easy to distinguish whether personal names were used as authorial figures, as titles of texts, or as both at the same time. This historiographical difficulty, in any case, highlights the extent to which individual names were used in connection with texts as a way to identify them and to give them validity. Labeling texts was a technique widely used by scribes, editors, and also translators, as the contributions by Elaine Leong, Alisha Rankin, Dror Weil, and Shireen Hamza in this volume also show.[32] Transformed into Trotula, the name of the healer Trota ended up not only being associated with the authorial figure of the famous compendium of women's medicine, but was ultimately to give name to a medical genre that offered treatments for the care of the female body, as we will see later.

Moreover, in Latin medicine her authority also traveled far beyond the aforementioned compendium. In medical and surgical works of the thirteenth century, Trotula is referred to as an authority on women's health. The physician Gerard of Berry (fl. 1220–30) and Abbé Poutrel (c. 1300), the author of a surgical text, mention her knowledge of cosmetics. Petrus Hispanus (d. 1227), while commenting on the *Isagoge* by Johanitius, also makes reference to her when he explains the physiological phenomenon of sexual pleasure. In the mid-thirteenth-century *Thesaurus pauperum*, a popular compilation of recipes often attributed to Petrus Hispanus, Trotula appears as an authority in some gynecological prescriptions taken from the *De sinthomatibus mulierum*, which, as we saw before, makes up an integral part of the *Trotula* compendium.[33]

Well established in the medical tradition, by the thirteenth century Trotula's figure leaps into literary works as well as into natural philosophy. Her name, already transcribed in different variants (Trotule, Torcule, etc.), appeared in a French translation with a commentary of the *Ars amatoria* of Ovid, in the *Dit de l'herberie* of the satirical poet Rutebeuf (d. c. 1285), and, most notably, in the prologue to *The Wife of Bath's Tale* by Geoffrey Chaucer (d. 1400).[34] Green has pointed out that it was in the textual traditions most distant from medicine where a negative image of her was created, as well as the texts linked to her figure. Those negative representations, originating to a large extent in the contextual association of the *Trotula* with the misogynist heritage of the pseudo-Albertian *Secreta mulierum*, developed in some of the most popular vernacular genres but eventually ended up having an impact on the learned Latin traditions.[35]

TROTA IN TRANSLATION

During the late middle ages, the vernaculatization of medicine brought about the production, adaptation, and translation of texts in circulation from the Latin into

[32] Elaine Leong, "Translating, Printing, and Reading *The Art of Distillation*"; Alisha Rankin, "New World Drugs and the Archive of Practice"; Dror Weil, "Unveiling Nature"; and Shireen Hamza, "Vernacular Languages"; all in the present volume (*Osiris* 37).

[33] Green, *Making Women's Medicine Masculine* (cit. n. 2), 84–5, 221.

[34] Benton, "Trotula" (cit. n. 2), 35; Monica H. Green, "'Traittié tout de mençonges': The *Secrés des dames*, 'Trotula,' and Attitudes towards Women's Medicine in Fourteenth- and Early Fifteenth-Century France," in *Christine de Pizan and the Categories of Difference*, ed. Marilyn Desmond (Minneapolis: Univ. of Minnesota Press, 1998), 146–78, on 160, 170.

[35] Green, *Making Women's Medicine Masculine* (cit. n. 2), 204–28; Green, "'Traittié tout de mençonges' (cit. n. 34).

languages more broadly used. In the Crown of Aragon, the growing audiences for medical books in Catalan appropriated and adapted earlier traditions of texts on women's health, and traces of the authority of Trota and the derivative forms of her name appear in the new corpus. Nevertheless, although the misogynist trends common in Europe were also current in Catalan literary texts, none of the negative views of Trotula that were popular in other vernaculars have as yet been found.[36] On the contrary, I will show testimonies of her acknowledgment in the Catalan setting, ranging from 1338 to 1479. During this period, a significant number of women practiced medicine, surgery, and midwifery in the Aragonese kingdoms: at least five were hired to work for the court between 1332 and 1421, and no fewer than twelve received royal licenses to protect their legal practice between 1374 and 1404.[37] This unveils an appreciation of their abilities as healers that is congruous with the circulation of textual forms of recognition of female authority in medicine.

In the Catalan context, labels, titles, and ascriptions of texts deriving from the name of Trota were also quickly associated exclusively with women's medicine, understood then as encompassing the domains of gynecology, obstetrics, and cosmetics. However, as highlighted above, Trota's practice—like that of many medieval women healers—was not reduced to women's concerns but covered broader areas of healthcare. There are no known translations in any language of the *Practica secundum Trotam*, or of the *De egritudinum curatione*, the twelfth-century Salernitan texts that fully or fragmentarily passed along her treatments. But there are examples that indicate the existence of Catalan translations of books that had been directly associated with her figure, as well as of their circulation in medical and lay circles. The Catalan linguistic tradition was not alone in displaying an active reception of the Latin *Trotula* texts, as there are over sixty examples of manuscripts of different versions in French, Irish, Dutch, Italian, Hebrew, English, and German.[38] Recently, an important Italian translation of the compendium has been edited, and the Hebrew and Dutch traditions have been the subject of studies that describe the textual configurations in those languages of the Latin renditions.[39] The focus of these analyses has been the evaluation and philological analysis of the contents and nature of the texts themselves, but they do not give significance or attention to whether the translations retain or obliterate female figures that had authorized them in earlier versions.

[36] On the misogynistic traditions in relation to medicine in the Crown of Aragon, see Antònia Carré, "La medicina com a rerefons cultural a l'Espill de Jaume Roig," in *Jaume Roig I Cristòfor Despuig: Dos assaigs sobre cultura i literatura dels segles XV i XVI*, ed. Antònia Carré and Josep Solervicens (Barcelona: Univ. de Barcelona; Vic: Eumo, 1996), 9–71; Michael Solomon, *The Literature of Misogyny in Medieval Spain: The Arcipreste de Talavera and the Spill* (Cambridge: Cambridge Univ. Press, 1998); and Jean Dangler, *Mediating Fictions. Literature, Women Healers, and the Go-Between in Medieval and Early Modern Iberia* (Lewisburg: Bucknell Univ. Press, 2001).

[37] Pere Guaita i Jiménez, *Dona i medicina a la Corona d'Aragó* (Manresa: Arxiu Històric de les Ciències de la Salut, 2010).

[38] Monica H. Green, "A Handlist of the Latin and Vernacular Manuscripts of the So-Called Trotula Texts. Part II: The Vernacular Texts and Latin Re-Writings," *Scriptorium* 51 (1997): 80–104; Green, *Making Women's Medicine Masculine* (cit. n. 2), 163–98.

[39] Carmen Caballero-Navas, "Un capítulo sobre mujeres: Transmisión y recepción de nociones sobre salud femenina en la producción textual hebrea durante la edad media," *MEAH: Sección hebreo* 52 (2003): 135–62; Orlanda S. H. Lie, "Women's Medicine in Middle Dutch," in *Science Translated: Latin and Vernacular Translations of Scientific Treatises in Medieval Europe*, ed. Michèle Goyens, Pieter de Leemans, and An Smets (Leuven: Leuven Univ. Press, 2008), 449–66; Rossella Mosti, "Una versione tardomedievale della *Trotula*: Il MS 532 della Wellcome Library di Londra. Edizione critica, analisi linguistica e glossario," in *Capitoli di Storia Linguistica della Medicina*, ed. Rosa Piro and Raffaella Scarpa (Milan: Mimesis, 2019), 105–64.

Unlike these approaches, my concern is to explore whether, in the course of translation, the texts lost, gained, or maintained marks of acknowledgment of female authority—a process which, as we have seen before, involved the association of healthcare texts with female personal names.

Both Trota and Trotula made their appearance in the new body of Catalan medical literature. A factitious manuscript, today held at the Riccardiana Library in Florence, contains a self-identified translation of *Trotula*, an important text for the entangled histories of translation and of female medical authority, showing a transference of the medieval figure from Latin medical culture into the vernacular.[40] It is copied in a disorganized miscellaneous codex that contains Italian, Latin, and Catalan medical treatises and collections of recipes, many in a fragmentary state or without headings. Nevertheless, a codicological inspection of the remains allows for a reconstruction of what the different medieval units may have looked like. The *Trotula* is included in a group of three texts written in the main by one single hand. However, an old foliation indicates that they had originally been part of a larger volume, probably containing other medical texts in Catalan, likely also translations from the Latin. The *Trotula* is the first of the extant texts of the group; it lacks the beginning, but on two of its pages the name of Trotula appears as a heading as well as at the end of the text, where five recipes in Catalan and Latin follow. The second bears the title *Con regnen les quatre humors en lo cor[s] de l'hom* (How the four humors reign in the human body), and it has no personal name or label associated with it. It consists of a single chapter on the administration of laxative medicines that undoubtedly comes from a work of a theoretical nature. On the same page, the *On pulses* by Philaretus begins, containing five further chapters that mirror about half of the Latin text very closely, finishing abruptly. The three works contain diverse annotations in the margins, including some recipes in Latin and in Catalan in the lower margins of the folios, written in the same hand that copied the body of the text.

The grouping of these three texts, in conjunction with other features of the manuscript, indicate that the Catalan *Trotula* belonged to a volume envisaged as a text for professional use. *On pulses* was part of the seven texts that formed the core group of the *Articella*, or *ars medicinae*, which constituted the very basis of university medical education.[41] The theoretical bent of the fragment on the administering of laxatives and the fact that it included one of the *Trotula* texts could indicate that the volume was made out of some form of the *Articella*. Although the *Trotula* was never part of the university curriculum, its presence in the context of academic medical training is not rare, given that versions of the *Trotula* have been identified in manuscripts containing the *Articella* collection, and its use by physicians and surgeons is widely attested.[42]

[40] Biblioteca Riccardiana, MS 2827. For a full description of this manuscript, see Montserrat Cabré, "Trota, Tròtula i *Tròtula*: Autoria i autoritat femenina en la medicina medieval en català," in *Els manuscrits, el saber i les lletres a la Corona d'Aragó, 1250–1500*, ed. Lola Badia, Lluís Cifuentes, Sadurní Martí, and Josep Pujol (Barcelona: Publicacions de l'Abadia de Montserrat, 2016), 77–102, on 86–93.

[41] Cornelius O'Boyle, *Thirteenth- and Fourteenth-Century Copies of the 'Ars Medicine': A Checklist and Contents Descriptions of the Manuscripts* (Cambridge, UK: Cambridge Wellcome Unit for the History of Medicine, 1998), 13–4.

[42] On versions of the *Trotula* copied in manuscripts containing the *Articella*, see ibid., 38, 40, 42, 74. On the uses of the compendium by medical practitioners, see Green, *Making Women's Medicine Masculine* (cit. n. 2), 74–84.

This initial core group of texts was extended to include three short alchemical and medical tracts as well as recipes in Latin and in Catalan on a wide range of issues, naming some male individuals as their sources.[43] Two of these recipes are particularly interesting as they include the reception date of 1453 and provide significant evidence of the exchange and circulation of knowledge involved in the production of these books: one was sent from Barcelona to Majorca by Joan Cabaspre, the son of a merchant; therefore, at some point of its elaboration the book may have been compiled in the kingdom of Majorca.

Trotula, therefore, featured in a compilation that was adapting a group of basic materials to the vernacular language for the use of non-university medical practitioners.[44] Lluís Cifuentes has shown the popularity of those vernacular miscellanies, especially among surgeons and barbers, in both the urban and rural settings.[45] It was elaborated by an unnamed male or female practitioner who undoubtedly had a certain level of training, since the text is translated from Latin. Although there is limited evidence on the possession of and engagement with the production of medical books by medieval women, in particular by women healers, nevertheless, the possibility of linking the book with the work of a female practitioner should not be completely discarded.[46] In the inventory of the properties of the *medicissa* Margarida de Tornerons, carried out in Vic in 1401, eleven books of medicine, astronomy, and surgery in Latin and Catalan are listed, one identified as a *Thesaurus pauperum*, where Margarida could read the attributions to Trotula of diverse gynecological recipes.[47]

The self-identified *Trotula* opens the codicological unit made up of the three texts and exhibits her name as the running title in red, and again at the end of the treatise itself. The text is not a translation of the whole compendium but of the *De curis mulierum*, one of the three texts of the ensemble.[48] As its beginning is missing, it is not possible to know if it was preceded by either of the other two, in particular

[43] The three texts are the alchemical epistle *Aque magistri Raymundi*, the *Ars operativa medica* ascribed to Raimundus Lullius, and the *Tractatus de bonitate memorie secundum magistrum Arnaldum de Vilanova*; for a full description see Cabré, "Trota, Tròtula i *Tròtula*" (cit. n. 40). The whole book has been called *Florència II*; for an evaluation in the context of its genre, see Lluís Cifuentes, "El receptari mèdic baixmedieval i renaixentista: Un gènere vernacle," in Badia et al., *Els manuscrits* (cit. n. 40), 103–60, on 124–32.

[44] Lluís Cifuentes i Comamala, "'Translatar sciència en romans catalanesch': La difusió de la medicina en català a la baixa Edat Mitjana i el Renaixement," *Llengua & Literatura* 8 (1997): 7–42; Cifuentes, "Estratègies de transició: Pobres i versos en la transmissió extraacadèmica del saber a l'Europa llatina tardomedieval," in *Translatar i transferir: La transmissió dels textos i el saber (1200–1500)*, ed. Anna Alberni, Lola Badia, and Lluís Cabré (Santa Coloma de Queralt: Obrador Edèndum, 2010), 241–63; Cifuentes, "Les miscel·lànies mèdiques medievals en català: Una proposta de classificació," in *Sabers per als laics: Vernacularització, formació, transmissió* (Corona d'Aragó, 1250–1600), ed. Isabel Müller and Frank Savelsberg (Berlin: Walter de Gruyter, 2021), 255–90. I am indebted to Lluís Cifuentes for sharing his paper with me in advance of publication.

[45] Cifuentes, "Les miscel·lànies mèdiques medievals" (cit. n. 44).

[46] Green, *Making Women's Medicine Masculine* (cit. n. 2), 129–45, has argued that records documenting the possession or loan of medical books by medieval women, healers or not, are very scarce.

[47] The library contains books in both Catalan and Latin, but the inventory states that this item was in Latin: "Ítem, ·i· libra cobert de pergamín, en paper, qui s'apella *Thesaurus pauperum*, e és en latín," Arxiu i Biblioteca Episcopal de Vic, Vic, Arxiu de la Cúria Fumada, ACF-3705 (inventaris anònims 1397–1403), fol. 3v. For a full description of the library, see Sciència.cat DB doc28, in *Sciència.cat*, coord. Lluís Cifuentes i Antònia Carré, Universitat de Barcelona, 2006, http://www.sciencia.cat, accessed April 23, 2021.

[48] The translation extends to chapters 132–64 of the standardized edition of the compendium edited by Green; *The Trotula* (cit. n. 1), 116–33.

De sinthomatibus mulierum, which usually went in first place, or even the *De ornatu mulierum*, although that one normally circulated in the third place of the triad. What we do know for sure, however, is that the *De curis* closed the Latin version that was used as a base for the translation because the Catalan retains the customary end: "The *Trotula* is finished, thank the Lord."[49]

The Riccardiana manuscript testifies to the revision of a translation previously made either by the editor emerging in the text or by another person:

> I have myself checked this treatise entitled *Trotula*, which either the original [from which the translation was made] fails or the original that I have checked is more copious, because all the additions that are below this point are taken from the original that I have checked.[50]

Explicitly noting some divergences, the editor added in Latin the omitted chapters found in the version s/he was consulting and that turned out to be longer than the Catalan translation.[51] And thanks to the voice emerging in the text, we can infer that the translation was made from a copy of one of the oldest abridged versions of the *Trotula* compendium.

The labor of the Catalan editor brings Trota back into the vernacular history of the text. The Latin manuscript used for the initial translation contained the name of Trotula, whereas the text used to collate it later on still retained that of Trota—*the Latin says Trota*, as s/he inserted in the text. Therefore, the paragraph places the collated version in the early phases of the development of the *Trotula*, when the *De curis* had not yet lost its principal historical actress. The fact that while annotating the divergence the editor gives preeminence to the spelling "Trotula" would suggest that this was already the culturally predominant form of the name through which both the text and the Salernitan healer were known when the translation and its revision were made.[52] At present, neither of them can be dated with precision; on philological grounds, however, the extant rendering can be placed as deriving from the Western region of the Catalan linguistic domain.[53]

Working with fourteenth- and fifteenth-century examples, Michael McVaugh and Antònia Carré have shown that the revision of previous translations was a technique used by some Catalan translators and copyists of healthcare texts, who acted, in fact,

[49] "Acabada és la *Tròtula*, a Déu gràcies," Biblioteca Riccardiana, MS 2827, fol. 172r.

[50] "Aquest tractat apellat *Tròtula* és stat per mi conprovat, lo qual o a l'original de aquest ffallie o a l'original que yo l'é comprovat és pus copiós, per què totes les addiccions qui ssón d'así avayll són tretas de l'original, que yo he conprovat aquest," ibid. The note follows the translation of chapter 164 of the standard Latin version of the *De curis*; Green, *The Trotula* (cit. n. 1).

[51] These are chapters 162 (omitted from the translation), 163, and 164 (previously translated after chapter 161), and then follow chapters 165, 176, 179, 183, 209, 212, 217, 222, 218, 225, 227, 228, and 229; Green, *The Trotula* (cit. n. 1).

[52] It is important to note here that the editor writes "Explicit suma Trotule" at the end of the paragraphs that s/he added, showing that their Latin version contained both forms of the name, Trota and Trotula; Biblioteca Riccardiana, MS 2827, fol. 178r.

[53] The text does not present traces of the primitive Catalan *scripta*, and this prevents us from dating the first translation before 1320–30. On the dating of Catalan texts, see Lola Badia, Joan Santanach, and Joan Soler, "Els manuscrits lul·lians de primera generació als inicis de la *scripta* llibrària catalana," in Alberni, *Translatar i transferir* (cit. n. 44), 61–90; Lluís Cifuentes I Comamala, "La scripta llibrària catalana primitiva als primers textos mèdics en català," in "*Qui fruit ne sap collir," Homenatge a Lola Badia*, ed. Anna Alberni, Lluís Cifuentes, Joan Santanach, and Albert Soler (Barcelona: Universitat de Barcelona, 2021), 1:157–70.

as editors of earlier versions.[54] Articles by Elaine Leong, Ahmed Ragab, Alisha Rankin and Dror Weil in this volume show that these practices were culturally widespread and belonged to a repertoire of methods used widely that influenced the frame of the final texts that were produced.[55] The glaring comments of the Catalan editor evince how these techniques molded the distribution of the authority upholding the refashioning of texts.

In the case of the Catalan *De curis*—understood then as *Trotula*—the immediate objective of the editor was to obtain a text that was as faithful as possible to a Latin original and that perhaps s/he disposed of only temporarily—and might even intend to translate fully later on. It is important to note that only the translation of the *De curis* was collated with a Latin original, and the other accompanying texts show no further investment on the part of the scribe or editor to ascertain their accuracy, indicating a clear interest in this precise text. It is as yet unclear whether the author of the initial translation and the person who later revised it are the same, as they were working in two different periods and with different Latin originals. Nevertheless, it is possible to identify the Latin texts that were used to elaborate both the initial Catalan translation and the second phase of revision to which the Riccardiana manuscript is witness. In spite of the fact that the *De curis mulierum* is the text that circulated less independently of the three that comprised the *Trotula*, we know that shortened versions were in use.[56] And it is one of those fragmentary versions that is to be found at the base of the original Catalan translation. One of the fragmentary extant versions belonging to the second stage of evolution of the independent text—which Green calls *De curis mulierum* 2—is made up of exactly the same chapters that were translated into Catalan in the Riccardiana text.[57] Independent of which specific Latin versions were used in the elaboration of the translation and its collation, the visibility of Trota in translation is remarkable. An English text on women's health bears as a title a variant of her name (*The book made by Rota*).[58] But the Catalan version is the only independent translation of the *De curis mulierum* known to date.[59]

[54] This is the case with Bernat de Berriac, who at the beginning of the fourteenth century consulted diverse Latin originals to make his Catalan version of the *Chirurgia* of Teodorico Borgognini, a project he carried out by adapting and correcting the translation of Guillem Corretger from the first three books and translating the rest of the work anew. See Michael McVaugh, "Academic Medicine and the Vernacularization of Medieval Surgery: The Case of Bernat de Berriac," in *El saber i les llengües vernacles a l'època de Llull i Eiximenis: Estudis ICREA sobre vernacularització*, ed. Anna Alberni, Lola Badia, Lluís Cifuentes, and Alexander Fidora (Barcelona: Publicacions de l'Abadia de Montserrat, 2012), 257–81. In the same way, the person who made the mid-fifteenth-century copy of Bernat Sarriera's translation of the *Regiment de sanitat* by Arnau de Vilanova collated the Catalan version they worked on with a Latin original; see Arnau de Vilanova, *Regiment de sanitat per al rei d'Aragó: Aforismes de la memòria*, ed. Antònia Carré (Barcelona: Universitat de Barcelona, 2017), 134–42; Antònia Carré, "Del llatí al català: El cas del *Regimen sanitatis ad regem Aragonum* d'Arnau de Vilanova," in Müller and Savelsberg, *Sabers per als laics*, 317–35 (cit. n. 44). I am indebted to the author for generously sharing her work with me before publication.

[55] Elaine Leong, "Translating, Printing, and Reading"; Ahmed Ragab, "Translation and the Making of a Medieval Archive"; Alisha Rankin, "New World Drugs"; and Weil, "Unveiling Nature"; all in *Osiris* 37.

[56] Green, *Making Women's Medicine Masculine* (cit. n. 2), 83.

[57] These are chapters 132–64; for the attestation, see Green, "Handlist Part I: Latin" (cit. n. 22), 151, no. 41.

[58] Green, *Making Women's Medicine Masculine* (cit. n. 2), 188–9.

[59] Green, "Handlist Part II: Vernacular" (cit. n. 38); Caballero-Navas, "Un capítulo sobre mujeres" (cit. n. 39); Lie, "Women's Medicine" (cit. n. 39).

THE AUTHORITY OF TROTULA IN CATALAN

In the body of medieval medical literature in Catalan, we have not found more traces of the historical Trota. Nevertheless, transformed into the figure of Trotula, her authority spread in Catalan medical culture. One route for its dissemination was via the fortunes of the translation of the *Thesaurus pauperum* (Treasure of poor men), which circulated under the name of Petrus Hispanus, and which, as we have seen before, attributed a handful of gynecological treatments to Trotula. The work was very popular in the late Middle Ages, and three manuscripts of one single Catalan translation have already been identified.[60] Two are extant in full and retain the attributions of the treatments to the different medical authorities presented in the original, acknowledging Trotula as the source for various procedures in two of the chapters devoted to female medicine.[61] Interestingly, one of the Catalan versions omits the mentions of Trotula, because the corresponding sections do not cite authorities for any of the recipes provided.[62] However, a mid-fourteenth-century version attributes five treatments to Trotula in the chapters "De embarguar el concebiment" (For avoiding conception) and "Qual cose ajude a concebre" (For helping to conceive).[63]

We know that Catalan texts identified with the name of Trotula were in use at the royal court and also circulated among medical practitioners in the fourteenth- and fifteenth-century Crown of Aragon. Some of the known instances come from book inventories and offer little information, making it difficult to ascertain not only the exact treatise referred to, but even the language of the text. For example, the surgeon Bernat Serra had a book entitled *Trotula* when he died in 1338, but this text was probably in Latin if we are to go by the complete learned collection of books that he possessed.[64] In any case, it attests to the circulation of Trotula as a figure with a presence in the libraries of medical practitioners. While compiling lists of books when taking stock of goods, some notaries transcribed sentences from the beginning and the end of the items that they were enumerating, but often they simply wrote short titles or names as a way to identify the texts. The lack of contextual evidence makes it difficult to discern the details in every case. However, neither notaries, scribes, nor readers labeled or referred to texts at random. Personal names of authorities were commonly used in association with texts but even more broadly with particular fields of expertise, offering value and legitimacy to what the books contained. This was not a gendered phenomenon, and it was not peculiar to women's medicine. The name of Palladius, a fourth-century author of a widely acknowledged treatise on agriculture, came to identify a variety of different treatises on the subject, which, although unrelated to his original work, circulated under his name in Catalan during the late Middle Ages—the so-called *Pal·ladis arromançats*.[65] Owing

[60] Cifuentes, "El receptari mèdic" (cit. n. 43), 118–20.

[61] The citations in the original Latin are found in chapter 44, *De impedimento conceptu*, and 45, *Ut mulier concipiat*, in Maria Helena da Rocha Pereira, *Obras médicas de Pedro Hispano*, 2 vols. (Coimbra: Universitatis Coninbrigensis, 1973), 1:265.

[62] Biblioteca de Catalunya, Barcelona, MS 864, fols. 27ra–rb, transcribed by Asunción Escudero Mendo, "Manuscritos de la Biblioteca de Catalunya de interés para la farmacia y las ciencias médicas: En particular el Tresor dels pobres" (PhD diss., Univ. of Barcelona, 1993), 655–6.

[63] Arxiu i Biblioteca Episcopal de Vic, MS 191, fols. LXXIIIV, LXXXVIV. I thank Lluís Cifuentes for providing me with a copy of this manuscript.

[64] "[26] Item, quendam alium librum, vocatum *Trotula*, scriptum in papiro," in Josep Hernando, *Llibres i lectors a la Barcelona del segle XIV* (Barcelona: Fundació Noguera, 1995), 1:136, no. 76.

[65] Lluís Cifuentes i Comamala, *La ciència en català a l'Edat Mitjana i el Renaixement* (Barcelona-Palma de Mallorca: Universitat de Barcelona–Universitat de les Illes Balears, 2001; 2nd rev. ed., 2006), 289–90.

to his recognized authority on a particular topic, his personal name came to authorize a whole area of knowledge. Similarly, in medieval Catalan texts, Trotula stars as a figure of authority in women's healthcare in a number of ways.

Most visibly, her name appears at the beginning of texts, quickly according their contents authoritativeness. In the library of Pere Company, an apothecary and citizen of Barcelona, there were various books of pharmacology and medicine, one of which is described in detail in a survey of his properties in 1428:

> Item, another book entitled *Tortule of Greece*, written on paper and with wooden covers lined in red leather. The first page of the book starts in black letters: "Here starts the book" and ends: "or lice on the head"; and on the last page of the book, which is unfinished, it ends with "contrary foods."[66]

The description does not allow for a precise identification of the text. It begins with the care of the hair, as was usual in treatises of practical medicine from head to toe, but cosmetic tracts also devote a good part of their contents to hair treatments—and as seen above, cosmetics are a substantial part of the texts associated with Trotula and the collective authority ascribed to women. The final sentence, or *explicit*, indicates that the text ended with diet recommendations, typical advice given in handbooks of practical medicine and regimens of health. Therefore, the text described here is not the compendium that circulated in Latin under the name of *Trotula*, but a further expression of the medical authority that had originally been recognized in Trota. The fact that next to her name appears a demonym—of Greece—is an evident sign that it is referring to a personal female figure giving authority to the text. The identification of that woman with Greek medicine—with the land of origin of Hippocrates and Galen—offers further validity to the authoritative figure. The association of Trotula with Greece is not common in medieval medical literature, but neither is it unique: Johannes Hartlieb, the translator of the most popular German versions of the *Trotula* compendium, made in the 1460s, also places the female authority figure in Greece but raises her status to that of a Greek queen.[67] As we saw before, instances of royal women validating medical knowledge are found in other linguistic traditions, most notably in the case of Cleopatra.

In the above examples, both a surgeon and an apothecary possessed texts that used the name of Trotula to identify and grant authority to a medical text. In neither case can it be ascertained exactly which texts they held since the extant descriptions do not provide sufficient information. The treatise owned by Bernat Serra could be the *Trotula* compendium. However, Pere Company's book contained a different and currently unknown text, likely a text on women's medicine. The association of the name of Trotula with texts on women's health that have no direct connection with the *Trotula* is not without parallel. One of the known copies of a Middle English text on obstetrics and gynecology was entitled *Liber Trotularis*, but its contents had no connection with the *Trotula* or the work of Trota; rather, it is associated with Muscio's gynecological

[66] "Item, ·I· altre libre cubert de posts de fust cubertes de pell vermella, scrit en paper, appellat *Tòrtule de Grècia*, la primera carta del qual libre comensa en letres negres: 'Ací comensa lo libre', e feneix: 'o lèmens del cap'; e la derrere carta del qual libre, no acabada, feneix: 'viandes contràries.'" I follow the transcription, slightly modified, made by Josep A. Iglesias i Fonseca, "Llibres i lectors a la Barcelona del s. XV: Les biblioteques de clergues, juristes, metges i altres ciutadans a través de la documentació notarial (anys 1396–1475)" (PhD diss., Univ. Autònoma de Barcelona, 1996), 416, doc. 107. For more details, see Sciència.cat DB doc10 (cit. n. 47).

[67] Green, *Making Women's Medicine Masculine* (cit. n. 2), 225.

works as well as with Gilbertus Anglicus's discourse on women's conditions in his medical handbook.[68]

There is a one further example of the use of the name of Trotula as an authoritative label on women's healthcare in Catalan texts. Probably dating from the third quarter of the fourteenth century, a text entitled *Trotula* was addressed to a lady of the Aragonese royal family, as the prologue of the work states:

> Here starts the book that speaks well and frees the Queen from any cure so that she might live healthily all her life; it has been made by Master Joan, pleasing the Infanta very much, and he has entitled it *Trotula*.[69]

The book is addressed to a "flower of Aragon" and was commissioned by a queen to a certain medical practitioner called Joan. The rhetoric of the prologue is courtly and contains a love poem through which the book speaks to a young lady, expressing hope that it will help her to keep her beauty and health. Exact confirmation of who these three characters were is lacking, but the book could have been made for the infanta Elionor of Aragon (1358–82) at the request of Queen Elionor of Sicily (1325–75, r. 1349). From the start, the author states that his work is a compilation from other texts rather than an original work. Master Joan never discloses his sources apart from a generic mention of Hippocrates and Galen when considering the retention of the menses. Nor does he reveal the language of the texts he is working with.[70] But he is very clear in his identification of the work with the unequivocal label of *Trotula*, as he refers twice to her name: "Here start the rubrics of *Trotula"* (Assí començan les rúbliques de *Tròtule*).[71] Nevertheless, the book does not contain any of the three texts that formed part of the old Latin *Trotula* compendium.

The compilation is a handbook for domestic use, a text that promises the woman who follows its advice that it will help her avoid having to consult a "physician except in case of plague or apoplexy."[72] Nevertheless, beyond the ladies of the court, master Joan's *Trotula* was of interest to medical practitioners. A copy was owned by a barber from Valencia, Miquel Domenge; by 1479 it had passed into the possession of the surgeon Jaume Boixadell in Barcelona. The extant description of the book shared by these practitioners does not mention its title, as it only transcribes the beginning of the dedicatory poem where there is no mention of *Trotula*, but it is distinctive enough to be identified as the same text.[73]

[68] For *Liber Trotularis*, see Monica H. Green and Linne R. Mooney, "The Sickness of Women," in *Sex, Aging, and Death in a Medieval Medical Compendium: Trinity College Cambridge MS R.14.52, Its Texts, Language, and Scribe*, ed. M. Teresa Tavormina, Medieval and Renaissance Texts and Studies 292, 2 vols. (Tempe, AZ: Arizona Center for Medieval and Renaissance Studies, 2006), 2:455–568, on 568n91.

[69] "Assí comença lo libre qui parla gint e desliure de tot adop de la regine, per la qual en tot son temps viurà sana; lo qual à fet mestre Johan a la infante molt agradant, al qual à mès nom *Tròtula*," Biblioteca Nacional de España, MS 3356, fol. 1ra.

[70] Montserrat Cabré, "From a Master to a Laywoman: A Feminine Manual of Self-Help," *Dynamis* 20 (2000): 371–93.

[71] Biblioteca Nacional de España, MS 3356, fol. 1rb.

[72] "Aiats vós donchs, madona, si a vós plau, aquest libre per ops de estar en vostra sanitat; car si voletz son mandament seguir, nuyl temps no aurets ops a ebeir metge, sinó és per aventura de plages e de feridura, les quals coses no pot hom de tot en tot esquivar. . ." ibid., fol. 27rb.

[73] "Ítem, ·i· altre libre scrit en paper de forma mijana, ab posts cubertes de cuyr burell o turrat, ab ·x· bolles e ·ii· gaffets de lautó, e comença: «Saluts a vós, flor d'Aragon, a qui don Déus rich e baron» et cetera," original edition, without identification, in Josep Maria Madurell i Marimon and Jordi Rubió i

Master Joan created a health manual from diverse texts or adapted a compilation that someone else, in Catalan or in another language, had put together from different, albeit still distinguishable, elements. Nonetheless, despite the lay audience and the fact that it is a different text, not only its name but also its thematic features directly link it to the Latin *Trotula*. The compilation contains an extensive first section devoted to cosmetics, which brings together materials that circulated under the name of *De ornatu*, although they do not correspond textually with those presented by the anonymous *De ornatu* that forms part of the Latin *Trotula*. This section is followed by the literal version of a gynecological text known in a Hebrew version and three French versions, *Des aides de la maire et de ses medicines* (On the aids of the womb and its medicines); in these latter, the name of the author is assigned as Jehan de Trabarmaco or Tarbamacho, which the Catalan version makes into Joan de Reimbamaco. Finally, the treatise ends with a short section on sexual medicine and a short regimen for health.[74]

The Catalan text that master Joan entitled *Trotula*, therefore, contains traces neither of the work of Trota nor of the Latin compendium that extended her fame. Now that the personal name was transformed into a label of a medical specialty, the choice to employ it was imbued with meaning: used as a rubric, it came to be understood as a genre of medical texts. As was the case with other authorities giving their name to specific domains of knowledge, *Trotula* still conveyed the idea of female authority in women's healthcare. In a similar manner, the name of Trotula in Latin had been saturated with the acknowledgment of female authority in women's medicine that had once been accorded to Trota.

CONCLUSION

Translation was an integral part of the process of the vernacularization of medicine underway in Western Europe during the late Middle Ages. Healthcare treatises were adapted and transported from Latin into Catalan, broadening the audiences that used them, which in the fourteenth and fifteenth centuries included a wider range of practitioners and laypeople alike. Together with the texts, notions of authority over the knowledge that they conveyed were transferred or suppressed. As an epistemic technology, translation involved the use of many techniques to intervene in the texts, including the association of personal names to the renderings in new languages as a method to authorize them.

In Latin medicine, female authority was particularly vulnerable because its acknowledgment was limited in two gender-specific ways. First, recognitions of anonymous women's authority over specific treatments, as individuals or as part of a distinctive collective, circulated widely but were invisible in the large bodies of medical literature. Second, women who were recognized individually as figures of authority were known only for their knowledge of women's health—even if their expertise was broader. The historical Trota of Salerno, through the evolved name of Trotula, stands as a prominent figure peculiar to Latin medical culture.

Balaguer, *Documentos para la historia de la imprenta y librería en Barcelona (1474–1553)* (Barcelona: Gremios de Editores y Libreros y de Maestros Impresores, 1955), 14, doc. 7. For full details, see Sciència.cat DB doc179, text number 4 (cit. n. 47).

[74] Cabré, "From a Master to a Laywoman" (cit. n. 70), 385–7. For the Hebrew version, see Carmen Caballero-Navas, *The Book of Women's Love and Jewish Medieval Medical Literature on Women* (London, 2004), 95, 160–2.

Late medieval Catalan medicine recreated and disseminated this female figure of authority, departing from the historicity of the Salernitan healer in a manner similar to that of the Latin traditions. Nevertheless, by reading as gender-significant the acts of identifying and labeling texts, a wealth of traces of a notion of female authority are still retained in treatises on women's health, showing that Trota and Trotula came to authorize an important part of medieval women's medicine in Catalan.

[Un]Muffled Histories: Translating Bodily Practices in the Early Modern Caribbean

*by Pablo F. Gómez**

ABSTRACT

This article focuses on the translating practices of people of African descent and Amerindians in the early modern Caribbean. The history of translation processes in the interconnected constellation of islands and shorelines that constituted the early modern Caribbean provides a framework for thinking about translation that exists outside analytical models based on notions of subalterities, center-periphery, or "contact zones." Narratives about translation based on imperial and colonial dynamics, or those that depend on the divides between nature and culture Europeans ideated in the seventeenth century, are not apposite for uncovering the history of medical translation in the Caribbean. By stepping outside of these narratives, this article shows how, in the early modern Caribbean, a multitude of people—Caribbean Amerindians, and migrants, mostly kidnapped people from Africa and their descendants—developed rich communal processes of translation around bodily matters.

Late one night during the first months of 1611, a despondent man knocked on the door of the Jesuit college in Cartagena de Indias (today Colombia) desperately asking for Father Alonso de Sandoval. The unnamed man wanted de Sandoval's help: one of his enslaved Africans had been unconscious for two days. De Sandoval rushed to the man's house to find the enslaved man "lying on his back with blank eyes, hands over his head, and a profuse amount of matter coming out of his mouth." After praying by his side, de Sandoval left the African captive for dead and put a reliquary and a bible at the head of the man's bed. The Jesuit priest came back to see the African man the next day on his way back from the harbor where he had been baptizing enslaved men and women arriving in a vessel "coming from Angola." To his surprise, de Sandoval found the African captive awake and talking in a language he recognized as Folupo, an ethnolinguistic group from West Africa. The African man was, in the terminology that Iberian slave traders used at the time, "a *bozal* [a muffled person]." This is a recently disembarked African person who did not speak any European language. Sandoval rushed

* Department of History and Department of Medical History of Bioethics, University of Wisconsin, 1300 University Ave., Room 1419, Madison, WI 53706, USA; pgomez@wisc.edu.

Osiris, volume 37, 2022. © 2022 History of Science Society. All rights reserved. Published by The University of Chicago Press for the History of Science Society. https://doi.org/10.1086/719228.

back to the college looking for a Folupo translator so he could properly baptize him. The enslaved man, according to de Sandoval, eventually recovered.[1]

This vignette brings into relief several essential themes common to histories of the region and, more generally, those of the Atlantic world—among them, the overwhelming presence of disease and death and the essential role of forced African immigrants in the shaping of translation practices in these historical spaces. The story also exemplifies how early modern Europeans conceived of Africans unable to speak European languages as being muffled (culturally and linguistically). This imposed muteness shaped the histories of African captives in the early modern Caribbean and their depictions in the historical record—an artificial muting of words and worlds that has persisted, more broadly, in most histories of translation related to the healing practices of Africans and their descendants in the Atlantic world.

In the cacophonous early modern Caribbean, translations of all sorts, physical, linguistic, ritual, performative, and medical, modeled social, ethnic, cultural, military, and political landscapes. Paradoxically, for all of Europeans' depictions of Africans as *bozales*, most of these transactions involved languages and mores other than European ones. Every day early modern Caribbean encounters around bodily suffering and its alleviation, such as those experienced by the unnamed Folupo captive, ensued through linguistic, material, and corporeal avenues that do not fit within the usual confines of most early modern historical narratives of knowledge-exchange related to the human body.

The history of the explosion of translation practices on a global scale that characterized the early modern era is inextricably linked to the violence that sustained the expansion of European imperial projects—including the rise of the transatlantic slave trade. Most studies of medical translation in the early modern era still depend on categories and analytical strategies that Europeans developed in the process of creating the globalized worlds of empire, capital, and colonialism. The need for translation and translators, after all, appeared in most cases in the context of the development of colonial projects. And it is on these episodes of translation practices that historians have focused. The best-known narratives about the histories of translation and circulation of early modern medical knowledge examine processes that were directly linked to the massive mobilizations of people, goods, and, more generally, biota, that Europeans unleashed during the early modern era in the name of commercial profit.[2] We have plenty of insightful works linking the emergence of scientific practices, anthropology, ethnology, and political economy, among others, to the history of slavery, racism, and

[1] Archivum Romanun Societatis Iesu, Vatican City (hereafter ARSI), N.R.G. et Quit 12 I, fols. 95v-96r. Unless otherwise noted, the translations in this article are mine.

[2] See, for instance, Harold Cook, *Matters of Exchange: Commerce, Medicine, and Science in the Dutch Golden Age* (New Haven, CT: Yale Univ. Press, 2008); Londa Schiebinger, *Plants and Empire: Colonial Bioprospecting in the Atlantic World* (Cambridge, MA: Harvard Univ. Press, 2007); James Delbourgo and Nicholas Dews, eds., *Science and Empire in the Atlantic World* (New York: Routledge, 2007); Antonio Barrera, "Local Herbs, Global Medicines: Commerce, Knowledge, and Commodities in Spanish America," in *Merchants and Marvels: Commerce, Science, and Art in Early Modern Europe*, ed. Pamela H. Smith and Paula Findlen (New York: Routledge, 2002); Kapil Raj, *Relocating Modern Science: Circulation and the Construction of Knowledge in South Asia and Europe, 1650–1900* (New York: Palgrave Macmillan, 2007); Simon Schaffer, Lissa Roberts, Kapil Raj, and James Delbourgo, eds., *The Brokered World: Go-Betweens and Global Intelligence, 1770–1820* (Sagamore Beach, MA: Watson Publishing International, 2009); and Pratik Chakrabarti, *Materials and Medicine: Trade, Conquest and Therapeutics in the Eighteenth Century* (Manchester, UK: Manchester Univ. Press, 2010).

colonialism of different sorts.[3] The very idea of the imperial nature of science's universalizing impulses has also been a point of departure for the literature on global histories of science and medicine for a while now.[4] In addition, the history of the obscuring of so-called indigenous knowledges in European works of natural history and natural philosophy has also been a central tenet of recent scholarship on scientific and medical translation during the early era of globalization.[5] These analyses of silencing are, however, still dependent on definitions of the study of "culture" and nature—also a product of European colonialism—that originated in the early modern era and located non-European ways of knowing the natural world outside the purview of histories of science and medicine.[6]

In other words, for all the advances that the global turn has produced, studies of early modern medical translation that engage with the work and lives of non-European historical actors have focused almost exclusively on episodes that link their histories (those of non-Europeans) to European medicine, natural history, or natural philosophy in one way or another. The literature is, thus, full of examples of how European processes of translation moved the knowledge of non-European groups to metropolitan spaces to be incorporated into projects of natural history or pharmacopeias.[7] Underlying these

[3] See, for instance, John Law and Wen-yuan Lin, "Provincializing STS: Postcoloniality, Symmetry, and Method," *East Asian STS* 11 (2017): 211–27; Warwick Anderson, "From Subjugated Knowledge to Conjugated Subjects: Science and Globalisation, or Postcolonial Studies of Science?," *Postcolon. Stud.* 12 (2009): 389–400; Jorge Cañizares-Esguerra, *How to Write the History of the New World: Histories, Epistemologies, and Identities in the Eighteenth-Century Atlantic World* (Stanford, CA: Stanford Univ. Press, 2001); Dipesh Chakrabarty, *Provincializing Europe: Postcolonial Thought and Historical Difference* (Princeton, NJ: Princeton Univ. Press, 2000); Arturo Escobar, *Territories of Difference: Place, Movements, Life, Redes* (Durham, NC: Duke Univ. Press, 2008), 373–88; and Suman Seth, "Colonial History and Postcolonial Science Studies," *Radical History Review* 127 (2017): 63–85.

[4] Among others, see Carla Nappi, "The Global and Beyond: Adventures in the Local Historiographies of Science," *Isis* 104 (2013): 102–10; Christopher A. Bayly, *Empire and Information: Intelligence Gathering and Social Communication in India, 1780–1870* (Cambridge, UK: Cambridge Univ. Press, 1999); David N. Livingstone, *Putting Science in Its Place: Geographies of Scientific Knowledge* (Chicago: Univ. of Chicago Press, 2003); James A. Secord, "Knowledge in Transit," *Isis* 95 (2004): 654–72; Kapil Raj, *Relocating Modern Science: Circulation and the Construction of Knowledge in South Asia and Europe, 1650–1900* (Basingstoke: Palgrave Macmillan, 2007); Lissa Roberts, "Situating Science in Global History: Local Exchanges and Networks of Circulation," *Itinerario* 33 (2009): 9–30; Neil Safier, "Global Knowledge on the Move: Itineraries, Amerindian Narratives, and Deep Histories of Science," *Isis* 101 (2010): 133–45; and Sujit Sivasundaram, "Sciences and the Global: On Methods, Questions, and Theory," *Isis* 101 (2010): 146–58.

[5] See, for instance, Robert N. Proctor and Londa Schiebinger, eds., *Agnotology: The Making and Unmaking of Ignorance* (Stanford, CA: Stanford Univ. Press, 2008).

[6] Among others, see Pablo F. Gómez, "Caribbean Stones and the Creation of Early Modern Worlds," *Hist. & Tech* 34 (2018): 11–20; Stephan Palmié, "When Is a Thing: Transduction and Immediacy in Afro-Cuban Ritual; Or, ANT in Matanzas, Cuba, Summer of 1948," *Comp. Stud. Soc. Hist.* 60 (2018): 786–809; Philippe Descola, *Beyond Nature and Culture* (Chicago: Univ. of Chicago Press, 2014); Michel-Rolph Trouillot, *Silencing the Past: Power and the Production of History* (Boston, MA: Beacon Press, 1995); Dipesh Chakrabarty, "The Muddle of Modernity," *Amer. Hist. Rev.* 116 (2011): 663–75; and Ashis Nandy, "History's Forgotten Doubles," *Hist. & Theory* 34 (1995): 44–66.

[7] Among many others, see Daniela Bleichmar, "Books, Bodies, and Fields: Sixteenth-Century Transatlantic Encounters with New World Materia Medica," in *Colonial Botany: Science, Commerce, and Politics in the Early Modern World*, ed. Londa L. Schiebinger and Claudia Swan (Philadelphia: Univ. of Pennsylvania Press, 2005), 83–99; James Delbourgo, *Collecting the World: Hans Sloane and the Origins of the British Museum* (Harmondsworth, UK: Penguin, 2017); Bleichmar, Paula de Vos, Kristin Huffine, and Kevin Sheehan, eds., *Science in the Spanish and Portuguese Empires, 1500–1800* (Stanford, CA: Stanford Univ. Press, 2009); Schiebinger, *Plants and Empire* (cit. n. 2); the essays in Roy MacLeod, ed., *Nature and Empire: Science and the Colonial Enterprise*, *Osiris* 15 (2000); or Junia Ferreira Furtado, "Tropical Empiricism: Making Medical Knowledge in Colonial Brazil," in Delbourgo and Dews, *Science and Empire* (cit. n. 2), 127–52; and, more recently, Matthew J. Crawford and Joseph M. Gabriel, eds., *Drugs on the Page: Pharmacopoeias and Healing Knowledge in the Early Modern Atlantic World* (Pittsburgh, PA: Univ. of Pittsburgh Press, 2019).

studies is an assumption about "the asymmetrical character" of the relationships between Europeans and (the majority) of historical actors of non-European descent involved in transactions around medical knowledge.[8] Not surprisingly, the analysis of printed texts (usually focused on the work of single individuals) has been at the center of translation studies that involve European centers of knowledge. Outside of Europe, the study of medical translation practices in the early modern era has been limited to areas with recognizable (for Europeans) textual traditions, such as China or Japan.

The limited number and nature of the extant sources we have to examine early modern Caribbean histories of translation is at the core of the difficulties historians have in their attempts to recover non-European voices from them. As a number of scholars have pointed out, these sources, including the Inquisition documents I examine in this work, were created through violent processes of colonial categorization. The voices, desires, and ideas of people of African descent appear in them (if they do) as ventriloquized.[9] They are channeled through categories of imperial organization that bound peoples and cultures. Disentangling these archives requires an engagement that resists viewing the archive as a space of death and erasure and instead follows a more subtle and tentative manner of approaching historical evidence, one that requires a disavowing of the dynamics around which historians of translation have usually construed their narratives.[10] Such a project allows for the discovery of traces, and admittedly only traces, of how people of Amerindian and African descent (most of them living in the world of slavery) translated the worlds of healing in the early modern era.

This article, thus, is not concerned with the history of how Europeans translated and incorporated, or not, ideas about health and healing coming from the New World or Africa. It also does not focus on how the traditions of non-European Caribbean people became obscured, labeled, or disappeared in linguistic, ontological, and epistemological processes of translation when appearing in European texts. By stepping outside of the narratives of these types of analyses, this article aims to avoid the kind of epistemological "fencing in" that has located early modern Caribbean people as enclosed in artificial labels and defined their knowledge-making practices as incommensurable.

The rest of this essay focuses on the practices of people of African descent and Amerindians in an attempt at "unmuffling" the history of medical translation in the early modern Caribbean. As the following pages show, the history of translation processes in the interconnected constellation of islands and shorelines that constituted the early modern Caribbean provides a framework for thinking about translation that exists outside analytical models based on notions of subalterities, center-periphery, or "contact zones." The Caribbean was not, obviously, only a place that provided data, ideas, experiences, and biota to be translated and taxonomized by European natural historians and

[8] Roberts, "Situating Science in Global History" (cit. n. 4), 10.

[9] Among others, see Saidiya Hartman, "Venus in Two Acts," *Small Axe* 12 (2008): 1–14; or Jennifer L. Morgan, "Accounting for 'The Most Excruciating Torment': Gender, Slavery, and Trans-Atlantic Passages," *History of the Present* 6 (2016): 184–207.

[10] Departing from the recognition that, as Michel de Certeau wrote three decades ago, our own western ways of knowing are imagined, it seems to me that the inevitable contradictions inherent in the epistemologies scholars use to study the histories of translation of early modern medicine in Caribbean spaces might be best approached by following scholars like Frantz Fanon, who insisted that "the real leap consists in introducing invention into existence." Michel De Certeau, *The Writing of History* (New York: Columbia Univ. Press, 1988); Fanon, *The Wretched of the Earth* (New York: Grove Press, 1963).

natural philosophers. This article shows how in the early modern Caribbean a multitude of people—Caribbean Amerindians, and migrants, mostly kidnapped people from Africa and their descendants—developed rich communal processes of translation around bodily matters. For the Caribbean, ideas about translation based on imperial and colonial dynamics, or those depending on the divides between nature and culture Europeans ideated in the seventeenth century for the creation of non-European others, do not suffice.[11]

A LAND OF MANY (AFRICAN) NATIONS

Cartagena de Indias was the main slave entrepôt in the Americas during the late sixteenth century and the first decades of the seventeenth century. The dynamics governing the slave trade in the city, thus, provide a rough picture of the demographics of the Caribbean—a region that was the point of entry for the majority of forced African immigrants arriving in the Spanish Americas up to the 1670s.[12] Around 70 percent of the population living in the Caribbean by the 1640s were of African descent, and most lived in what we today call the Spanish Caribbean. Most of the kidnapped Africans arriving in the Caribbean during the seventeenth century came from West Central Africa (mainly from the kingdom of Kongo and Angola); a second larger group came from West Africa (predominantly from the Senegambia delta in "upper Guinea," and the Bight of Benin), with smaller percentages arriving from East Africa and the Atlantic islands.[13]

Upon the arrival of "*armazones*" (shiploads) of enslaved Africans to Cartagena's port, to use the terminology of contemporary slave traders, Jesuits like Alonso de Sandoval hurried to the city docks to tend to health needs and baptize captives within the ships. The priests informed "themselves about the *naciones* [nations] from where the Africans came, how many sick people arrived in [the ships], and what risk [the captives] had of dying."[14] Next to the vessels, Jesuits would commonly find hundreds of men, women, and children lying in agony in Cartagena's sweltering heat; the slave traders discarded them as being "soul in mouth, bones in a sack."[15] Ostensibly, the college's primary mission and the justification for their expansion in the New Kingdom of Granada had been the baptism and care of African captives, even if these in extremis practices were controversial. Jesuits' interest in health care, however, went hand in hand with their proselytizing mission, and Cartagena's college even had an infirmary specifically devoted to the treatment of sick enslaved people.[16]

[11] For a summary of this history, see Philippe Descola, *Beyond Nature and Culture* (Chicago: Univ. of Chicago Press, 2014). See also William F. Hanks and Carlo Severi, "Translating Worlds: The Epistemological Space of Translation," *HAU: Journal of Ethnographic Theory* 4 (2014): 1–16.

[12] See, among others, Archivo General de Indias (hereafter AGI), Seville, Spain, Santa Fe 73, "Carta que el Capitan Duarte de Leon Marques escribio al Rey." For data on the demography of the slave trade, see http://www.slavevoyages.org. The historiography on this area is extensive; for recent studies on African captives' origins in the seventeenth-century Caribbean, see David Wheat, "The First Great Waves: African Provenance Zones for the Transatlantic Slave Trade to Cartagena de Indias, 1570–1640," *Journal of African History* 52 (2011): 1–22.

[13] David Wheat, *Atlantic Africa and the Spanish Caribbean, 1570–1640* (Chapel Hill: Univ. of North Carolina Press, 2016).

[14] ARSI, N.R.G. et Quit 12 I, fol. 94r.

[15] See, for instance, the report in AGI, Santa Fe 449, "Madrid, 3 de julio de 1700."

[16] José Fernández, *Apostólica y penitente vida de el V.P. Pedro Claver de la Compañia de Iesús . . .* (Zaragoza: Diego Dormer, 1666). Specifically, for the role of baptism in the establishment of Jesuit missions in the Caribbean, see Alonso de Sandoval's classic, *De instauranda Aethiopum salute; el mundo de la esclavitud negra en América*, ed. Angel Valtierra (Bogotá: Empresa Nacional de Publicaciones, 1956).

Slave trafficking boats, in which up to 90 percent of all captives could die of diseases ranging from dysentery to typhus, smallpox, measles, and many others, were famously insalubrious.[17] These spaces of violent and forced movement of peoples, languages, cultures, and ideas certainly provide a stark point of contrast to older narratives that viewed them in terms of circulation and exchange of knowledge, and "discovery" about early modern medicine. These spaces of death and suffering linked the production of both racialized concepts of human bodies and the conceptual geographies that emerged out of the encounters between Africans and Europeans.[18]

Of the seventy or so languages reportedly spoken by Africans in mid-seventeenth-century Cartagena, Alonso de Sandoval could clearly distinguish at least ten: Angola, Arda, Carabalí, Banulo, Bañol, Mandinga, Biojo, Bran, Biáfara, and Folupo. Seven of these languages were spoken in West Africa, one in Angola, and two in other regions in West Central Africa.[19] Many of these languages shared similar linguistic roots. This was particularly true of Upper Guinean languages. Around the Senegambia delta, for instance, some polyglot groups (most prominently the Zapes) spoke Lindagoza, Zozo, Pelicoya, Baga, and Boloncho, among other languages.[20]

Worried about how to understand newly arrived enslaved Africans properly, Jesuit priests faced the fact that, as one of the missionaries put it, "the blacks had so many and so varied languages, that it is impossible to recognize them . . . They come [being] so *bozales* that they do not understand a single word of [Castilian]."[21] To communicate with arriving enslaved Africans, and "according to time and need," Jesuits brought with them "skilled Ladino *lenguas* [tongues/languages]" when meeting arriving slave ships. These *lenguas* were also enslaved Africans who spoke Castilian and served as translators for different African languages.[22] Bartolome (referred to only by his first name), for instance, was a famous *lengua* working for the Jesuits; he spoke fifteen different African languages.[23] Addressing Pedro Claver, another Jesuit famous for his work with enslaved people in Cartagena, superiors in Rome wrote that "in answer to your letter of March 21, 1623 . . . I charge the [rector of the college in Cartagena] that under no circumstances should he occupy [in any other type of job] the blacks that you have bought for [working as] lenguas."[24] But sometimes, not even these skilled translators sufficed.

Speaking to this fact, the rector of the Cartagena college reported to Rome in the 1610s that "one of the most difficult jobs that we have in this ministry is the large variety of languages that come in each *armazón*. [So much so] that it often happens that [we] spend two or three days searching for a translator." On one occasion, the rector explained, the priests had baptized all the enslaved people that "came in an *armazón*"

[17] See, among others, Joseph. C. Miller, *Way of Death: Merchant Capitalism and the Angolan Slave Trade, 1730–1830* (Madison: Univ. of Wisconsin Press, 1988), 105–39; and Linda Newson and Susie Minchin, "Slave Mortality and African Origins: A View from Cartagena, Colombia, in the Early Seventeenth Century," *Slavery and Abolition* 25 (2004): 18–43.

[18] Katherine McKittrick, "Plantation Futures," *Small Axe* 17 (2013): 1–15.

[19] Sandoval, *De instauranda* (cit. n. 16), 46; ARSI, N.R.G. et Quit 12 I, fols. 95v-96r.

[20] Nicolás del Castillo Mathieu, *Esclavos negros en Cartagena y sus aportes léxicos* (Bogotá: Instituto Caro y Cuervo, 1982), 161. For a recent, and astute, examination of the work of Black translators in Cartagena, see Larissa Brewer-García, *Beyond Babel: Translations of Blackness in Colonial Peru and New Granada* (Cambridge, UK: Cambridge Univ. Press, 2020), 116–63.

[21] ARSI, N.R.G. et Quit 12 I, fol. 93v.

[22] Ibid., fol. 94r.

[23] ARSI, N.R. Gran et Quit 14, No. 34; ARSI, Vitae 111, Claver.

[24] ARSI, N.R.G. et Quit 12 I, fol. 240r.

except for two "of Zape cast" who spoke such an "impenetrable language, that after a month," and working with available translators, the only word that the Jesuit had understood was "*boloncho*" [Bolongo], the name of an ethnolinguistic group from the Senegambia delta. Eventually, the priests located a "Zape black man to whom they asked what kind of language was Bolongo" and, more important, "if he knew of somebody in this land who spoke it." The unnamed enslaved man answered that the Zape languages were divided along the lines of different "castes and languages" and included "cololo, zemele, limba, bloom, cumba, bucca, landasoca, zoso, peli, coya, baba, and bolongo, which is the one that he [the Zape captive] was being asked about." He also said that a Black *ladina* [acculturated African] woman "who lived a quarter of a *legua* [2.6 miles] outside the city spoke bolongo."[25] The woman was promptly located to help with the rite. It is clear, thus, that already by the beginning of the seventeenth century, there was a universe of African languages being spoken in the Caribbean, and that the issue of translation of bodily related matters was one of pressing importance for a region that was shaped, precisely, by the commerce of human bodies, and in which the main translators were themselves enslaved. Caribbean processes of translation, this story illustrates, most commonly ensued through the communal practices of the survivors of forced transatlantic passages.

COMMUNAL TRANSLATIONS

A number of studies have appeared in the past decade that demonstrate the crucial role of communities in shaping ideas about healing in the early modern Black Atlantic.[26] This is also evident in the examples of African words related to healing and knowledge makers that appear in the few extant contemporary dictionaries of African languages coming from the early modern Americas. In a compilation of Angola grammar coming from Brazil and published in 1697, the Jesuit missionary Pedro Dias translated into Portuguese the meanings that enslaved people in Brazil, the majority of them from West Central Africa, ascribed to terms related to people involved in healing or transmission of knowledge. *Mulungi*, for instance, was the term Brazilian Angolans used for "keepers of the verb." This term for repositories of communal knowledge was explicitly related to a social responsibility for actively safeguarding social and cultural traditions and comes from the verb *Lunda*, to keep. West Central Africans commonly referred to healers as *Ginganga* (sing. *Nganga*), a word which, as has been explored in multiple studies of the African diaspora, referred to individuals with religious functions involved in the healing of social ailments.[27] For early modern people of African descent, disease (or at least some types of disease) was a sign of the disequilibrium not of individuals' bodies, but of communities. These aggregates of beings included, for Afro-Caribbean people, a vast panoply of material and immaterial ancestors across human and nonhuman landscapes. Disease diagnosis and treatment, thus, required practices of communication and translation that included all these variegated groups of manifold origins.

[25] Ibid., fols. 94v-95r

[26] See, for example, Pablo F. Gómez, *The Experiential Caribbean: Creating Knowledge and Healing in the Early Modern Atlantic* (Chapel Hill: Univ. of North Carolina Press, 2017); or James H. Sweet, *Domingos Alvares, African Healing, and the Intellectual History of the Atlantic World* (Chapel Hill: Univ. of North Carolina Press, 2011).

[27] Pedro Dias, *Arte da lingua de Angola* (Lisbon: Na officina de Miguel Deslandes, 1697), 32. For a recent, and fascinating, study of the world of healing in Atlantic Africa, see Kalle Kananoja, *Healing Knowledge in Atlantic Africa: Medical Encounters, 1500–1850* (Cambridge, UK: Cambridge Univ. Press, 2021)

Antonio Congo (from West Central Africa), for instance, became one of the most powerful healers in the northern New Kingdom of Granada, after traveling for a decade from the Kingdom of Kongo to the Netherlands, England, and Spain, and then settling in Cartagena.[28] Like Antonio, dozens of ritual practitioners were also fluent in European languages and cultural mores. Uncertainty about the meaning of words and actions was an essential part of their world. Africans and their descendants living in the early modern Atlantic basin had been exposed to European and indigenous languages for more than a century already. Many, after traveling around the Atlantic with their masters, or by themselves, spoke European languages, including the pidgin versions of Portuguese and Spanish that were the common language of commerce in West and West Central Africa and the Caribbean. Yet the movements of ideas and terms between these languages, and the ruptures in the transmission of knowledge common to all sorts of translations, more often than not happened through communal linguistic and material conduits outside of recognizable European channels.

Many early modern Caribbean health practitioners of African origin were able to communicate in several African languages while tending to the needs of their clientele, but plenty of them, even after living in the Caribbean for decades, did not speak Castilian or other European languages proficiently.[29] Indeed, successful healers like Luis Biáfara, Mateo Arará, and Luis Yolofo from West Africa, or Pedro Congo from West Central Africa, required translators when facing Inquisition or governmental authorities—in some instances needing even two interpreters during their trials, as was the case for the West African healer Francisco Mandinga in 1664.[30] That these healers requested translators during their encounters with the Inquisition might have been part of a strategy for deflecting the accusations that had brought them in front of the Holy Office. By portraying themselves as *bozales*, healers could inhabit, for their own benefit, the trope of the unknowing, untranslatable, muzzled African that Europeans had imposed on them. For healers living in Black Caribbean communities located in the interconnected urban and rural spaces of early modern Caribbean islands and the shorelands of northern South America and Central America, it was essential to be polyglot and clever travelers of African worldviews. At the same time, pretending not to know European languages, laws, and customs could be an advantage

Communal translations were a matter of everyday life in early modern Caribbean locales, even for surviving Amerindian communities that had a long history of displacement, migration, and knowledge exchange before the arrival of modern Europeans. Language discordances were the norm in most of the Amerindian areas of the Caribbean shores in the New Kingdom of Granada. When explaining the animosities between different groups in the Indian town of Chenú (located east of Cartagena), for instance, Fray Diego Becerra wrote in 1612 that "there are [constant] brawls and fights" between the Amerindians "because they are from different tongues." We have to remember that Amerindian populations in many Caribbean locales were composed of the survivors of civil wars and epidemics whom Spaniards had relocated into "Indian towns" and settlements established under forced-labor systems like *encomiendas* and *repartimientos*.[31]

[28] Archivo Historico Nacional de España, Madrid, Spain (hereafter AHN), Inquisición, L. 1023, fol. 480v.

[29] Gómez, *Experiential Caribbean* (cit. n. 26).

[30] AHN, Inquisición, L. 1022, fols. 292r-292v.

[31] See Julián Bautista Ruiz Rivera, *Los indios de Cartagena bajo la administración española en el siglo XVII* (Bogotá, Colombia: Archivo General de la Nación, 1996); Gómez, *Experiential Caribbean* (cit. n. 26), chap. 1.

Ritual practitioners of African descent repeatedly reported to the Inquisition about their exchanges of information with their Amerindian counterparts and the translation of their healing practices. Some of these Amerindian ritual practitioners were renowned, including one unnamed healer whom Jesuit priests said lived in the town of Tubara and "was famous all across the land."[32] In Inquisition testimonies related to Black Caribbean ritual practitioners, Caribbean people of all origins recognized these *indios herbolarios* as masters of the same natural world inhabited by the spiritual entities and energies of a social nature identified by their African and Afro-descended counterparts. By the 1610s, another Amerindian healer, Luis Andrea, was among the more renowned practitioners around Cartagena and healed people of all origins. We have precious little information about the ways in which Luis Andrea communicated with his patients and translated his cures. We do know, however, that people in his own community admonished him: "You only want to cure Spaniards. They [the Spaniards] will cut [off] your head. Instead, you should be curing Indians who are good people." As members of his group predicted, the retribution Luis Andrea faced for his translation of healing practices was, if not losing his head, banishment from the province and prohibition from curing people.[33] In reality, however, Spanish authorities' sentences were difficult to enforce. Movement of people and mores was the norm in a region that was Spanish only in the imagination of imperial authorities in Europe.

Healers of African descent like Domingo de la Ascension, Antonio Congo, and Pedro Congo traveled around the Cuban and Northern New Kingdom of Granada's countryside learning about plants from Amerindian practitioners like Luis Andrea.[34] Such processes of exchange and translation were even more pronounced in the constellation of Maroon settlements populated by cosmopolitan communities of Africans, Creoles, Spaniards, and Amerindians that proliferated in places such as la Maria and el Limon, in the New Kingdom of Granada; or in Panama or Nicaragua; or around Santiago de Cuba.[35] Practitioners of African descent such as Mateo Arará and Francisco Mandinga also commonly borrowed healing and diseasing bodily techniques from each other and shared information about their own experiences and different plants and other curative elements. They also used, or translated, if you will, European medical knowledge, including humoral techniques and therapeutic elements, and incorporated them into their practices.[36]

An additional characteristic of the communal dimensions of the knowledge-making and translation processes in Black Atlantic locales is evident in the practices of Antonio Congo, who performed rites to communicate with the dead "Caciques [Amerindian rulers]" that populated the underground of his home in Cartagena de Indias during the 1680s.[37] In their explorations of the natural landscape of the Caribbean, healers of African descent did not restrict themselves to the physical classification of the plants, animals, and minerals they encountered. As the West Central African healer Pedro Congo told inquisitors, such explorations also involved communication and the sensing of local

[32] ARSI, N.R. et Quit, 12 I, fol. 172r.
[33] AHN, Inquisición, L. 1020, fol. 3–21r.
[34] AHN, Inquisición, L. 1020; AHN, Inquisición, L. 1021; AHN, Inquisición, L. 1022; AHN, Inquisición, L. 1023.
[35] Gómez, *Experiential Caribbean* (cit. n. 26), chaps. 2–3.
[36] Ibid., chaps. 4–6.
[37] AHN, Inquisición, L. 1023, fol. 480r.

(presumably Amerindian) spiritual entities. Although these complex embodied and immaterial processes of knowledge circulation might seem beyond the purview of traditional histories of medical translation, they had profound consequences for the making of the healing landscapes of the region.

The experience of the West African ritual practitioner Francisco Mandinga demonstrates the working conditions for healers in the region. In the 1660s and 1670s he was called repeatedly by Caciques (Amerindian chiefs) and Spanish *encomenderos* (beneficiaries of *encomienda* forced-labor settlements) to the towns of Turbo and Coloso, south of Cartagena de Indias, to heal Amerindians afflicted by unnamed pestilence. Francisco would spend several days in the communities he visited, interrogating individuals and diagnosing the social dynamics of the group. He also communicated with spiritual entities in the Amerindian settlement to provide a diagnosis and prescribe a cure. His diagnosis and therapeutic methods, in turn, also needed to be linguistically, materially, and performatively translated in ways that were understandable to his patients. Francisco's methods included the identification of both material evidence of social pollution (such as bundles of hair, bones, and insects, among others, in bodies or buried in the abodes of the sick people) and specific individuals involved in causing communal unrest, as the cause of the disease outbreak. At the same time, bundles crafted by ritual specialists could be used as protection against spiritual attacks.[38] And, as discussed below, Afro-Caribbean ritual practitioners explained to patients that such bundles, when appearing as tumors in bodies (or expelled as vomit), were evidence of witchcraft. That he was repeatedly called to perform cures in several Amerindian *encomiendas*, some of them hundreds of miles from Cartagena, where he lived, only speaks to the power of Francisco Mandinga's healing methods and, crucially, to the effectiveness of his translating practices.[39]

Another expert translator of bodily matters was Mateo Arará, an African ritual practitioner coming from Allada in West Africa who worked in the Caribbean shores of the New Kingdom of Granada during the 1640s. Mateo traveled the Caribbean countryside, healing in mining towns like Morocí, a place populated by Amerindians, and enslaved Africans and people of African descent. Similarly to Francisco, Mateo spent several days during his healing procedures speaking with members of the communities of forced laborers in these settlements. He also performed rites with diagnosing tools that included a walking mat and a dancing "little horn" with which he identified the cause (and culprits) of the illnesses plaguing his patients and the "best herbs to cure Christians."[40] All of the medical translation processes described above, and most of those in which practitioners like Mateo and Francisco were involved, were based on communal engagement and involved the participation of dozens of people.

The communal histories of translation of Caribbean healers are in direct contrast with the ones that dominate the literature on early modern medical translation and depend largely on the practices of individuals toiling and publishing alone. These histories also confound the notion that medical translation occurred mainly between "contrasting,

[38] For a discussion of "bundles" as protective devices in the Andes, see Margaret Brown Vega, "Ritual Practices and Wrapped Objects: Unpacking Prehispanic Andean Sacred Bundles," *Journal of Material Culture* 21 (2016): 223–51.

[39] AHN, Inquisición, L. 1021; AHN, Inquisición, L. 1022. For the history of *encomienda* in the New Kingdom of Granada, see María Teresa Molino García, *La encomienda en el nuevo reino de Granada durante el siglo XVIII* (Seville: Escuela de Estudios Hispano-Americanos, 1976). See also Ruiz Rivera, *Los indios de Cartagena* (cit. n. 31).

[40] AHN, Inquisición, L. 1021, fols. 340r–341r.

bounded languages."[41] Traditional approaches to the history of translation, as Susan Gal argues, individualize a process that "is invariably the work of a variegated social world," one which, I would add, also involved processes of a spiritual and performative nature not usually seen as part of the processes of knowledge exchange.[42]

A fundamental, and often bypassed, aspect of the communal nature of early modern Caribbean medical translation processes is that they were infused with power creation dynamics. It might seem odd to focus on the importance of medical translation for the enactment of social and political power-making strategies in a history of communities of enslaved Africans and displaced Amerindian communities under the putative rule of a European empire. When thinking about the political importance of ritual and healing practices of Atlantic Black and Amerindian groups, historians have framed these practices in terms of cultural survival and resistance against the evils of capitalism, colonialism, and racism.[43] There is no point in denying the importance of ritual practices in the struggle for the physical, social, and cultural conformation and reinvention of communities of African and Amerindian descent in the Atlantic. But such a portrayal risks missing the political and power-making resonances of practices of translation as uniquely reactionary against European political, cultural, or social systems.

Communal processes of medical translation in the early modern Caribbean were fundamentally defined by strategies of political and social negotiation within the communities in which they were deployed. In the process of translating physical, spiritual, and social characteristics of diseases, and therapeutic practices, people like Antonio Congo, Francisco Mandinga, and Mateo Arará carved a space of elevated social and political hierarchy for themselves and their bodily practices.[44] African healers occupied prominent political positions in their societies of origin and aimed at replicating such status in New World locales. Their social and political importance, and the recalcitrant persistence of their ideas in the imaginary of bodies and disease in the broader Caribbean region, even in the face of prosecution by ecclesiastical and judicial authorities, is testimony of how successful they were in establishing such power. In this sense, as much as those created by Europeans, Black Caribbean healers' translating practices also created boundaries of authority, healing hierarchies, and registers. They modeled "performance" as related to healing and denoted ideas about what terms and material evidence could transmit.[45] These struggles were not related only to the naming of diseases or the peddling of herbal cures. Instead, through their involvement in these processes of medicinal translation, African healers claimed social and political positions around fraught ontological issues. Not surprisingly, many of these healers ended up in jail or exiled. At stake was not so much how to precisely translate words or ideas, but rather how to make sure that one's way of classifying and thinking about the world had political and social relevance.

[41] Susan Gal, "Politics of Translation," *Annual Review of Anthropology* 44 (2015): 225–40, on 229.
[42] Ibid.
[43] For example, Vincent Brown, *The Reaper's Garden: Death and Power in the World of Atlantic Slavery* (Cambridge, MA: Harvard Univ. Press, 2010).
[44] This is because "the direction and purpose of translation matter in creating boundaries," as Gal argues; Gal, "Politics of Translation" (cit. n. 41), 231. If this is true of collaborations between "agents of colonial power" and "subaltern populations," the focus of Gal's discussion here, it is equally applicable to encounters between non-European populations, between Blacks and Amerindians and their descendants, for instance. Power and its delimitation were not uniquely a European matter.
[45] I am borrowing here from anthropologists and philosophers of medicine who argue that the denotation of terms for engaging in medical encounters is modeled by practice. See Gal, "Politics of Translation" (cit. n. 41), 234; and Annemarie Mol, *The Body Multiple: Ontology in Medical Practice* (Durham, NC: Duke Univ. Press, 2003).

MATERIAL TRANSLATIONS

Some of the most common ideas about diseases and tools for healing and diseasing in the Caribbean were composites of communal translations between communities of Amerindians and African descent. In his treatise on surgery first published in Spain in 1628, for example, the Cartagenero surgeon Pedro López de León wrote about the effectiveness of local Caribbean treatments for social ills that materialized in what I call "bundles of disease." These bundles of disease were evidenced in patients vomiting balls of hairs, bones, and worms, or in the form of "*apostemes*" (tumors/abscesses), which signified spiritual attacks ("poisoning") by ritual practitioners. The bundles depended on beliefs about the materialization of spiritual attacks that most likely had an Amerindian origin (even if the origin had strong resonances in contemporary West and West Central Africa). These bundles of disease appear repeatedly in testimonies related to the practices of healers of African descent in the Caribbean and circulated widely in early modern Spanish and Portuguese Atlantic locales. The West Central African healer Francisco Angola, for instance, chewed on herbs that he spread in the wounds of his patients before sucking on them and taking out of their bodies "small sticks, stones, hairs, and other similar things."[46] In Cuba, Antón Carabalí took out "small balls" from the foreheads "of sick people and worms and hairs from their bod[ies]."[47] There are dozens of similar examples in the historical record. Black Caribbean ritual practitioners adopted the notion of the bundle as a signifier of disease and developed techniques to treat poisoning that could be evidenced by the identification and extraction or expulsion of these bundles.

Attesting to the widespread importance of the idea of bundles of disease, the surgeon Pedro López de León explained in his book how *ambire*, a tobacco-based compound substance used by Amerindian ritual practitioners in the northern New Kingdom of Granada around the city of Santa Marta, "makes the sick person expel the curse through his mouth and intestines little by little. I myself have seen how some people [after using the remedy] throw out little bones of toads through their mouths"—animals' bones being a common component of these bundles.[48] These Afro-Caribbean bundles were crucially embodied modes of translation that made visible the world of disease and functioned in the modeling of communal understandings of social ills for Caribeños of all extractions.

Natural historians and physicians and surgeons like López de Leon, as discussed in the growing literature on the history of medicinal trade in the Atlantic, gathered information about the natural world from Amerindians and populations of African descent in the New World.[49] The example of *ambire* is just one of many ways in which the practices of Black and Amerindian healers were inscribed in the published record. These

[46] AHN, Inquisición, L. 1023, fol. 349v.

[47] AHN, Inquisición, L. 1020, fol. 297r.

[48] López de León, *Prática y teórica de las apostemas en general, question y practicas de cirugía, de heridas llagas y otras cosas nuevas y particulares*... (Christóbal Gálvez: Calatayud, 1685), 200. *Ambire* also shows up in inventories of medicines purchased in Cartagena by Portuguese slave traders; see Linda Newson and Susie Minchin, *From Capture to Sale: The Portuguese Slave Trade to Spanish South America in the Early Seventeenth Century* (Boston, MA: Brill, 2007), 274–5.

[49] See, for instance, Crawford and Gabriel, *Drugs on the Page* (cit. n. 7); or Renate Dürr, "Early Modern Translation Theories as Mission Theories: A Case Study of José de Acosta 'De procuranda indorum salute' (1588)," in *Cultures of Communication: Theologies of Media in Early Modern Europe and Beyond*, ed. Helmut Puff, Ulrike Strasser, and Christopher Wild (Toronto: Toronto Univ. Press, 2017), 209–27.

acts of translation, obviously, depended on Europeans' reliance on the importance of specific types of communication. Non-European linguistic, material, performative, and embodied channels were relevant for Europeans only insofar as they framed specific epistemological and linguistic parameters and questions. This type of translation revolved around European vernaculars or Latin and was highly scripted and formulaically prescribed in predesigned questionnaires and ritualistic formulas. The ways natural historians and physicians used to organize categories and analytical practices from non-Europeans depended on similar processes of translation into European linguistic and epistemological models—and were framed around similar dynamics of patronage and competition.[50]

Except for the brief mention of *ambire* in López de Leon's treatise, most of the practices of Caribbean Black and Amerindian communities (with notable exceptions for those associated with tobacco) did not make it to the published record of early modern compendia of materia medica coming from the New World.[51] Scholars have espoused several reasons for Europeans' failure to incorporate African or Afro-Caribbean healing plants into the therapeutic arsenal of the early modern West. Among these theories are the lack of commercial possibilities for most substances used by health specialists of African descent and, prominently, early modern Europeans' characterization of most African knowledge about the natural world as tainted by superstition and linked to ritual practices that could not easily circulate (or be translated, if you will). More broadly, the existent literature on the topic has focused on how, when incorporating them into their pharmacopeias, Europeans made sure to void African (and Caribbean) therapeutic elements of the indigenous knowledge and traditions behind their usage.[52]

Caribbean Amerindian and Black communities' therapeutic elements do, however, appear repeatedly in inventories of the medicines that local pharmacy owners provided to slave traders—including records coming from Panama and Cartagena, where substances like *ambire*, the balsam de Tolu, and *caraña* are priced in bills for medical services.[53] We also know of the existence of dozens of botanicals with bodily effect that ritual practitioners of African descent used with great success in the region during the seventeenth century.[54] Some, like the herb *limpiadientes*, are briefly cited in unpublished (at the time) manuscripts by European physicians like Juan Mendez Nieto at

[50] See Michael Wintroub, "Translations: Words, Things, Going Native, and Staying True," *Amer. Hist. Rev.* 120 (2015): 1185–217; Harold Cook and Timothy D. Walker, "Circulation of Medicine in the Early Modern Atlantic World," *Soc. Hist. Med.* 26 (2013): 337–51; Jürgen Renn, "The History of Science and the Globalization of Knowledge," in *Relocating the History of Science*, ed. Theodore Arabatzis et al. (Boston, MA: Boston Studies in the Philosophy and History of Science, 2015), 241–52; or Secord, "Knowledge in Transit" (cit. n. 4).

[51] Most of the Amerindian knowledge from the New World appearing, even if obscured, in printed texts in Europe during the sixteenth and seventeenth century came from New Spain (Mexico), most famously in the work of Nicolás Monardes and its multiple translations; Nicolás Monardes, *Dos libros. El uno trata de todas las cosas que traen de nuestras Indias Occidentales, que sirven al uso de medicina . . .* (Seville, 1565); and Monardes, *Segunda parte del libro, de las cosas que se traen de nuestras Indias Occidentales, que siruen al vso de medicina* (Seville, 1571). See also Antonio Barrera-Osorio, *Experiencing Nature: The Spanish American Empire and the Early Scientific Revolution* (Austin: Univ. of Texas Press, 2006).

[52] See Benjamin Breen, "The Flip Side of the Pharmacopeia? Sub-Saharan African Medicines and Poisons in the Atlantic World," in Crawford and Gabriel, *Drugs on the Page* (cit. n. 7), 143–59.

[53] See, among others, ARSI, Opera Nostrorum 17; AGI, Contaduria, 263; AGI, Contaduria, 264A; ARSI, Brasil, 58; AGN, Colonia, Hospitales, 34, fols. 592–617.

[54] See, for example, AHN, Inquisición, L. 1021; AHN, Inquisición, L. 1022; AHN, Inquisición, L. 1023; AGN, Colonia, Negros y Esclavos; AGI, Contaduria, 243; AGI, Justicia, 74.

the end of the sixteenth century.[55] Moreover, some of the therapeutic elements described above, as well as ideas about bundles of disease, also appear in the surviving records of the Jesuit college and correspondence sent back to Rome. The Jesuit pharmacy in Cartagena amassed what was perhaps the largest pharmacy in the Caribbean.[56] In addition to these medicinals, Black ritual practitioners developed and used bodily procedures, healing tools, and curative rites that became the most common therapeutic practices across the region throughout the seventeenth century.[57]

In other words, it is evident that even though Black and Amerindian knowledge and information about medical goods in the Caribbean did not make its way to Europe, it was actively exchanged and translated among people of all origins. Surviving manuscript records speak to the rich processes of incorporation and translation of medicinal substances from the region. Such exchanges, however, mostly ensued through translations between health practitioners working outside European institutions and languages.

I have explored elsewhere how a materialistic approach to the history of healing tools used by early modern Caribbean people allows us to interrogate assumptions about Caribbean worlds' ontologies, and how thinking with objects helps historians of science and medicine to understand the limitation of the types of epistemological assumptions they have used to deal with these histories.[58] Material culture elements, such as the famous ritual bags that circulated through the early modern Atlantic, also functioned as effective and common, if often unacknowledged, methods for the translation of beliefs around ideas about health, disease, and healing.

The best known of these artifacts, as explored in multiple studies, are *Mandinga* bags and Obeah pouches.[59] These assemblages of pieces of scripture, relics, and animal, vegetable, and mineral elements appear in the historical record throughout the Atlantic, in Brazil, Portugal, West Africa, and Spain, among other places. People used them for therapeutic purposes and protection against spiritual attacks. Similar objects circulated in seventeenth-century Caribbean locales. For example, Antonio de Salinas, a sixty-year-old free Black fisherman who descended from "blacks from Guinea" and had previously lived in Nicaragua and Guatemala, explained in 1666 that a bag an Inquisition official had found in his possession protected him from bullets and the attacks of animals. The contents of the sack, he said, were also "very beneficial for the body."[60] Among the elements that inquisitors found inside the bag were "powders wrapped in two small papers"; icons of saints and the virgin; "a stick that was seemingly part of a cross stuffed with relics"; several pieces of red and blue fabric; a "*havilla*, which is a contra against snakes"; corn; slivers of roots and leaves; a leaf of tin in the shape of a cross; pieces

[55] See Juan Méndez Nieto (1611), *Discursos medicinales* (Salamanca: Universidad de Salamanca, Junta de Castilla y León, 1989).

[56] Archivo General de la Nación de Colombia, Bogotá, Colombia, Colonia, Hospitales, 34, fols. 592–617. Examples of the Jesuits' incorporation of indigenous materia medica are legion. See, for instance, the impressive compilation of medicinal preparations crafted in 1766 in Rome using the techniques contributed by Jesuits from Brazil, Macao, India, and Portugal (ARSI, Opp. NN. 17).

[57] Gómez, *Experiential Caribbean* (cit. n. 26).

[58] For a discussion of early modern Black Caribbean material culture and issues of ontology and epistemology, see Gomez, "Caribbean Stones" (cit. n. 6).

[59] Benjamin Breen recounts how the prohibition of similar pouches in Jamaica in the 1720s barred enslaved healers from "making use of any blood, feathers, parrots beaks, dog's teeth, alligator's teeth, broken bottles, grave dirt, rum, egg, shells or any other material relative to the practice of obeah or witchcraft"; Breen, "Flip Side of Pharmacopeia" (cit. n. 52). See also Farris Thompson, *Face of the Gods: Art and Altars of Africa and the African Americas* (New York: Museum for African Art, 1993).

[60] AHN, Inquisición, L. 1023, fols. 402r–403r.

of bread; "an old bull of the Holy Crusade"; a written prayer that begins "very miraculous prayer and very helpful for the soul and body" and finishes "with the will of Jesus"; a "written prayer [protecting the wearer] from drawing or being injured or taken prisoner"; and "another that begins: 'I entrust myself to the just Jesus and his mother whose son is.'"[61]

Contemporary natural historians and physicians located power objects like those used by Juan de Salinas and their Black creators within categories that contained the realms of the magical and unnatural. These powerful bags were part of what for European natural historians was a category of Atlantic objects that defied epistemic categorization within learned frameworks, other than that of witchcraft. These "fetishes" were foundational for the imagination of the untranslatable/irrelevant characteristics of non-European bodily practices and knowledge.[62] Juan de Salinas's bag is, thus, a strange chimera: An object crafted out of materials coming from multiple healing/spiritual cultures. Roots, Virgin Mary paintings, and corn kernels wrapped in a rough fabric that encompassed a world. Because of their promiscuous nature, these artifacts held wondrous powers. Afro-Caribbean bags functioned, then, as "boundary objects" existing in and between ages, oceans, and disciplines. As much as they translated, they also served as an index of otherness through which Europeans epistemologically organized powers and obscured knowledge within the defined categories, organizational practices, and authoritative models of *historia*, botany, and natural history.[63]

De Salinas's sack, however, is also a vehicle that allows us to think about materialized translations of communal practices with social, political, and bodily effects. These materialized translations in the form of power objects or bundles of disease, their effects, and the important role they occupied in the ideation of early modern Caribbean notions of health and disease are an example of the fact that "not merely lexically different ways" of naming something are responsible for how communities think about material objects and their effects on bodies and the natural world. This is of particular significance for the analysis of elements such as purges, bundles of disease, or crucifixes, and our study of their role as epistemic and ontological conductors in early modern Caribbean communities of patients and healers.[64] In the Caribbean, health specialists

[61] Ibid., fols. 401v-403r.

[62] For the usage of the concept of the fetish as an epistemological device of differentiation between what were ostensibly similar European and non-European objects and practices, see Bruno Latour, *On the Modern Cult of the Factish Gods* (Durham, NC: Duke Univ. Press, 2011); and Latour, "Fetish—Factish," *Material Religion* 7 (2015): 42–9.

[63] For usages of the multiple slippages of boundary objects, see Susan Leigh Star and James R. Griesemer, "Institutional Ecology: 'Translations' and Boundary Objects: Amateurs and Professionals in Berkeley's Museum of Vertebrate Zoology, 1907–39," *Soc. Stud. Sci.* 19 (1989): 387–420; and Kathryn DeLuna, "Marksmen and the Bush: The Affective Micro-Politics of Landscape, Sex and Technology in Precolonial South-Central Africa," *Kronos* 41 (2015): 37–60. For *historia* practices, see Gianna Pomata, "Praxis Historialis: The Uses of Historia in Early Modern Medicine," in *Historia: Empiricism and Erudition in Early Modern Europe*, ed. Pomata and Nancy Siraisi (Cambridge, MA: MIT Press, 2005), 1–38.

[64] Gal, "Politics of Translation" (cit. n. 41), 232. For a recent overview of the role of objects as conduits of translation and "the interplay between human agents and material objects," see Maeve Olohan, "History of Science and History of Translation: Disciplinary Commensurability?," *The Translator* 20 (2014): 9–25. See also Andrew Pickering, "Decentering Sociology: Synthetic Dyes and Social Theory," *Perspectives on Science* 13 (2005): 352–405; and, obviously, Bruno Latour, *Reassembling the Social: An Introduction to Actor-Network-Theory* (Oxford: Oxford Univ. Press, 2005). These processes muddle differences between terminology agents and objects. See, for example, Bruno Latour, *The Pasteurization of France* (Cambridge, MA: Harvard Univ. Press, 1988).

crafted a type of healing material culture and languages that were defined by their changeable, indexable qualities.[65]

As Todd Ochoa explains when discussing the power of *prendas-ngangas-enquisos* in his study of Palo practitioners in Cuba, complex objects like de Salinas's bag allowed for "marvelously ambivalent transfers of social and conceptual force."[66] When in the appropriate hands, these objects could be a channel for "forces that solidify one version of reality, just as it can create a tear in the screens of petrified, fated thought and for an instant make fluid thinking possible."[67] Antonio de Salinas's bag worked as a translator of a plethora of social understandings of corporeality intimately interconnected. In the early modern Caribbean, these reality-congealing materials also included a variety of ritual and medicinal tools: feathers, drums, rattles, walking and talking Black figurines, dancing horns, and animated mats, among others. They represent the manner in which Caribbean people incorporated and translated into their own understandings of the world different medical cultures of European, Amerindian, African, and, increasingly, creole origins. These objects are also evidence of how translation efforts involving interactions between human and material objects "condense and express branching histories of social relations" of crucial importance for the history of healing practices.[68]

CONCLUSION

As the preceding pages show, in the early modern Caribbean, non-European historical actors, mainly Amerindians and enslaved Africans and their descendants, were at the center of productive communal processes of translation around health, healing, and, more generally, bodily matters. Existent frameworks for thinking about medical translation that depend on analyses of imperial and colonial dynamics; or divides between nature and culture, and religion and science; or those that emphasize individual projects of translation, are not adequate for examining most of the translation practices around bodily matters that ensued in the early modern Caribbean. Looking at translation through the lenses of the early modern Caribbean provides new possibilities for the analysis of translation practices elsewhere that do not depend on traditional geographical and epistemological boundaries (both outside EuroAmerica, as well as within it).

Historians of medicine have rarely considered the types of exchanges that occurred between Africans and Amerindians in the Americas to be part of the processes of knowledge circulation that fueled the proliferation of translation practices during the early modern era. Like the other articles in this collection, this article discusses evidence that emphasizes the importance of looking at practices of translation that go beyond "key moments" of translation and their (usually European) protagonists as they appear in most existent histories of medicine and science. The articles in this volume also emphasize, crucially, the value of incorporating practices that historians of medical translation have ignored, particularly those occurring both in non-European settings (or

[65] Gómez, *Experiential Caribbean* (cit. n. 26), chap. 6.

[66] Todd R. Ochoa, *Society of The Dead: Quita Manaquita and Palo Praise in Cuba* (Berkeley: Univ. of California Press, 2010), 8.

[67] Ibid., 7.

[68] Gómez, *Experiential Caribbean* (cit. n. 26), chap. 6. See also Stephan Palmié, "Thinking with Ngangas: Reflections on Embodiment and the Limits of 'Objectively Necessary Appearances,'" *Comp. Stud. Soc. Hist.* 48 (2006): 852–86.

that did not involve Europeans) and in the immaterial, spiritual, and ritual contexts detailed in the contributions by Alberts and Mukharji in this volume.[69]

Most histories of translation of early modern science and medicine in the Atlantic world revolve around a set of parameters defined by the inclusion of imagined bounded systems of knowledge, identified under labels such as Nahuatl, Congo, Arará, or Taino within European epistemologies. The more sophisticated of these studies would not claim that there are direct continuities between seventeenth-century historical actors like Mateo Arará and contemporary Cuban *Babalawos* on the basis of vague cultural/religious traditions ahistorically brewing in the region (although there are, of course, examples of this sort of work). Still, most historians do locate them in a common analytical space defined by ideas about Africanness, slavery, and, of course, culture and religion, among others.[70] These analyses commonly proceed through a reconstruction of how indigenous knowledges were either (partially) incorporated within projects of natural history and pharmacopeias in Europe or how they disappeared from the historical record.

Existent methodologies for the analysis of medical translation practices in the Atlantic depend, thus, on the ideation of an analytical space where historians can identify colonial relationalities and cultural syntheses of distinct imagined traditions—what scholars have labeled "hybrid," "mestizo," or even "entangled" knowledges.[71] Consequently, scholars place the emphasis of these types of Atlantic histories on the study of how knowledge moves (or not) within the confines of European tenets, and gets translated (or lost) as it moves from one fictitiously discrete epistemological realm to the next one.[72]

Unlike existent histories of transatlantic knowledge-movement, the histories of translation in the Caribbean allow for the elucidation of new possibilities for thinking not only about different modes and nodes for medical translation practices but also their consequences. The translations of Caribbean enslaved people and Amerindians were at the center of bodily transactions of all sorts in the region and were fundamental for the creation of powerful healing traditions. In the process of translating words, sounds, smells, pain, and ecstasies, these Caribbean actors congealed material and immaterial worlds. They depended on communal dynamics that birthed localized political and epistemic hierarchies. Histories of translation, thus, do not always have to look at knowledges, objects, or gestures spanning hemispheres. In the early modern Caribbean, productive practices of translation most commonly moved words, rituals, and objects across imperceptible boundaries separating individuals and communities living in close proximity (in many cases under the same roof).

[69] See Tara Alberts, "Translating Alchemy and Surgery"; and Projit Bihari Mukharji, "Casting Blood Circulations," both in *Osiris* 37.

[70] For instance, see Robert A. Voeks, *Sacred Leaves of Candomblé: African Magic, Medicine, and Religion in Brazil* (Austin: Univ. of Texas Press, 1997); Heather M. Kopelson, "'One Indian and a Negroe, the First Thes Ilands Ever Had': Imagining the Archive in Early Bermuda," *Early Amer. Stud.* 11 (2013): 272–313; or Sweet, *Domingos Alvares* (cit. n. 26).

[71] For a recent discussion, see Marcy Norton, "Subaltern Technologies and Early Modernity in the Atlantic World," *Col. Latin Amer. Rev.* 26 (2017): 18–38.

[72] See, for instance, the recent collection of essays on knowledge-making practices in the Iberian world: Helge Wendt, ed., *The Globalization of Knowledge in the Iberian Colonial World* (Berlin: Max Planck Research Library for the History and Development of Knowledge–Open Access, 2016). See also Timothy D. Walker, "The Medicines Trade in the Portuguese Atlantic World: Acquisition and Dissemination of Healing Knowledge from Brazil (c. 1580–1800)," *Soc. Hist. Med.* 26 (2013): 403–31; or Matthew J. Crawford, *The Andean Wonder Drug: Cinchona Bark and Imperial Science in the Spanish Atlantic World, 1630–1800* (Pittsburgh, PA: Univ. of Pittsburgh Press, 2016).

At the same time, and as other articles in this volume demonstrate (for instance, Benjamin Breen's piece), Caribbean histories of translation demonstrate the importance of examining these processes as taking place through different avenues—linguistic, literary, performative—and through material culture. Thinking with the world of translation in the early modern Caribbean also allows us to focus on the ambiguity of the transfers that ensued in all early modern medical transactions (as discussed in Hansun Hsiung's piece in this volume), and the multiple channels, embodied, material, performative, and also linguistic, through which these transactions flowed. It invites us to consider the histories of medical translation in the early modern world as windows into processes of "perpetual indeterminacy."[73]

The seemingly fragmented and isolated lands of the early modern Caribbean can be productive points for ideating methods that allow for the elucidation of the evanescent, yet indispensable, kinds of histories, geographies, temporalities, and translations of the people at the receiving end of the processes of colonialism and imperialism of the early modern era. This approach transcends histories of subalterns, or models of knowledge circulation that still think about center and peripheries or divides between religion/science, nature/culture. Engaging fruitfully with the borderless characteristics of early modern Caribbean histories of translation should make us think not about the inappropriateness of its extant sources but rather about the ebullient and inescapably messy nature of the world spurned by slavery and colonialism—dynamics characterizing scenarios of knowledge transfers elsewhere. The history of Caribbean Black and Amerindian translations of bodily knowledge should invite us to recognize not alien worlds but instead the possibilities that created ours.

[73] See Benjamin Breen, "Pyric Technologies and African Pipes"; and Hansun Hsiung, "'Use Me as Your Test!,'" both in *Osiris* 37. The quoted phrase is from Ochoa, *Society of the Dead* (cit. n. 66), 18.

Translating Surgery and Alchemy between Seventeenth-Century Europe and Siam

*by Tara Alberts**

ABSTRACT

René Charbonneau (1643–72) traveled to Siam (modern-day Thailand) in 1676 as a lay auxiliary to a band of French missionary priests. It was intended that he would practice his skills as a surgeon to treat missionaries and the indigent poor in the mission hospital established in Ayutthaya. An examination of the few letters that survive in his hand reveal fascinating insights into some of the treatments he used. Details emerge that suggest how much healers like Charbonneau were transformed as they moved into new contexts: how they were able to experiment with new techniques and cures, and how they were often required to reconceptualize their approaches for new audiences. This article explores Charbonneau's life and career in seventeenth-century Ayutthaya to reassess the "foreign" healers in the history of Thai medicine and to explore the methodological complexities of reconstructing the medical world of Ayutthaya.

In 1730 an ecumenical funeral was held in Ayutthaya, the royal capital of Siam (modern-day Thailand). Present were representatives of various Catholic communities—Portuguese, French, Vietnamese, Spanish, and Siamese—the Protestant head (*Opperhoofd*) of the Dutch factory and his assistant, and various "people of other religions."[1] The man being laid to rest was René Charbonneau (ca. 1646–1730), a "venerable elderly gentleman" of eighty-four, whose passing was "regretted by all the Christians and even the pagans."[2] Charbonneau was born in Les Herbiers, Poitou, France, the son of a tenant farmer.[3] By the time of his death he had lived in Siam for over fifty years. Fluent in Thai and knowledgeable about the country, he was often asked for advice by locals and visitors.[4]

* Department of History, University of York, Heslington, York, YO10 5DD, UK; tara.alberts@york.ac.uk

[1] Bhawan Ruangsilp, *Dutch East India Company Merchants at the Court of Ayutthaya: Dutch Perceptions of the Thai Kingdom, c. 1604–1765* (Leiden: Brill, 2007), 45.

[2] Paul Aumont, "Mémoires de M. Aumont," in *Histoire de la mission de Siam: Documents historiques*, ed. Adrien Launay, 2 vols. (Paris: P. Téqui, 1920), 1:266. Charbonneau was also spelled "Cherboneau" and "Cherbonneau." Unless otherwise noted, the translations in this article are mine.

[3] Bernard Martineau to superiors in Paris, Phuket (Junk Ceylon), 26 November 1686, printed in Joseph Grandet, *Les Saints Prêtres Français du XVIIe siècle . . . Troisième Série. Prêtres Angevins* (Angers: Germain et G. Grassin, 1898), 359.

[4] Bénigne Vachet, "Mémoires de M. Bénigne Vachet," in Launay, *Histoire* (cit. n. 2), 1:317.

Osiris, volume 37, 2022. © 2022 History of Science Society. All rights reserved. Published by The University of Chicago Press for the History of Science Society. https://doi.org/10.1086/719229.

He was probably known to the Siamese by the noble title *ok phraya* (ออกพระยา) and addressed as *chaokuhn* (เจ้าคุณ - "lord"): honors granted by the former king, Phra Narai (r. 1656–88), when he appointed Charbonneau governor of Phuket in around 1682.[5] Many of the French also recognized his ennobled status, referring to him as *Sieur* René. In 1715 a Catholic missionary had written to the French governor of Pondicherry recommending Charbonneau as someone "highly esteemed by all the foreigners in this kingdom, who very often consult with him about their business, and appoint him as arbiter in their quarrels."[6] "The Siamese also respect and honor him greatly," he continued. "They always treat him like one of their old mandarins, who has been responsible for the government of a province, and who throughout all the time of his governorship pleased [everyone] greatly."[7]

Charbonneau had arrived in Siam in 1676 in a much more humble role: as house servant (*domestique*) or valet to a group of priests from the French missionary society, the Missions Étrangères de Paris (MEP). He had been selected because of his piety and his knowledge of surgery.[8] The missionaries intended to make medical care a key part of their proselytization strategy in Siam, where "missionaries apply themselves to the cure of illnesses of the body in order to come more easily to that of the soul."[9] They had established a small cottage hospital in Ayutthaya where Charbonneau was to serve as a surgeon, bandaging, bloodletting, and handing out simple remedies.[10] Yet his designation as a "*chirurgien*" in the French sources masks the increasing complexity of his role and the transformation that he underwent.

This article explores Charbonneau's life as a healer in Ayutthaya and reconstructs his trajectory from servant and surgeon to provincial governor. In Section 1, I set the scene, mapping out the medical, scientific, and cultural world of Ayutthaya. I address how any attempt to reconstruct histories of seventeenth-century Siam is hindered by silences and disarticulations in the archive. Because of the destruction of Ayutthaya in 1767 during a Burmese invasion, only a small number of contemporary Thai-language sources related to medical practices survive.[11] To catch sight of many aspects of life in the capital, historians therefore often turn to European sources of the period. These accounts provide many details of Siamese medical practices, but they appear only in translation: sketched perceptions derived from fleeting experience.

[5] Simon de la Loubère, *Du Royaume de Siam*, 2 vols. (Paris: Veuve Jean-Baptiste Coignard, 1691), 1:177. Nicolas Gervaise describes the "*ocpraya*" as the "second order of nobility"; Gervaise, *Histoire naturelle et politique du Royaume de Siam* (Paris: Claude Barbin, 1688), 122. On the five ranks of Thai civil nobility, see Robert B. Jones, "Thai Titles and Ranks," *Cornell Southeast Asia Programme Data Paper* 81 (1971): 127. For the titles of previous governors of Phuket, see Dhiravat na Pombejra, "Towards an Autonomous History of Seventeenth-century Phuket," in *New Terrains in Southeast Asian History*, ed. Abu Talib Ahmad and Tan Liok Ee (Athens: Ohio Univ. Press, 2003), 292–3, which suggests that previous governors had the rank of *phra* (พระ), the third rank. Other histories of Phuket give Charbonneau's name and title as *okphraya khlang Charbonno* (ออกพระยาถลาง ชาร์บอนโน). See also G. E. Gerini, "Historical Retrospective of Junkceylon Island," *J. Siam Soc.* [hereafter *JSS*] 2 (1905): 121–268, especially 146–50, 162; and Michael Smithies, "Les gouverneurs français de Phuket, Bangkok et Mergui au XVIIe siècle," *Aséanie* 7 (2001): 59–78; see 63–78.

[6] Mgr De Cicé to M. Hébert, 1715, in Launay, *Histoire* (cit. n. 2), 2:102.

[7] Ibid.

[8] Called "domestique" in Jacques de Bourges, *Relation du voyage de Monseigneur l'Evêque de Beryte* (Paris: Denis Bechet, 1668), 168; "valet" in Loubère, *Du Royaume de Siam* (cit. n. 5), 351.

[9] "Autres nouuelles enuoiées de Siam en 1677," Archives de la Mission Étrangères de Paris [hereafter AMEP], Paris, 859:26.

[10] "Lettre de M. de Berithe escritte de Siam à M. d'Heliopolis au mois d'Octobre 1667," AMEP, 857:224.

[11] Somchintana Thongthew Ratarasarn, *The Principles and Concepts of Thai Classical Medicine* (Bangkok: Thai Kadai Research Institute, 1989), 29–30.

The resultant mistranslations have reverberated throughout the historiography. I explore Charbonneau's place in traditional narratives of Thai medicine and argue for a reassessment of the self-fashioning and translation practices of "foreign" experts like Charbonneau as they were remade by the Ayutthayan context.[12]

In Sections 2 and 3 I turn to the range of treatments used by Charbonneau. I show how, with careful contextualization, European sources can be mined for evidence of how concepts, materials, and techniques were translated in Siam. In these sections I trace both how the practices of "foreign" healers like Charbonneau were modified in translation, and how new contexts created the space for the transformation of healers themselves. As a healer in Ayutthaya, Charbonneau did not restrict himself to his former (and in France, regulated) "art" of *chirurgie*, "which teaches diverse manual operations on the human body for the healing of injuries, wounds, fractures etc."[13] He was quick to adopt wider responsibilities—researching and compounding medicines, and training others in internal medicine and surgery. Most important, he strove to learn new techniques, especially those that were deemed valuable to local audiences: how to use unfamiliar materia medica; new approaches to diagnosis and cure; and "alchemical" techniques for the purification of metals and preparation of medicinal substances.

Of course, the intimate, personal relations between healer and patient only rarely leave an archival trace. We are fortunate that we can catch occasional glimpses of Charbonneau in action. His posterity is formed in institutional archives, with all the consequences this has for the survival of only certain aspects of his story: the few letters in his hand that survive are from his time in the service of the Missions Étrangères.[14] The experiences of patients themselves, and their understandings of his cures, are, of course, even more obscure. How might we reconstruct the translations in which Charbonneau was engaged, given the light footprint he has left in the written record? Reading across multiple sources, I consider how patients may have understood techniques and materials utilized by practitioners like Charbonneau.

In tracing Charbonneau's travels, I follow in the wake of a number of historians who have used microhistories to disrupt or add nuance to conclusions drawn through more broad-brush mappings of global circulations.[15] As John Paul Ghobrial has argued, a close focus on the "moving stories" of such individuals "can reveal new geographies that we do not see otherwise" and can help to "unsettle, to surprise, and to oblige us to revise our traditional views."[16] I trace how local contexts shaped Charbonneau's

[12] On the complexity of Thai concepts of "foreignness" in this period, see Edward van Roy, *Siamese Melting Pot: Ethnic Minorities in the Making of Bangkok* (Singapore: ISEAS Publishing, 2017); and John Smith, "State, Community, and Ethnicity in Early Modern Thailand, 1351–1767" (PhD diss., Univ. of Michigan, 2019). On the broader literature concerning the figure of the mediator, broker, or go-between, see the introduction to this volume, *Osiris* 37.

[13] *Dictionnaire de l'Académie française*, 1st ed. (Paris: Jean Baptiste Coignard, 1694), 1:187.

[14] Charbonneau's letters were preserved as some sections provided publishable information about the missions. Parts of a letter written in October 1677 (see n. 89 in the present article), for example, were excerpted in *Relation des missions et des voyages des evesques vicaires apostoliques, et de leurs ecclesiastiques és Années 1676 & 1677* (Paris: Charles Angot, 1680), 183–4.

[15] See especially Natalie Zemon Davis, *Trickster Travels: A Sixteenth-Century Muslim between Worlds* (London, 2008); Sanjay Subrahmanyam, *Three Ways to be Alien: Travails and Encounters in the Early Modern World* (Waltham, MA: Brandeis Univ. Press, 2011); Jonathan Gebhardt, "Microhistory and Microcosm: Chinese Migrants, Spanish Empire, and Globalization in Early Modern Manila," *J. Medieval Early Mod. Stud.* 47 (January 2017): 167–92; and John Paul Ghobrial, ed., "Global History and Microhistory," Issue supplement 14, *Past & Present* 242 (2019).

[16] John-Paul Ghobrial, "Moving Stories and What They Tell Us: Early Modern Mobility between Microhistory and Global History," in Ghobrial (cit. n. 15), 243–80, on 279.

translations and, analogously, how he himself was transformed. I show how these translations and transformations created opportunities to lay claim to new forms of authority and value.

1. LOST LEAVES AND (RE)DISCOVERED GEMS: CONSTRUCTIONS OF MEDICINE IN AYUTTHAYA

Sources for the History of Thai Medicine

The destruction of Ayutthaya by the Burmese in 1767 included the wreckage of the kingdom's intellectual capital, with books and manuscripts burnt or scattered.[17] During the reign of Rama I (r. 1782–1809), there began a concerted attempt to reconstruct the lost leaves of Ayutthayan intellectual heritage by recovering as far as possible texts of law, Buddhism, and medicine.[18] Texts that had survived, often in provincial temples, were gems which transmitted the essence of the lost golden age. Further texts were rediscovered in the nineteenth and early twentieth centuries. "Traditional" medical practices became a key area of investigation.

During the reigns of the Chakri kings Mongkut (r. 1851–68) and Chulalongkorn (r. 1868–1910), in parallel with the introduction of "modern" or "western" (ฝรั่ง - *farang*) medicine, programs of textual recovery and revision were undertaken to identify historical medical texts.[19] A canonical corpus of urtexts—representative of a supposedly national tradition—was established. The seventeenth-century *Tamra Phra Osot Phra Narai*, discussed below, for example, was "rediscovered" by Krom Luang Wongsa, the chief physician of the royal court, at this time when "traditional medicine was fiercely challenged by Western medical knowledge."[20] These texts were also brought to a wider public: an edition of this *Tamra* was printed in 1917 by Prince Damrong Rajunubhab (1862–1943).[21] Necessarily, these compilations often involved the deconstruction of texts and their translation—out of the formats of palm-leaf manuscripts and *samut khoi* (สมุดข่อย - folding books of paper made from *Streblus asper* bark) and into

[17] See Chris Baker and Pasuk Phongpaichit, *A History of Ayutthaya: Siam in the Early Modern World* (Cambridge, UK: Cambridge Univ. Press, 2017), 253–8.

[18] See especially Jean Mulholland, "Thai Traditional Medicine: Ancient Thought and Practice in a Thai Context," *JSS* 67 (1979): 82–3; James Nelson Riley, Fred L. Mitchell Jr., and Dan Bensky, "Part Two - Thai Manual Medicine as Represented in the Wat Pho Epigraphies: Preliminary Comparisons," *Medical Anthropology* 5 (1981): 155–94; and Nopphanat Anuphongphat and Komatra Chuengsatiansup, "Krom Luang Wongsa and the House of Snidvongs: Knowledge Transition and the Transformation of Medicine in Early Modern Siam," in *Histories of Health in Southeast Asia: Perspectives on the Long Twentieth Century*, ed. Tim Harper and Sunil S. Amrith (Bloomington: Indiana Univ. Press, 2014), 19–43, on 21.

[19] Riley, Mitchell, and Bensky, "Thai Manual Medicine" (cit. n. 18), 157–60; C. Pierce Salguero, *Traditional Thai Medicine: Buddhism, Animism, Ayurveda* (Prescott, AZ: Hohm Press, 2007), 11–8.

[20] Anuphongphat and Chuengsatiansup, "Krom Luang Wongsa" (cit. n. 18), 21.

[21] *Tamra Phra Osot Phra Narai* [King Narai's medical treatise] (Pranakhon: Sophinpipattanakorn, 1917). Through this printed version, the *Tamra* became a popular text to be distributed as part of a cremation volume given out to mourners. See Jean Mulholland, *Medicine, Magic and Evil Spirits: Study of a Text on Thai Traditional Paediatrics* (Canberra: ANU Press, 1987), 10–12; Junko Iida, "The Invention of Medical Tradition in Thailand: Thai Traditional Medicine and Thai Massage," in *Translating the Body: Medical Education in Southeast Asia*, ed. Hans Pol, C. Michaele Thompson, and John Harley Warner (Singapore: NUS Press, 2017): 273–94; and Salguero, *Traditional Thai Medicine* (cit. n. 19) 7–9.

modern print and paper; out of the poetic obscurity of premodern Siamese. The colonial context of the nineteenth and twentieth centuries inevitably shaped the ways in which "traditional" Thai medical systems were reconstructed from these texts and set against or alongside "modern" or "Western" frameworks.[22] For example, the texts often set out the principles of Thai medicine in a way that emphasized synergies with explanatory frameworks found in Ayurvedic medicine to emphasize the equal intellectual legitimacy and scholarly heritage of the Thai system through reference to an ancient and recognized system.[23]

These drives for textual rediscovery have had a number of consequences for historians. In many cases we now have access only to nineteenth- and twentieth-century recensions of earlier texts. The principles by which copyists established the "correct" version of texts are often opaque. The emphasis on a stable national tradition has often meant that many other aspects of continuing knowledge complexes were replaced by these texts made canonical by accidents of survival and rediscovery. Owing to a lack of surviving evidence, for example, it is difficult to know how far the theories and practices of the *mo ratsadon* (หมอราษฎร - doctors who treated commoners) and the *mo luang* (หมอหลวง - those who practiced royal medicine) overlapped. All the surviving texts dating back to early periods relate to the latter. Necessarily, in this article I therefore focus mainly on royal medicine, which certainly came to inform the practices of many different types of practitioners throughout central Siam. However, we should be alert to the diversity of approaches among practitioners and patients outside of the royal court, which remain obscure in many sources.

Because of the limited nature of the Thai-language material, scholars continue to draw on contemporary European accounts of premodern Siamese medicine. Early modern European accounts were often rather schematic and tended to see Siamese medicine as one practice, characterized by vague generalities. This was partly owing to difficulties in gaining access to texts that would have illustrated the underlying principles: recipe compendia, treatises, and other materials used by healers were often professionally privileged, and jealously guarded.[24] No seventeenth-century European accounts refer to having been able to study such documents. There was a wide range of healers in Ayutthaya, Siamese and foreign, all with their own traditions of knowledge transfer. Secrecy and the gradual induction of tyros were important parts of medical training. Moreover, training in many skills was necessarily "hands on." Learning some practices—acupuncture in China, surgery in Europe, the "manual science" of massage in Thai medicine—could not be done without translating between intellectualized and haptic understanding.[25] As scholars studying Thai healing traditions in later periods have emphasized, in addition to information encoded in writing, healers would often need to

[22] Chatichai Muksong and Komatra Chuengsatiansup, "Medicine and Public Health in Thailand: From an Elitist View to Counter-Hegemonic Discourse," in *Global Movements, Local Concerns: Medicine and Health in Southeast Asia*, ed. Laurence Monnais and Harold J. Cook (Singapore: NUS Press, 2012), 226–45.

[23] Jean Mulholland, *Herbal Medicine in Paediatrics: Translation of a Thai Book of Genesis* (Canberra: ANU Press, 1989), 12.

[24] Viggo Brun and Trund Schumacher, *Traditional Herbal Medicine in Northern Thailand* (Bangkok: White Lotus, 1994), 42–6; Salguero, *Traditional Thai Medicine* (cit. n. 19), 8–9; Jana Igunma, "Aksoon Khoom: Khmer Heritage in Thai and Lao Manuscript Cultures," *Tai Culture* 23 (2015): 1–8, on 7.

[25] Loubère describes massage; see *Du Royaume* (cit. n. 5), 1:242. See also Riley, Mitchell, and Bensky, "Thai Manual Medicine" (cit. n. 18), 161.

learn additional theory, ingredients, techniques, and incantations that would be passed on in person only.[26]

Perhaps the most influential contemporary account of Siamese medicine is that of Simon de la Loubère (1642–1729), who traveled with a French embassy to Siam in 1687–8. He set the tone for later depictions, writing that "Medicine cannot merit the name of a Science amongst the Siamese."[27] He outlined what he saw as the four main deficiencies of Siamese medicine. First, he considered that healers lacked crucial technical skills, exemplified by their ignorance of surgery. "Their chief ignorance," he began, "is to know nothing of Surgery, and to have need for Europeans, not only for Trepans and for all other difficult surgical operations, but even for simple blood-letting."[28] Second, Thai medicine was presented as unsystematic, exemplified by a lack of "principles" and a failure to understand certain mechanisms of cure. "They do not trouble themselves to have any Principle of Medicine, but only a number of recipes, which they learnt from their Ancestors, and in which they never change a thing."[29] Third, knowledge was often patchy and misapplied, emblemized by Thai approaches to and uses of chemistry/alchemy: "They are ignorant of Chemistry, even though they love it passionately, and though many among them boast of possessing the most rare secrets. Siam, like all the rest of the Orient, is full of two sorts of people in this regard: imposters and dupes."[30] Finally, he criticized what he saw as a lack of curiosity, which meant that these deficiencies were never remedied. He bemoaned an "ignorance" of anatomy, for example.[31]

His depiction has informed most subsequent descriptions of seventeenth-century medicine in the kingdom in both the European and Thai-language scholarship.[32] Consequently, certain understandings of the differences and boundaries between "Siamese" medicine and "Western" medicine—each treated as immutable, definable categories—have colored many attempts to explore the interplay between the variety of healers operating in Ayutthaya. Because of this framing, European medical practitioners in seventeenth-century Ayutthaya are often presented in general histories of Thai medicine as introducing "Western" (or even "modern") medicine to Siam.[33] Charbonneau, for example, is depicted as an "agent of modernisation," standing as a symbol of an initial Thai encounter with various reified categories of knowledge: "European" surgery, "French" medicine, "Western" science.[34] In these narratives

[26] For example, see Louis Golomb, *An Anthropology of Caring in Multiethnic Thailand* (Urbana: Univ. of Illinois Press, 1985); and Viggo Brun, "Traditional Thai Medicine," in *Medicine Across Cultures: History and Practice of Medicine in Non-Western Cultures*, ed. Helaine Selin (Dordrecht: Kluwer Academic Publishers, 2003), 115–32.

[27] Loubère, *Du Royaume* (cit. n. 5), 1:239.

[28] Ibid.

[29] Ibid., 1:240.

[30] Ibid., 1:244.

[31] Ibid., 1:239.

[32] See, for example, Mulholland, *Medicine* (cit. n. 21), 7–9; and Santisuk Sobhanasiri, "Prawattisat kan phatthana kan phaet phaen thai kan phaet phuenban lae kan phaet thanglueak nai prathet thai," in *Dan kan phaet phaen thai kan phaet phuenban lae kan phaet thanglueak*, ed. Vichai Chokevivat, Suwit Wibulpolprasert, and Prapoj Petrakard (Bangkok: DTAM, 2010), 1–65.

[33] On narratives of "modernization" in Thai historiography, see especially Muksong and Chuengsatiansup, "Medicine and Public Health" (cit. n. 22).

[34] See, for example, Somrat Charuluxananan and Vilai Chentanez, "History and Evolution of Western Medicine in Thailand," *Asian Biomedicine* 1 (2007): 97–101, on 98; and Jean Affie, "Sous le règne de Rama V (1868–1910), l'adaptation du Siam à la modernité occidentale," in *Vietnam: Le moment moderniste*, ed. Gilles de Gantès and Phuong Ngoc Nguyen (Aix-en-Provence: Presses

Narai, in his patronage of figures such as Charbonneau, is often compared to the nineteenth-century reformist monarchs Mongkut and Chulalongkorn as utilizing "Western" knowledge to introduce "scientific" principles to improve public health. However, if we carefully contextualize Charbonneau against the intellectual world of seventeenth-century Ayutthaya, a much more nuanced picture emerges.

Knowledge and Exchange in Seventeenth-Century Ayutthaya

The city of Ayutthaya is a compelling site from which to explore medical, scientific, and intellectual exchange. In recent years a number of historians have demonstrated the value of rewriting histories of premodern circulations from decentered perspectives, and of deemphasizing narratives that seek to find origins for today's modernities.[35] Ayutthaya offers a range of such new vistas. Peripheral now in many narratives of global exchange, by the late sixteenth century, it was a crucial node: networks of trade, pilgrimage, and diplomacy connected the city to communities and markets around Southeast Asia, China, Japan, India, the Middle East, and Europe.[36] There was "no city in the Orient," wrote a French diplomat in 1686, "where one sees so many different nationalities [. . .] and where so many different languages are spoken."[37] Ayutthaya was a place where "translation, interpretation, and multilingual conversations were part of everyday life," something that is reflected in much of the surviving literature of the period.[38]

The seventeenth century in particular was a time of great openness to foreign trade and ideas, especially under the reign of Narai.[39] Royal patronage of domestic and foreign learned men, skilled in medical, astronomical, mathematical, literary, sacred, and divinatory scholarship and arts, was expansive. Observers noted with interest how many "foreigners" were employed in various branches of government: as mercenaries; as engineers and city planners; and as ritual specialists, astronomers, and physicians. European commercial companies and religious societies supplied artisans and experts—"some skilled, some useless," in the words of the historian Bhawan Ruangslip—to gain favor at court, including physicians, glassblowers, chandlers, metal workers, gardeners, and mining experts.[40] The presence of the missionaries in Ayutthaya was also owing to royal interest in foreign lands.

---

universitaires de Provence, 2009), 26–41. See also Ian Hodges, "Western Science in Siam: A Tale of Two Kings," *Osiris* 13 (1998): 80–95; and Thanet Aphornsuvan, "The West and Siam's Quest for Modernity: Siamese Responses to Nineteenth-Century American Missionaries," *South East Asian Research* 17 (2009): 401–31.

[35] See, for example, Heather Sutherland, "Southeast Asian History and the Mediterranean Analogy," *J. Southeast Asian Stud.* 34 (2003): 1–20; and Natalie Zemon Davis, "Decentering History: Local Stories and Cultural Crossings in a Global World," *Hist. & Theory* 50 (2011): 188–202.

[36] See especially Baker and Phongpaichit, *History of Ayutthaya* (cit. n. 17).

[37] Alexandre Chaumont, *Relation de l'ambassade de Mr le Chevalier de Chaumont à la cour du Roy de Siam* (La Haye: Isaac Beauregard, 1733).

[38] Baker and Phongpaichit, *History of Ayutthaya* (cit. n. 17), 207. See also Christopher Joll and Srawut Areem, "Thai Adaptations of the Javanese *Panji* in Cosmopolitan Ayutthaya," *Southeast Asian Studies* 9 (2020): 3–25.

[39] See especially Dirk van der Cruysse, *Siam and the West, 1500–1700* (Chiang Mai: Silkworm Books, 2002); David K. Wyatt, *Thailand: A Short History*, 2nd ed. (New Haven, CT: Yale Univ. Press, 2003), 88–103; and Muhammad Ismail Marcinkowski, *From Isfahan to Ayutthaya: Contacts between Iran and Siam in the 17th Century* (Singapore: Pustaka Nasional, 2005).

[40] Ruangsilp, *Dutch East India Company* (cit. n. 1), 142; Keith Branigan and Colin Merrony, "The Gardens of the Royal Palace at Ayutthaya," *JSS* 87 (1999): 17–31, on 18–9; Van der Cruysse, *Siam and the West* (cit. n. 39), 236–5, 382, 391.

Yet this interest in "foreign" knowledge and skills was not—as some visitors assumed—born of a sense that local expertise was somehow inferior. Rather, it was an important part of the wider fashioning of the monarch as an idealized Buddhist monarch. Thai Buddhist concepts of kingship celebrated the figure of the *cakkavatti* monarch—the "wheel turner" world emperor—in whose realm multitudes choose to come and shelter.[41] He was judicious and learned. At a time when trade and diplomatic links with distant lands were increasingly important for the region's political economy, his kingly virtues were made manifest by the sight of foreigners bringing gifts, pledging themselves to his service, and adding to the intellectual riches of his realm.

This period was more broadly a golden age of cultural, literary, and intellectual production, encouraged by the court to reflect its own glories.[42] The court and nobility sponsored an increased production of *samut khoi* and manuscripts in temples, many of which had their own palm plantations for manuscript production.[43] Alongside and often bound with Buddhist scriptures, temple archives housed materials on topics including astronomy, alchemy, divination, and medicine, including pharmacopoeias and collections of healing prayers.[44] Learning was highly lauded among courtiers, with value placed on the possession of treatises or manuals (ตำรา - *tamra*) of various "principles" or "sciences" (ศาสตร์ - *sat*, from the Sanskrit *sāstra*): of medicine, astrology, divination, alchemy, cosmology, and warfare. Scholars sought to translate, explain, and interpret (แปร *prae* or แปล - *plae*) just as those versed in alchemy (แปรธาตุ - *praethat*) sought to explain and transmute (*prae)* the elements (*that*).[45]

Facility with elegant literary forms and elaborate courtly registers that used a number of Sanskrit and Pali loanwords was also increasingly important at court.[46] In around 1672 the court astrologer (*Phra Horathipbodi* - พระโหราธิบดี) completed the first normative guide to the prosody of the Thai language, the *Chindamani* (จินดามณี).[47] The work takes as its allegorical title the name of the "wishing gem," or "jewel of thought," which was sought by alchemists, analogous to the "philosophers' stone" described in other traditions.[48] The metaphorical jewel offered in this case was itself an intricate work of translation, setting out how Sanskrit and Pali verse forms could be used with the Thai language.[49]

[41] Baker and Phongpaichit, *History of Ayutthaya* (cit. n 17), 207.

[42] Ibid., 149–50; Charnvit Kasetsiri, *The Rise of Ayudhya: The History of Siam in the Fourteenth and Fifteenth Centuries* (Kuala Lumpur: Oxford Univ. Press, 1976), 7–10.

[43] Jana Igunma, "Southeast Asia II: The Mainland," in *The Oxford Companion to the Book*, ed. Michael F. Suarez and H. R. Woudhuysen, vol. 2 (Oxford: Oxford Univ. Press, 2010), 390–3; John Guy, *Palm-Leaf and Paper: Illustrated Manuscripts of India and Southeast Asia* (Melbourne: National Gallery of Victoria, 1982), 66–7.

[44] See Justin McDaniel, *Gathering Leaves and Lifting Words: Histories of Buddhist Monastic Education in Laos and Thailand* (Seattle: Univ. of Washington Press, 2008), 85–6.

[45] The definitions are from Jean-Baptiste Pallegoix, *Dictionarium linguae Thai, sive siamensis interpretatione latina, gallica et anglica illustratum* (Paris: Imperatoris impressum, 1854). In modern Thai the distinction between แปร, *prae* (to turn, change, convert, become), and แปล, plae (to translate, interpret, to mean), is more clearly drawn, and "to translate, explain, mean," is *plae*. See Domnern Garden and Sathienpong Wannapok, *Thai-English Dictionary* (Bangkok: Amarin, 1999).

[46] B. J. Terweil, "The Introduction of Indian Prosody among the Thais," in *Ideology and Status of Sanskrit: Contributions to the History of the Sanskrit Language*, ed. Jan E. M. Houben (Leiden: Brill, 1996), 307–26, on 322.

[47] This was probably begun in the sixteenth century; Terweil, "Introduction" (cit. n. 46).

[48] Derived from the Sanskrit *Cintāmaṇi* (thought-gem); Monier Monier-Williams, *A Sanskrit-English Dictionary* (1899; repr., Delhi: Motilal Banarsidass, 1986), 398.

[49] Terweil, "Introduction" (cit. n. 46), 314–6.

In written reports produced for the king, detail, empirical verification, and elegant prose were important. The preparation of detailed reports was a necessary prelude to any decision being made.[50] Diplomats were instructed to record in forensic detail everything they saw in foreign lands.[51] Royal physicians and pharmacists investigated materia medica brought into the city and kept detailed records of the treatment of the royal family.[52]

Recipes and treatment approaches suggested by foreign medical experts like Charbonneau were recorded in pharmacopeia such as the *Tamra Phra Osot Phra Narai* (King Narai's Medical Treatise).[53] Prepared by officials in the royal pharmacy, the *Tamra* described the causes of certain illnesses and recipes for their cure, prepared for members of the royal court by Siamese, Chinese, Indian, and "western" (*farang*) physicians during the reigns of Narai and Petratcha. Royal medical texts and pharmacopeia (ตำรับ - *tamrap*) had much in common with Buddhist sermon scriptures and were often written using the Pali-Thai-Sanskrit hybrid ritual language, the Khmer *khom* script, and the palm leaves typical of such texts.[54]

This was also practical knowledge. In 1695 we see the Brahman ritual specialists (พราหมณ์ - *phram*) of the court being called on to examine a list of medical herbs "in the Brahman language," and to set it out again in "an Indian tongue" so that supplies could be ordered from Surat via the Dutch.[55] Record-keeping extended to matters of public health. In 1682 Narai heard that the French missionaries were distributing remedies during a smallpox epidemic. He ordered his court physicians to accompany them and to provide an exact report of houses visited and successes achieved.[56] A similar report was also requested from the missionaries.

This is the context in which we should understand Charbonneau was operating: in a polylingual, cosmopolitan city; amid an explosion of scholarly production; under the patronage of a royal court which valued learning that could be practically applied. Reading Charbonneau's story against the broader context of Ayutthayan intellectual culture we find new ways to interpret this moment of exchange. Turning now to examine in more detail the variety of treatments that Charbonneau offered, we see a range of responses to the treatments he attempted to translate. Most important, we see how this context could reshape "foreign" experts, translating them, their knowledge, and their techniques into new frames.

[50] See Ian Hodges "Time in Transition: King Narai and the Luang Prasoet Chronicle of Ayutthaya," *JSS* 87 (1999): 33–44, on 36–8.

[51] See, for example, Kosa Pan, *The Diary of Kosa Pan: Thai Ambassador to France, June-July 1686*, trans. Visudh Busayakul, ed. Dirk van der Cruysse and Michael Smithies (Chiang Mai: Silkworm Books, 2002).

[52] For example, the *Tamra Phra Osot Phra Narai* (cit. n. 21), discussed in more detail below. For medical note-taking at court in a later period, see Leslie Anne Woodhouse, "A 'Foreign' Princess at the Siamese Court: Princess Dara Rasami: The Politics of Gender and Ethnic Difference in Nineteenth-Century Siam" (PhD diss., Univ. of California, Berkeley, 2009), 105.

[53] *Tamra Phra Osot Phra Narai* (cit. n. 21); C. Phichiansunthon, M. Chawalit, and W. Chiruang, *Kham atthibai tamra Phra Osot Phra Narai* [An explanation of Phra Narai's Medical Treatise] (Bangkok: Amarin, 2001).

[54] Mulholland, *Medicine* (cit. n. 21), 15–6.

[55] Hendrik E. Niemeijer, trans., "Letter from the Phrakhlang on behalf of the King of Siam Phetracha (r. 1688–1703) to the Supreme Government, 12 February 1689," in *Harta Karun: Hidden Treasures on Indonesian and Asian-European History from the VOC-Archives in Jakarta* (Jakarta: Arsip Nasional Republik Indonesia, 2014), document 19.

[56] "Journal de la Mission," AMEP, 878:189.

2. CHARBONNEAU THE SURGEON

Charbonneau's value to the missionaries who brought him to Siam lay in his knowledge of surgery. As we have seen, ignorance of surgery was viewed by Europeans as a chief deficiency of Thai medicine. The surgeons of most Europeans' imagination were not learned men. In many missionary letters, surgery is grouped with other manual skills: the missionary Pierre Langlois, writing to the directors of the Parisian seminary in 1674, for example, recommended, "If you wish to bring some lay Frenchmen with you [. . .] Do not bring one who cannot read and write and who knows not either surgery, tailoring, fur-making or carpentry."[57] In France, professional bodies and guilds fought to establish and maintain distinctions between the spheres of physicians, apothecaries, and surgeons. While the first university chair of surgery had been created in Paris in 1673, surgery was still largely considered a lowly profession, concerned with manual treatments that would be inappropriate for physicians, much less clerics, to carry out themselves.[58]

Yet on the missions, the missionaries acknowledged, surgeons were often more highly regarded. On his voyage to Siam, the missionary François Pallu (1626–84) reported how a surgeon he encountered was "listened to like an oracle," winning respect from many.[59] So it proved for Charbonneau. His traveling companion, missionary le Roux, described how Charbonneau's skills won friends on the ship from Surat to Siam. "God in his mercy has made it so that they all love us," he wrote. "Brother René has helped with surgery the majority of these different Nations, which has contributed in no small measure to winning their affection for us."[60]

The missionaries saw certain surgical practices as crucial to their medical provision. Owing to a lack of trained surgeons, Pallu urged that missionaries should acquire "a little surgery, including the use of a knife or razor, which is not too difficult to learn."[61] Louis Laneau (1637–96), head of the mission in Siam, agreed: "All missionaries in these lands must be doctors (*medecins)*," he wrote in 1684. "Each should have a surgeon's case—not one of those large expensive ones, merely one supplied with good lancets, good scissors, a razor to cut flesh, one or two needles for sewing, and a spatula."[62] Familiar surgical practices were joined to a range of other medical approaches. Mirroring the Buddhist monks of Siam, Laneau introduced lessons in basic surgery and medicine in the Ayutthaya seminary.[63] These were to be taught by auxiliaries like Charbonneau, but also by local Siamese and Chinese practitioners. Seminarians were taught to bandage and suture wounds, to mix remedies, to administer purges, and to let blood. They were also taught the rudiments of pulse medicine by a Chinese physician, which they used to determine when moribund babies would die, so they could be quickly and secretly baptized without the knowledge of their

[57] Pierre Langlois to Paris seminary, 1674, AMEP, 857:311.

[58] Of course, such distinctions were in practice often blurred. See Laurence Brockliss and Colin Jones, *The Medical World of Early Modern France* (Oxford: Oxford Univ. Press, 1997); and Christelle Rabier, "La disparition du barbier chirurgien: Analyse d'une mutation professionnelle au XVIIIe siècle," *Annales* 65 (2010): 679–711.

[59] Pallu to Superiors, in Launay, *Histoire* (cit. n. 2), 1:68.

[60] Quoted in Bourges, *Relation* (cit. n. 8), 181–2.

[61] François Pallu to Paris Seminary, Surat, 1663, in Adrien Launay, *Lettres de Monsigneur Pallu* (Paris: Les Indes Savantes, 2008).

[62] Laneau to Seminary, 22 November 1684, AMEP, 852:130.

[63] On the wat (temple) as a place of medical training, see Mulholland, *Medicine* (cit. n. 21), 20.

parents.[64] Missionaries went as far as to push against canon law injunctions that prevented clergy from performing certain procedures, attempting (unsuccessfully) through letters seeking "clarification" from the Roman Congregation de Propaganda Fide, to redefine "surgical incisions" in such a way that would permit them to perform, in extremis, major amputations without incurring ecclesiastical penalties.[65]

The library of the Ayutthayan seminary contained a range of medical treatises that missionaries could borrow. Charbonneau also brought over a number of medical books.[66] The missionary Étienne Paumard (1640–90) copied notes from these books "into a little compendium" while travelling to Siam with Charbonneau.[67] Paumard "finally became a doctor himself by dint of deepening his study of these things he had contemplated," leavening his medical notes with spiritual insights, and learning from experience: "he neglected nothing in order to become skilled."[68] Crucially, apart from Charbonneau and his books, the sources of Paumard's knowledge were found in Ayutthaya: "to these European sciences he joined those of the country. He informed himself very carefully about the manner in which illnesses are treated and the remedies used here."[69] Paumard, unlike Charbonneau, never became proficient in Thai. It is tempting to imagine Charbonneau acting as an intermediary to assist his apprentice in learning from unnamed local experts.

To some extent, then, the knowledge complexes of medicine and surgery as they were increasingly understood in France were disarticulated, with certain techniques abstracted and acquired as supplementary tools. These complexes were not "French" or "European": we see individuals transforming practices as they move into Siam. Most interesting, we see Charbonneau and other healers learning hybrid practices from scratch after they have left Europe.

To consider how certain practices might have been understood by the Siamese, I will focus now on one of the most common treatments offered by surgeons in Europe—bloodletting. For European observers such as Loubère, it was striking that this practice seemed to be a relatively recent innovation in Siam. For Loubère this epitomized Siamese "ignorance" of surgery. "Nowadays," he asserted, "the Siamese use bloodletting, as long as they have a European surgeon, and sometimes in place of bloodletting they use scarification cupping (*les ventouses scarifiées*) and leeches."[70] Clearly, cupping and leeches did not qualify, as far as Loubère was concerned, as "surgery": for him, bloodletting meant venesection or phlebotomy, where a larger quantity of blood was drawn.

The presence of a large variety of healers in Ayutthaya, trained in a range of Chinese, Ayurvedic, and Muslim practices that had their own traditions of bloodletting, suggest other routes by which some of these treatments may also have entered the Siamese medical marketplace.[71] It seems that Charbonneau was practicing during a

[64] Aumont, "Mémoires," in Launay *Histoire* (cit. n. 2), 2:64.
[65] See my *Conflict and Conversion: Catholicism in Southeast Asia, 1500–1700* (Oxford: Oxford Univ. Press, 2013), 114–5.
[66] Vachet, "Mémoires" (cit. n. 4), 326. Sadly, the titles of these are for the most part now lost.
[67] Ibid.
[68] Ibid.
[69] Ibid.
[70] Loubère, *Du Royaume* (cit. n. 5), 1:243.
[71] See, for example, D. C. Epier, "Bloodletting in Early Chinese Medicine and Its Relation to the Origin of Acupuncture," *Bull. Hist. Med.* 54 (1980): 337–67; Neil Pemberton, *Leech* (London: Reaktion Books, 2012); and Naseem Akhtar Qureshi et al., "History of Cupping (Hijama): A Narrative Review of Literature," *Journal of Integrative Medicine* 15 (2017): 172–81.

brief window of time in which European-style bloodletting joined these other bleeding practices to enjoy a degree of modishness in Ayutthaya in the seventeenth century. The missionaries offered bloodletting to the indigent sick in their hospital, and to patients they visited at home. They successfully petitioned Narai and his successor Petratcha (1688–1703) to allow them to set up mass treatment programs involving bloodletting and the administration of purgatives as measures to tackle severe outbreaks of epidemic disease in 1682 and 1696.[72]

We have very little evidence to ascertain how these practices were understood by ordinary people. However, we have one interesting account of bloodletting in court, recounted by Claude de Forbin (1656–1733), a member of a French embassy to Siam. According to Forbin, Narai returned one day from an elephant hunt feeling very ill. After four days, his doctors were all agreed that he needed to be bled, presumably on the advice of one of his European physicians. There were by this time, in addition to Charbonneau, at least two other Frenchmen whose medical advice was listened to in court: Daniel Brochebourd, a French Huguenot from Sard who was a surgeon for the Dutch East India Company; and Étienne Paumard, the missionary who began his medical training under Charbonneau.[73] However,

> There was a difficulty with this remedy: because the Siamese regard their King like a divinity, they dare not touch him. The matter was taken before the council, and a mandarin suggested that a great curtain be set up, through which the king could put his arm, so that the surgeon could bleed him without knowing that it was the king.[74]

Forbin was greatly displeased, and proposed instead that

> If bleeding was absolutely necessary, there was a French surgeon, being from a country where we bleed without difficulty kings and princes when necessary, he could be used and that I was sure His Majesty would not regret the confidence placed in him. The king approved of my advice and there was no cause to repent of it as the prince recovered his health.[75]

If Forbin's account is accurate, it is striking for a number of reasons. First, performing this treatment on a member of the royal family involved breaking a number of taboos. Most forms of surgery involved physically touching the body, while bloodletting necessarily entailed penetrating the skin to bring about a shocking gush of blood. Indeed, the German physician Englebert Kaempfer, who spent some time in Siam in the 1680s, contrasted the "savage employment" of a "sharp point of deadly steel, dripping with blood" in "Western" surgery, "with its grim attitude towards human beings," to gentler, more compassionate treatments of "the East"—acupuncture and moxibustion.[76]

[72] See Tara Alberts, "Experiments in Dealing with Epidemics in Seventeenth-Century Siam," *Asian Medicine* 16 (2021): 153–76.

[73] On Brochebourd, see Dhiravat na Pombejra, "Ayutthaya as Cosmopolitan Society: A Case Study of Daniel Brocheboorde and his Descendants," in *Court, Company and Campong: Essays on the VOC presence in Ayutthaya*, ed. D. N. Pombejra (Ayutthaya: Ayutthaya Historical Study Centre, 1992), 25–42.

[74] Claude de Forbin, *Le voyage de Forbin à Siam: 1685–88* (Paris: Zulma, 1998), 57–8.

[75] Ibid.

[76] Englebert Kaempfer, *Exotic Pleasures. Fascicle III: Curious Scientific and Medical Observations*, trans. and ed. by Robert W. Carrubba (Carbondale: Southern Illinois Univ. Press, 1996), 109–10. Of course, there were other practices of bloodletting in Chinese medicine, but a much smaller

But the sacral authority of a semidivine monarch in Siam was reflected in the systems that emphasized the otherness and untouchability of his body.[77] Since the reign of Narai's father, Prasat Thong (r. 1629–56), protocol concealed the king from view: he seldom appeared outside the palace, and when he did, his subjects were obliged to avert their eyes.[78] The physicians given care of his body performed elaborate rituals before offering treatment, and approaching his person was only done when necessary. The *Tamra* notes with reverence the names of special unguents prepared for use when the king's massage doctors (*mor nuad*) performed their treatments.[79]

These othering restrictions extended to language norms that wrapped the king in mystery. His name was not revealed until his death, "for fear, they say, that it be profaned by the indiscreet tongue of some impious subject."[80] Similarly, a Thai king's blood was not blood. Like the rest of his body, it was distinguished by the elevated register used to discuss it. "Royal language" (ราชาศัพท์ - *rachasap*) introduced an additional layer of translation: those who wished to discuss or address royalty used vocabulary derived from Pali and Sanskrit, prefaced by the particle *phra* (พระ - noble, holy). His feet, for example, were not feet (เท้า - *thao*), they were feet (พระบาทยุคล - *phrabatyukhon*), which was also the term used for Buddha's footprints. He had an armpit (พระกัจฉะ - *phrakatcha*) rather than an armpit (รักแร้ - *rakrae);* passed excreta (พระบังคน - *charangon*) rather than feces (อุจจาระ - *utchara*), and became sick (ประชวร - *prachuan*), not sick (ป่วย - *puai*).[81]

Royal blood (พระโลหิต - *phralohit*) had special significance. Should it be necessary to execute someone with royal blood, he should be placed in a velvet sack and beaten to death with a sandalwood cudgel, ensuring that no drop of his blood fell on the ground. Indeed all blood was powerful and potentially dangerous.[82] Elaborate rituals were required to take the "inauspiciousness, evil, danger and misfortune" out of the city should bloodletting—through violence or childbirth—occur within the royal palace.[83] Rebels against royal authority were to be executed without letting it touch the realm.[84] This rule is reflected in the sentence imposed on the heroine of the seventeenth-century folk epic *The Tale of Khun Chang Khun Phaen*: "Cleave open her chest with an axe without mercy," ordered the king. "Don't let her blood touch my land. Collect it on banana leaves and feed it to dogs."[85]

---

quantity of blood was drained. See Henry McCann, *Pricking the Vessels: Bloodletting Therapy in Chinese Medicine* (London: Singing Dragon Press, 2014), 17, 49–57. See also Hansun Hsiung, "Use Me as Your Test!," in the present volume (*Osiris* 37).

[77] Baker and Phongpaichit, *History* (cit. n. 17), 143–7.

[78] Ibid. See also Alan Strathern, "Sacred Kingship under Narai of Ayutthaya: Divinisation and Righteousness," *JSS* 107 (2019): 49–77.

[79] *Tamra* (cit. n. 21), prescriptions 64–69. See also the discussion in Ratarasarn, *Principles* (cit. n. 11), 252.

[80] Gervaise, *Histoire* (cit. n. 5), 78, 241–2. See also Baker and Phongpaichit, *History* (cit. n. 17), 148.

[81] On royal language as medical jargon, see Ratarasarn, *Principles* (cit. n. 11), 287–8.

[82] For a fascinating contemporary example of the polysemic meanings of blood, see Erik Cohen, "Contesting Discourses of Blood in the 'Red Shirts' Protests in Bangkok," *J. Southeast Asian Stud.* (2012): 216–33.

[83] Chris Baker and Pasuk Phongpaichit, *The Palace Law of Ayutthaya and the Thammasat* (Ithaca, NY: Cornell Univ. Press, 2016), 108.

[84] Chris Baker and Pasuk Phongpaichit (trans. and eds.), *The Tale of Khun Chang Khun Phaen* (Chiang Mai: Silkworm Books, 2010), 801n30.

[85] Ibid.

While some underpinning theories of royal medicine found in the *Tamra* and other treatises of royal medicine resonate (albeit slightly misleadingly) with European concepts of the humors, blood did not have the same qualities, nor did blood loss have the same effects. The body element *lom* ลม (air/breath, often translated as "wind") circulates the body, putting blood, which embodies all four of the elements, in motion and promoting the smooth circulation of the four types of body element (ธาตุ - *that*).[86] Piercing the skin, if done incautiously, can risk a loss of *lom*, a subsequent imbalance of the elements, and a failure of their proper circulation.[87] Similarly, in a comparative consideration of bloodletting in Europe and China, Shigehisa Kuriyama has demonstrated how different emphases on dangers posed by corruption versus dispersal, or on retention versus loss (of energies, vitalities, fluids, etc.), led to divergent understandings of the prophylactic and curative potential of bloodletting in the two medical systems.[88]

Given these differences in understandings of blood it is intriguing that bloodletting was adopted briefly at court and practiced more widely on patients in the capital. Its brief adoption suggests, perhaps, the success of experts' attempts to explain the efficacy of the procedure, and the extent of their own expertise, which justified the breaching of taboos and medical norms. Or it may rather suggest the appeal of novel, exotic, or even outlandish approaches in moments of crisis: something not "fully" translated, exactly, but effective precisely because of its lingering "foreignness" and strangeness.

It is also unsurprising that the heyday of European-style bloodletting was brief and limited. Outside of the court, European surgeons continued to offer the treatment—especially to fellow Europeans—but they always seemed to struggle to find adequate equipment. In his letters, Charbonneau requests needles, scalpels, scissors, and syringes for cleaning wounds and delivering clysters.[89] Later letters from the superiors of the Ayutthaya mission requested that all who traveled to Siam would be furnished with sets of lancets for bloodletting, suggesting that they had still not found suitable local materials or artisans to fashion these tools.[90] It seems that European bloodletting was not adopted widely: eighteenth- and nineteenth-century accounts continued to complain that the Siamese were "ignorant" of it.[91] The polyglot Thai dictionary printed by the MEP missionary Jean-Baptiste Pallegoix in 1854 also suggested a limited role for "surgeons" of the bloodletting sort in the Thai medical landscape. A "surgeon" was a หมอ บาดแผล—*mo batphlae*—that is, a doctor (*mo*) who attends to wounds (*bat phlae*), while his scalpel was merely a มีด ผ่า ฝี (*mit pha fi*), a knife to lance abscesses: his incisions were not primarily to drain blood.[92]

[86] As was the case with European translations of the concept of *qi* in Chinese medicine as analogous to Galenic notions of *aer*, or of *pneuma/spiritus*, translating the concept of *lom* has often caused difficulties for outside observers. See Ratarasarn, *Principles* (cit. n. 11), 79–80, 184.

[87] Ibid., 80.

[88] Shigehisa Kuriyama, "Interpreting the History of Bloodletting," *J. Hist. Med. Allied Sci.* 50 (1995): 11–46.

[89] Charbonneau to Houdan, 8 October 1677, AMEP, 857:411.

[90] For example, Laneau to Seminary (cit. n. 62).

[91] François Henri Turpin, *Histoire civile et naturelle du royaume de Siam* (Paris: Chez Costard, 1771), 139; D. B. Bradley, "Siamese Theory and Practice of Disease," *Bangkok Calendar* (1865): 53–93, on 85–6.

[92] Pallegoix, *Dictionarium* (cit. n. 45), 441, 439. The American physician and missionary Dan Beach Bradley reported in 1865 that most Siamese physicians avoided such practices, and "dare not open a boil with a lancet or even a needle"; Bradley, "Siamese Theory" (cit. n. 91). This language had changed by the time George MacFarland, who was medically trained, published his *Thai-English Dictionary* (Stanford,

The limited translation of bloodletting in the seventeenth century was perhaps analogous to the period of popularity enjoyed by moxibustion in late seventeenth- and early eighteenth-century Europe.[93] We can speculate that where it was tried, the purpose, mechanism, and consequences of European bloodletting—indeed its meaning—was perhaps similarly disarticulated and mistranslated in Siam.

3. VIRTUOUS ELIXIRS

If translations of the scalpel skills of the traditional French surgeon failed to "stick," what of Charbonneau's other knowledge? His letters give us some insight into diseases he encountered and the range of treatments he employed, and perhaps also the perceived preferences of his patients—what would "translate" well. He requests two expensive electuaries: *teriacque* (theriac) and *comfection des jacinte* (confection of hyacinth).[94] Theriac, which combined a large number of drugs, including opium, with snake flesh, honey, and minerals, featured in a large number of cures for pestilential illnesses and was a panacea against all sorts of poisoning. Confection of hyacinth, a remedy "invented by the moderns," was made from a long list of spices and precious minerals, including hyacinth, red coral, terra sigillata (medicinal clay), myrrh, pearls, and emeralds.[95] It remedied palpitations of the heart, illnesses of other "noble parts," and melancholy, venereal, and contagious diseases; it also restored those recovering from long illnesses.[96]

He also requested a number of common remedies such as *l'emplastre de vigo com mercurio* (Vigo's Plaster), suggesting that "Madam Miramion will have the goodness to have this made by the nuns."[97] Vigo's Plaster contained various aromatics, resins, gums, mercury, a concoction made from frogs, and a wax base; it was used "to soften and dissipate cold humors," for cysts, nodules, and to ease pain.[98] Into the nineteenth century it remained an important treatment for the tertiary symptoms of syphilis, such as tumors, nodes, pustules, and muscle damage, and for skin eruptions caused by smallpox, described by Loubère as the "real plague" of Siam.[99] Other common remedies that he requested were topical: *ongand admirable* (admirable unguent of Nicodemus),

CA: Stanford Univ. Press, 1944), in which "surgery" was ศัลยศาสตร์, *sanlaya saat*, "that branch of the healing art that relates to injuries, deformities and morbid conditions that require to be remedied by operations" and derived from the Sanskrit ศัลย, *Sanlaya*, a dart, javelin, lance, spear, iron-headed weapon: "anything causing pain or torment, such as extraneous substances lodged in the body as thorns, splinters, calculi in the bladder" (794). In contemporary Thai, *sanlaya* and การผ่าตัด – *gaan phaa dtat* – the act of cutting open, are used; Garden and Wannapok, *Thai-English Dictionary* (cit. n. 45), 523, 325.

[93] See, for example, Margaret D. Garber, "Domesticating Moxa: The Reception of Moxibustion in a Late Seventeenth–Century German Medical Journal," in *Translation at Work: Chinese Medicine in the First Global Age*, ed. Harold J. Cook (Leiden: Brill, 2020), 134–56.

[94] Charbonneau to Houdan (cit. n. 89), 412.

[95] Jean de Renou, *Les oeuvres pharmaceutiques*, 2nd ed. (Lyon: Nicolas Gay, 1637), 631.

[96] Ibid.; Thomas Corneille and Bernard de Fontenell, *Le dictionaire universel des arts et des sciences. . . . Tome premier, A-L* (Paris: P. G. Le Mercier Fils, 1732), 578.

[97] Miramion was Louis XIV's morganatic wife. By "les bones seurs" [*sic*], Charbonneau presumably means the nuns of the Congregation of St. Cyr.

[98] Over the course of the eighteenth century, recipes began to feature more aromatic ingredients and fewer frogs. See "Emplastre de Vigo," *Dictionnaire botanique et pharmaceutique* (Paris: Laurent le Conte, 1738), 194.

[99] See, for example, M. Ricord, "Lectures on Venereal and Other Diseases Arising from Sexual Intercourse: Delivered in the Summer of 1847, at the Hôpital du Midi, Paris," *Lancet* 51 (1848): 571–3; Loubère, *Du Royaume* (cit. n. 5), 1:146.

containing myrrh, eaglewood, and sarcocollier for wounds, ulcers, and fistulas; and the inexpensive *ongand diuin* (divine unguent) for wounds, swellings, stones, ulcers, plague, skin and eye conditions, toothache, rheumatism, and paralysis.[100] As a postscript Charbonneau also requests rather optimistically that his correspondent ask "some good doctor of your acquaintance the method of curing leprosy; there is a lot of it in this country."[101] Missionaries in Ayutthaya offered treatment to the indigent poor for wounds, fevers, and skin conditions as a form of charity: it is possible that initially, Charbonneau intended to employ these remedies in the seminary's clinic, or perhaps even to win favor at court. We know that around this time the MEP were hoping to effect a cure—by medicine or miracle—of the king's paralyzed brother.[102]

Comparing these remedies with contemporary Thai texts such as *Tamra Phra Osot Phra Narai*, we can speculate how some of them might "translate" into the context of Siamese royal medicine. The requested drugs all have applications for a range of symptoms for which the *Tamra* also offered cures, often in a similar form (unguents, liniments, draughts), and especially for pestilential illnesses, skin conditions, wounds, digestive complaints, "general weakness," and paralysis. The large number of ingredients in remedies such as theriac and confection of hyacinth echoes the wide range of substances used in most Thai prescriptions; indeed, there is some overlap between some of the *materia medica* used—including especially myrrh, sandalwood, eaglewood, pepper, camphor, honey, and opium.[103]

Confection of Hyacinth is also comparable to a number of prescriptions in the *Tamra* that were designed to restore the balance in the body elements.[104] The importance of balance and of maintaining circulation of the elements around the body is a crucial principle underpinning the system of medicine in the *Tamra*.[105] Although the concept of *that* in Siamese medicine is very different, on a superficial level, were there perhaps enough parallels with European concepts of balancing the humors for European physicians to be able to talk to their Siamese colleagues about restoring balance in the patient?

This is not to say that the "foreignness" of a remedy, the principle underpinning it, or the physician prescribing it was any bar to incorporation in court medicine. "Translations" were not required to efface exotic origins or even incommensurate etiological underpinnings. A number of remedies in the *Tamra* are attributed to non-Siamese court physicians, including two prescriptions attributed to "western" physicians (*mor farang*). These featured "unusual" ingredients alongside more commonly used Siamese *materia medica*. Recipe number 22, for example—a diuretic—calls for potassium nitrate, or saltpeter (ดินประสิวขาว - *dinprasio khao*), alongside *yira* (ยี่หร่า - shrubby basil, *ocimum gratissimum*, widely used in African, Indian and Southeast Asian medical systems) and tea leaves.[106] Nitrate of potassium, a subject of much speculation and study in

[100] *Manier de se servir de l'Onguent Divin, & ses vertus* [pamphlet], (n.d.), 1; *Remede universel pour les pauvres gens et leur bestiaux* (Paris: La Veuve Denis Langlois, 1680).
[101] Charbonneau to Houdan (cit. n. 89), 412.
[102] On Chao Phraya Apithot, see Gervaise, *Histoire* (cit. n. 5), 246. This is not to suggest that there is any evidence these specific remedies were used for this purpose, or were presented at court.
[103] On the large number of ingredients, see Scott Bamber, "Medicine, Food, and Poison in Traditional Thai Healing," *Osiris* 13 (1998): 339–53, on 342.
[104] See, for example, the opening section of *Tamra* (cit. n. 21), which addresses the diagnosis and consequences of when the fire element leaves the body.
[105] Ratarasarn, *Principles* (cit. n. 11), 63–70.
[106] *Tamra* (cit. n. 21), prescription 22.

Europe, was known to Thai scholars (it was extracted from bat feces) but does not seem to have been commonly used in the period, featuring in only one other recipe in the *Tamra*.[107] Recipe number 79, a wax-based cure for suppurating wounds, included ironwood, camphor, frankincense, and coconut oil alongside the exotic Mediterranean resin mastic.[108] The substance also featured in Persian (*mastagi*) and Indian (*mastaki*) pharmacopeia, so it was not necessarily a "European" import: indeed the Thai transliteration, มาตะกี - *mataki*, perhaps favors one of these other routes.[109] However, it does not feature in other recipes.

Charbonneau also experimented with remedies he had never made before. He requested that an interlocutor in Paris send him information about "metals and minerals and how to purify them, for the herbs are all different to in Europe, [and] how to make furnaces and to place them, the manner of the fire."[110] The unfamiliarity of local materia medica thus stimulated an attempt to adapt his practices and to develop new skills. Some of this reinvention was at the direct request of the Siamese. In 1677, for example, Charbonneau described attempting to make *aurum potabile* (drinkable gold), a particularly storied cure, using alchemical methods. He reported that this was at Narai's own request.[111]

European accounts suggest that alchemy was a widespread interest in the kingdom with a range of aims, not least in purifying metals and creating elixirs to prolong life indefinitely. Palm leaf manuscripts combining Buddhist prayers, medical recipes, and alchemical techniques suggest a widespread, intertwined interest in these topics among learned men. "Many Siamese," reported Turpin, "dissipate their fortune on the search for the philosophers' stone."[112]

As we have seen, Siamese knowledge of alchemy/chemistry was often disparaged by Europeans.[113] "Imposture introduced chemistry here," ran one typical account, "but this science, which uncovers the veritable riches of nature, is only cultivated by rogues, who make their contribution to the credulity of imbeciles."[114] Loubère was similarly scornful about the previous king's alleged waste of "two million" on the search for the "philosophers' stone."[115] However, to the Siamese, quests for treasures—metaphorical and tangible—were proper for a Buddhist king. Under a meritorious monarch the hidden riches of the kingdom would be uncovered, embodied in

[107] Ibid., prescription 62. See also Ratarasarn, *Principles* (cit. n. 11), 161; and David Cressy, *Saltpeter: The Mother of Gunpowder* (Oxford: Oxford Univ. Press, 2013).
[108] *Tamra* (cit. n. 21), prescription 79.
[109] See Paul Freedman, "Mastic: A Mediterranean Luxury Product," *Mediterranean Historical Review* 26 (2011): 99–113.
[110] René Charbonneau, 1 Dec. 1677, AMEP, 861:41–2
[111] Ibid. On the history of drinkable gold and its relation to the philosophers' stone, see especially Lawrence C. Principe, *The Secrets of Alchemy* (Princeton, NJ: Princeton Univ. Press, 2012), 113–5, 181–7; and Renzo Console, "Pharmaceutical Uses of Gold from Antiquity to the Seventeenth Century," in *A History of Geology and Medicine*, ed. C. J. Duffin, R. T. J. Moody, and C. Gardner-Thorpe (London: Geological Society, 2013), 171–92.
[112] Turpin, *Histoire* (cit. n. 91), 141.
[113] While recent scholarship has attempted to reassess alchemical metallurgy and spagyric medicine in China and India (for example, Nathan Sivins, "Research on the History of Chinese Alchemy," in *Alchemy Revisited*, ed. Z. R. W. M. von Martels [Leiden: Brill, 1990]: 3–20), alchemical practice and knowledge in Siam have been almost entirely neglected. Further scholarship into the texts, practices, and uses of *prae thaat* is much needed.
[114] Turpin, *Histoire* (cit. n. 91), 141.
[115] Loubère, *Du Royaume* (cit. n. 5), 1:244.

its deposits of metals, minerals, jewels, gemstones, and Buddhist relics. Scholars would similarly reveal the jewels of (lost) knowledge. Narai's accumulated merit was manifest in discoveries and achievements under his aegis.[116]

Charbonneau recognized that it would not suffice to mix gold flakes into an electuary, as found in many recipes.[117] Narai was seeking a true, alchemical *aurum potabile*, which was reputed in Europe and Siam to have near miraculous curative properties. Charbonneau knew little to nothing about alchemy: as part of his reinvention, he would attempt to learn. Thanks to one of his letters, we catch a rare glimpse of someone attempting to translate the instructions of a vernacular text, published for the edification of surgeons and others unskilled in Latin, into practice.[118] He requested rather breathlessly that his correspondent send instructions

> of the manner of making [potable gold] and purifying it into the state in which it can be taken, written out fully in good French and not in Latin and not in terms of chemistry as I am not versed in that art . . . I already have a copy of Glaser's *De la Chimie*, but it treats the matter in too elevated a manner for me. Nevertheless, we tried to do it from the book and we succeeded a little with tin.[119]

Glaser's text describes various methods of purifying and then making preparations out of gold, through fulmination, calcination with mercury, or dissolution in the aptly named "royal water" (*aqua regia*—nitrohydrochloric acid).[120] These proved harder than expected. The missionary Charles Sevin wrote to Paris that while they had the necessary ingredients, the glass retorts they acquired in Siam shattered before reaching the necessary temperatures.[121] The French hoped to be able to demonstrate their technique for the king, perhaps mirroring the alchemists who entertained European royalty, but Charbonneau reports that they were not yet confident enough to do this.[122] From the 1680s onward, missionaries therefore made frequent requests that laymen with knowledge "of alchemy" be sent to Siam.

In December 1682 Narai asked again for drinkable gold. By this time Charbonneau had left the service of the mission, but the MEP had been joined by an "extremely capable" Swiss physician-chemist, and so they replied confidently that they would meet his request.[123] The king ordered built in the seminary's enclosure "the laboratory and furnaces necessary for this operation, and ordered for us to be provided with the necessary gold, men, and wood."[124] The following February, "we placed in the hands of

[116] Richard D. Cushman, trans., and David K. Wyatt, ed., *The Royal Chronicles of Ayutthaya* (Bangkok: The Siam Society, 2000), 272. See also Patrick Jory, *Thailand's Theory of Monarchy: The Vessantara Jataka and the Idea of the Perfect Man* (Albany: State Univ. of New York Press, 2016).

[117] Nicolas Lefèbvre condemned the practice of mixing gold leaf into recipes as "an abuse introduced to Pharmacy by the Arabs," as nothing in the body would be able to break it down; Lefèbvre, *Traicté de la Chymie* (Paris: T. Jolly, 1660), 2:795.

[118] Charbonneau, 1 Dec. 1677 (cit. n. 110), 41–3.

[119] Ibid.

[120] Christophe Glaser, *Traite de la Chymie: Enseignant par une brieue et facile methode toutes ses plus necessaires preparations* (Paris: Chez l'Autheur, 1663), 70–88.

[121] Charles Sevin to MEP directors, AMEP, 851:190.

[122] On alchemy as spectacle in the European courts, see, for example, Pamela H. Smith, *The Business of Alchemy: Science and Culture in the Holy Roman Empire* (Princeton, NJ: Princeton Univ. Press, 2016).

[123] "Journal de la Mission" 1682–84, AMEP, 878:488.

[124] Ibid., 488.

the king's physicians a vial of liquid gold, very well made, with which the king was said to be very satisfied, having never received the like in this country, although he had desired it for a long time due to the reputation of this liquor."[125] The missionaries were not themselves convinced of its medicinal effects, reporting that the substance "at heart does not have all the virtues attributed to it, if we are to believe the most sincere physicians."[126] This was not, then, an attempt to provide the most efficacious European remedy available, but rather to match and surpass Siamese alchemists on their own terms.

In attempting to create remedies valued by the Siamese, Charbonneau engaged in a multifaceted process of translation. This was not a straightforward story about the transplantation of "Western" technology and knowledge into Siam. As Charbonneau translated between medical systems he knew from Europe and those he encountered, he refashioned his skills and acquired new knowledge. Indeed, Ayutthaya transformed Charbonneau, who reinvented himself to meet Siamese expectations of valued expertise and skill.

CONCLUSION: RENÉ'S REINVENTIONS

To conclude, we will follow René out of the service of the MEP as he refashioned himself as a man of substance and came to be perceived as a man of skill, virtue, and merit in Ayutthayan society.[127]

His departure from the service of the mission is often depicted in later accounts as another instance in which this good Catholic followed the direction of superiors. He would have preferred to remain in service, reported the missionary Paul Aumont, but followed the wishes of Laneau, "who had not the slightest suspicion about his conduct, but who was persuaded that he would render more service to the mission being married."[128] However, contemporary records paint a picture of a self-willed operator and suggest that his departure was disruptive for missionaries in the seminary.

In 1681 Claude Gayme, for example, adopted a scandalized tone when he reported that Charbonneau, "having left off being a brother, is on the road to being a father," having married at the end of August the previous year "a twenty-five year-old widow of two husbands. After having been such a hypocrite for a long time, he begged [Laneau] so much that he allowed him to do it. . . . He has set himself to commerce, and I think that he will soon lose a good part of the best thing about him (I mean to say his devotion)."[129] Charbonneau's departure seems to have been particularly upsetting to Paumard, who beseeched his confrères to pray that he, Paumard, remain constant in his own vocation. "You know already," he explained, unable to resist a pun, "that [for] René Cherbonneau, the fire dying away, nothing remains but charcoal [*Cherbon*], which keeps of the fire of virtue but a weakened and diminished heat and flame. He is married in Siam."[130] Charbonneau's new life seemed enviable, Paumard

[125] Ibid.

[126] Ibid.

[127] On the concepts of virtue and merit, and their relationship to skill, see Lucien M. Hanks, "Merit and Power in the Thai Social Order," in *Everyday Life in Southeast Asia*, ed. Kathleen M. Adams and Kathleen A. Gillogly (Indiana Univ. Press, 2011), 89–100.

[128] Aumont, "Mémoires" (cit. n. 2), 265–6n1.

[129] M. Gayme to MEP directors, 18 January 1681, Bantam, AMEP, 859:137.

[130] Étienne Paumard to Brisacier, 7 November 1682, Tenasserim, AMEP, 859:183.

continued, "for its inconstancies and, principally, because of my cowardice, and incapacity, and extraordinary unworthiness for the service of God."[131]

Yet while he was temporarily diminished in the eyes of his former superiors, at court Charbonneau's merit seemed ever more evident. With his apparent familiarity with a range of valuable skills and his apparent access to esoteric knowledge, Charbonneau rose to prominence. He won the respect of many at court with his courtesy and appropriate demeanor.[132] His skills and propriety suggested capabilities in other areas, which Narai put to the test: in the early 1680s Charbonneau was ordered to construct a fortress on the frontier of Pegu.[133] Pleased with the result, Narai elevated him, appointing him as governor of the important and strategic tin-producing province of Phuket, where he was required to administer the business of government.[134] This was an important appointment. An English visitor to the province in the 1670s had noted that "the natives entitle [the governor] Radja, vizt. Kinge, as indeed he is a vice-kinge to the great King of Syam."[135] A number of his predecessors had come to violent ends in the difficult posting, but Charbonneau was generally well respected.[136] Such appointments remade the individual. They came with new clothes, a new name and title, paraphernalia of office, and a new lexicon of honorifics with which one would be addressed for the rest of one's life.

Charbonneau's time as governor of Phuket was so successful that he may have excited the jealousy of Constantine Phaulkon, the mercurial Greek minister who had risen to a position of importance in the court.[137] By other accounts, Narai himself had recalled Charbonneau: having been informed "of the manner he had governed his people, more as a father than a master, he had been so pleased that he had recalled him in order to give him a much better governorship than this one."[138] He was also considered for the role of official translator and advisor to the first Siamese embassy to France, which set sail in 1684, only to be lost at sea.[139]

In 1688 Charbonneau played a hitherto unnoticed role in the "revolution" following the death of Phra Narai, which reflected his shifting identities between biddable surgeon and independent man of substance. Around the beginning of May, Phra Narai had fallen gravely ill, and machinations to secure the succession intensified. Constance Phaulkon, correctly fearing for his own future, begged the commander of a French garrison in Bangkok to send reinforcements. General Desfarges, the leader of the French squadron, demurred, sending an officer, Beauchamp, to explain the situation to Phaulkon: there were many sick among the troops, he explained, playing for time, and they had only one surgeon, so they could not send the requested soldiers. Phaulkon countered that he would send them Charbonneau. But the latter "said to Mr Constance . . . that there was no way that he could take on that role, which made him [Phaulkon] very angry."[140] Charbonneau was unmoved by the anger of Phaulkon,

[131] Ibid., 183.
[132] Ruangsilp, *Dutch East India Company* (cit. n. 1), 157.
[133] Loubère, *Histoire* (cit. n. 5), 1:177.
[134] Ibid.
[135] Thomas Bowrey, cited by Pombejra in "Towards an Autonomous History" (cit. n. 5), 277.
[136] Ibid., 293–4.
[137] See Baker and Phongpaichit, *History* (cit. n. 17), 161–4.
[138] Martineau to superiors (cit. n. 3), 359–60.
[139] Vachet, "Mémoires" (cit. n. 4), 317–8.
[140] Beauchamp, 17 November 1689, Archives Nationales de France, Paris, C1 25, fols. 73–82, on 75r.

and Beauchamp returned to Bangkok. His refusal to assist Phaulkon's faction may not have been decisive in the revolution that followed, but it poses an interesting counterfactual: Would Narai's successor Petratcha have succeeded in taking the throne if the French had backed Phaulkon at this point?

Charbonneau was doubtless prudent in distancing himself from the doomed Phaulkon. He weathered the storm of the revolution, remaining free and unmolested while the majority of the French were imprisoned. Phetratcha also respected him enough to entreat the Dutch factory to protect him against anti-French sentiment. Charbonneau was credited with keeping the French prisoners alive. "It is almost he alone that has supported all these prisoners over such a long time," wrote Laneau, valuing the cost to him at over 750 écus.[141] Charbonneau's fortune seems to have come at least in part from trade.[142] In 1695, when he acted as a witness to the appointment of new mission procurators, Charbonneau the servant and surgeon had all but disappeared: he signed as "a French merchant."[143]

Yet his medical knowledge continued to be an important part of his identity until the end of his life. In 1720, for example, the missionary Roost had fallen ill, first with "a type of apoplexy, or transport of the brain, which made the seminarians think he was dead," then with "a type of fluxion that fell all of a sudden on both his eyes, and rendered him almost blind." However, "with the remedies which the famous M. Charbonneau gave him [. . .] he escaped these two illnesses."[144]

Until his last, Charbonneau was a regular, extraordinary sight on the waterways of Ayutthaya, travelling in state to church in a long, narrow boat rowed by a dozen girls. His wife, a Portuguese-Eurasian woman whom he had met in Ayutthaya, had had responsibility for the young female orphans who came into the care of the French missionaries. After the death of his wife and daughter, Charbonneau took over the care of these young women, and he always had around twelve to fifteen of them living in his home. According to one MEP missionary who met him in the 1720s, despite these unusual arrangements, and the spectacle of his team of orphan rowers, "his reputation was so well established amongst the idolaters, that none would ever be found who entertained the slightest suspicion about him."[145] Charbonneau's carefully cultivated skills proved his merit, brought him authority and status, allowed him some extraordinary access, and put him beyond reproach.

This article has followed Charbonneau as he plied his trade, operating between (at least) two languages and cultures. The context of Ayutthaya shaped his translations and transformed him in the process. Charbonneau's career thus allows us to think abstractly about the movement of bodies of medical knowledge into new contexts: about how the rearticulation of theories and concepts about health and the body also

[141] Laneau to M. de Seignelay, 15 January 1690, Siam, reproduced in Launay, *Histoire* (cit. n. 2), 1:265; Laneau, *Lettre de M. l'Evêque de Metellopolis Vicaire Apostolique de Siam Au Superieur & aux Directeurs du Seminaire des Missions Etrangers* (Paris: Charles Angot, 1690), 8.

[142] On his trade see, for example, the proceedings issued by Marianne Lagroye before the Conseil souverain de la Compagnie des Indes in Pondicherry to recover a sum of 800 pagodes owed to her late husband from a contract involving Charbonneau and the French Compagnie, 22 August 1720, *Procès-verbaux des délibérations du Conseil souverain de la Compagnie des Indes* (Puducherry: Société de l'histoire de l'Inde française, 1911–14), 1:252–3.

[143] Laneau, 20 October 1695, in Launay, *Histoire* (cit. n. 2), 2:371.

[144] Cicé to Seminary, 17 Feb 1721, in ibid., 2:85.

[145] Aumont, "Mémoires" (cit. n. 2), 266.

involved the fragmentation and rearticulation of associated material complexes, embodied skills, practitioner's identities, and frameworks of meaning. The process of reconstructing Charbonneau's shifting personas, and of imagining what his medical practices *meant* to different audiences in Siam, allows us to create new narratives about the history of Thai medicine, and of intercultural exchange in the early modern period.

"Use Me as Your Test!": Patients, Practitioners, and the Commensurability of Virtue

*by Hansun Hsiung**

ABSTRACT

Cutting deeply into a patient's body posed a problem for medical deontology in premodern East Asia, defined by the Confucian virtue of "humaneness" and a preference for noninvasive cures. How did Japanese physicians reconcile "humaneness" with their interest in invasive European surgical techniques? This essay offers answers through the tale of Kan (1743–1804) and her physician, Hanaoka Seishū (1760–1835). Inspired by the writings of the German physician Lorenz Heister (1683–1758), Hanaoka attempted to remove a cancerous tumor from Kan's breast in 1803—the first reliably documented operation of its kind in East Asia. In the process, Hanaoka outlined a new reasoning by which the *testing* of *un*tested foreign techniques could be construed as "humane." While scholarship on the translation of European medicine in East Asia has focused on epistemic shifts, I argue that translation was also about the renegotiation of ethical relations, reconfiguring patient-practitioner roles and boundaries of the morally permissible.

First one sister died, and then another: so it was that Kan had already surmised her illness, when she, at the age of sixty, felt in her breast a lump "the size of a bean." The time was the early summer of 1803. A year and a half later, having visited numerous practitioners close to her village of Gojō, she found herself at wits' end. Breast cancer, they told her, could not be cured. Yet as she made her tour, a sliver of hope emerged. It was reported that a certain physician in the neighboring Kii Province, Hanaoka Seishū (1760–1835), had of late displayed particular interest in "treat[ing] extreme illnesses." Deciding not to "idly await death," Kan instead chose to pursue the rumor and, in the late autumn of 1804, set out on the journey days west to visit Hanaoka.[1]

* School of Modern Languages and Cultures, Durham University, Elvet Riverside, New Elvet, Durham, DH1 3JT, U.K.; hansun.hsiung@durham.ac.uk.
Sincere thanks to the attendees at the Translating Medicine conferences of 2017 (Berlin, London), and successive writers' workshops led by Tara Alberts, Sietske Fransen, and Elaine Leong. Special gratitude is owed to co-contributor Daniel Trambaiolo, whose perceptive comments aided greatly in the reshaping of this paper. Late in this article's development, Ludmilla Jordanova contributed a valuable discussion of how to write a history of ethics not as doctrine, but as practice.

[1] Hanaoka Seishū, *Nyūgan chiken roku* [Record of a test treatment for breast cancer] (1805), in Kure Shūzō, *Hanaoka Seishū sensei oyobi sono geka* (Tokyo: Ōzorasha, 1994), 260–8, 260. Unless otherwise noted, the translations in this article are mine. Also, terms from texts composed in literary Sinitic will be glossed according to contemporary Mandarin (C) and Japanese (J) pronunciations.

Osiris, volume 37, 2022. © 2022 History of Science Society. All rights reserved. Published by The University of Chicago Press for the History of Science Society. https://doi.org/10.1086/719230.

The resulting consultation, as dramatized by Hanaoka himself, allegedly unfolded as follows. I quote at length, for in this dramatization lies the translational tangle—between languages and linguistic registers, across verbal and visual representations—that stands at the center of this article.

> I examined her. As depicted in the images at left, her left breast was greatly swollen. . . . This was indeed breast cancer. The *Waike zhengzong* [Orthodox lineage of surgery] says, "Of those who are afflicted by breast cancer, one hundred of a hundred always die." . . . I said [to her] in confidence, "Although the ancients said that breast cancer was an ailment impossible to cure, it is probably not true that a hundred out of a hundred will always die. If one uses appropriate methods to save them, then out of ten, one or two should live. But I have not yet tried this, so it is difficult to venture a statement on the matter. The ancients do not speak of it. . . . What is one to do, as there seems to be no technique [for this] in Japanese and Chinese experience? Once, there was something I saw in a book of the red-hairs [the Dutch]. I dedicated myself to this, and reflected for many years. By the might of the gods, I have come to some conclusions. But the tumor has already ruptured your skin, and the color of your flesh is black and purple. It will be hard to operate on this. . . . So for the time being I am declining treatment."
>
> The sick woman replied, "Master, do not refuse! I know well that breast cancer is difficult to treat. In the past, my sisters already suffered from this. They went to village doctors, and all of them declined to treat it. The tumors broke through the skin, and they died. . . . By saying that one will not offer treatment, one just extends the days to no end, prolonging the pain. If one treats it, it would likely hasten death, but would alleviate the pain. Rather than extend the days and prolong pain, hasten death and alleviate pain. . . . If you truly purpose to make an attempt, then use me as your test (C. *yan*; J. *ken*)! . . . Master, do not hesitate. I tell you that I am well prepared.[2]

These words find their earliest occurrence in a manuscript entitled *Nyūgan chiken roku* (Record of a test treatment for breast cancer), dated to the opening months of 1805, and likely written in Hanaoka's own hand.[3] Accompanying the text was a portrait of Kan (see fig. 1). Copied, recopied, modified dozens of times in manuscript, then excerpted selectively for print in 1847, these images and words, in their various configurations, have come to be firmly enshrined in the annals of Japanese medicine. They have become so, however, not on account of Kan's own struggles, but because of the nature of her subsequent treatment. For in attempting the surgical removal of Kan's breast tumor, Hanaoka also provided the first documented instance of the use of a full-body anesthetic in world history, over forty years before Morton's celebrated demonstration of ether in the amphitheater of Massachusetts General Hospital.

The secondary literature concerning this episode has thus understandably focused foremost on two tasks. First, it has attempted to tease out the ingredients of Hanaoka's anesthetic, kept an unwritten proprietary secret among his disciples, and referred to primarily by the name *mafutsusan*, an allusion to the drug employed by the storied Han dynasty physician Hua Tuo (ca. 140–208).[4] Second, the literature has attempted to identify the "red-hair" book mentioned by Hanaoka—a Dutch translation of Lorenz

[2] Ibid., 260–7.

[3] On debates over scribal identity, see Akitomo Matsuki, *Seishu Hanaoka and His Medicine: A Japanese Pioneer of Anesthesia and Surgery*, 2nd ed. (Hirosaki: Hirosaki Univ. Press, 2011), 83–100.

[4] The leading candidates are aconitum, datura, and hemp. See ibid., 47–51, 53–68; Tomio Ogata, "Seishu Hanaoka and His Anaesthesiology and Surgery," *Anaesthesia* 28 (1973): 645–52; and Masaru Izuo, "Medical History: Seishu Hanaoka and His Success in Breast Cancer Surgery under General Anesthesia Two Hundred Years Ago," *Breast Cancer* 11 (2004): 319–24.

Figure 1. *Portrait of Kan, in* Nyūgan chiken roku *(1805). The tag itself indicates, in reverse order, her age (60), and that she was the mother of the dyer Rihei from the station town of Gojō in Washū. Courtesy of the National Library of Medicine.*

Heister's (1683–1758) *Chirurgie* (first ed. 1718 or 1719)—and determine the precise extent to which the book informed Hanaoka's surgical practice compared with influential Chinese works such as the *Waike zhengzong*.[5] Framing debates over the latter has been the broader historiography of "Dutch Learning" (*rangaku*). Stationed on reclaimed land off the coast of Nagasaki, the Dutch were the only European power to enjoy official

[5] William Johnston, "Of Doctors, Women, and the Knife of Hope: The Surgical Treatment of Breast Cancer in Early Modern Japan," in *History of Ideas in Surgery, Proceedings of the 17th International Symposium of the Comparative History of Medicine—East and West*, ed. Yosio Kawakita, Shizu Sakai, and Yasuo Otsuka (Tokyo: Ishiyaku EuroAmerica, 1997), 153–80; Achiwa Gorō, *Kindai ishigaku ronkō* (Kyoto: Shibunkaku Shuppan, 1986), 87–116; Achiwa, *Kindai Nihon no igaku* (Kyoto: Shibunkaku Shuppan, 1982), 139–60.

trade relations with Japan until the closing decades of the Tokugawa period (1600–1867). Encounters with imported texts and Dutch East India Company (VOC) ship surgeons thus served as a portal into European knowledge. This portal, according to earlier accounts, alerted Japanese physicians to the "transmitted errors" of Chinese medicine, paving the way for the wholesale introduction of "modern Western science" later in the nineteenth century.[6] More recent scholarship, in contrast, has attempted to recover the continued role of Chinese medical and scientific knowledge in the Tokugawa world.[7]

A concern with the interaction of Chinese and Dutch texts in the Japanese medical world forms one key thread of my article. Yet as this volume emphasizes, translation is rarely ever an exclusive matter of the textual, emerging instead at the intersection of multiple communicative and representational practices, linguistic and beyond.[8] By beginning with the above transcript of Kan's travails, I argue that textual interactions between Chinese, Dutch, and Japanese medicine must be reinterpreted in light of three additional dimensions of translation appearing in the same passage. First, there is translation between word and image, marked by the presence of Kan's portrait next to its verbal record. Second, the passage itself comprises a translation of Kan's voice, inscribing the colloquial Japanese that a commoner such as Kan would have spoken into the erudite literary Sinitic that was Hanaoka's preferred register of writing. Third, we are confronted at the very end by the thorny Chinese character *yan* 驗 (J. *ken*), translated most frequently both as "experience" and "experiment," although rendered here, for reasons that will become clear, as "test."[9] I contend that rather than being isolated phenomena, these dimensions of translations—between Chinese and Dutch medical works into Japanese, between word and image, between patient's vernacular and physician's cosmopolitan, and between "experience" and "test"—were linked by a shared problem. The problem: how to translate virtue. The site where it unfolded: the relation between patient and practitioner.

Virtue, especially in its "epistemic" capacities, has for well over a decade now played a central role in our understanding of the construction and maintenance of scholarly personae—as scientists, as philosophers, as physicians.[10] Turning to the early modern Japanese context, we find the centrality of virtue all the more pronounced. East Asian

[6] Genpaku Sugita, *Kyōi no gen* (1775), in *Nihon shisō taikei: Yōgaku jō*, ed. Numata Jirō, Matsumura Akira, and Satō Shōsuke (Tokyo: Iwanami Shoten, 1976), 231; Genpaku Sugita, *Dawn of Western Science in Japan: Rangaku kotohajime*, trans. Ryōzō Matsumoto (Tokyo: Hokuseido Press, 1969), 30; Marius B. Jansen, "Rangaku and Westernization," *Mod. Asian Stud.* 18 (1984): 541–53.

[7] See, for instance, Daniel Trambaiolo, "Translating the Inner Landscape," in this volume (*Osiris* 37); and also Hansun Hsiung, "The Problem of Western Knowledge in Late Tokugawa Japan," in *The New Cambridge History of Japan, Volume 2: Early Modern Japan*, ed. David L. Howell (Cambridge, UK: Cambridge Univ. Press, forthcoming).

[8] See Tara Alberts, Sietske Fransen, and Elaine Leong, "Translating Medicine, ca. 800–1900"; and Daniel Trambaiolo, "Translating the Inner Landscape"; both in this volume (*Osiris* 37).

[9] For an early usage of *yan/ken* to signify "test" or "having been tested" rather than "experience" or "experiment" in the Chinese medical context, see Donald Harper, *Early Chinese Medical Literature: The Mawangdui Medical Manuscripts* (London: Kegan Paul, 1998), 66; and Yan Liu, "Understanding Efficacy (Yan) in Tang China," Asian Medicine Zone, published September 1, 2019, http://www.asianmedicinezone.com/chinese-east-asian/understanding-efficacy-yan-in-tang-china/.

[10] Lorraine Daston and H. Otto Sibum, "Introduction: Scientific Personae and Their Histories," *Sci. Context* 16 (2003): 1–8; Conal Condren, Stephen Gaukroger, and Ian Hunter, introduction to *The Philosopher in Early Modern Europe: The Nature of a Contested Identity*, ed. Condren, Gaukroger, and Hunter (Cambridge: Cambridge Univ. Press, 2006); Hunter, "The History of Philosophy and the Persona of the Philosopher," *Mod. Int. Hist.* 4 (2007): 571–600; Herman Paul, "What Is a Scholarly Persona? Ten Theses on Virtues, Skills, and Desires," *Hist. & Theory* 53 (2014): 348–71.

medicine at large, to borrow Qiu Renzong's epithet, was a form of "applied Confucianism."[11] This was emphatically true for Japan during the Tokugawa period, when the elevation of physicians into the ruling class of Neo-Confucian status elites also subsumed medical practice under Neo-Confucian moral imperatives of virtuous self-cultivation, in particular the cultivation of "humaneness" 仁 (C. *ren*; J. *jin*).[12]

Investigating the translation of Dutch surgery in Japan thus requires conceiving of translation not only as an epistemic procedure but as a site of ethical negotiation. "The spread of Western medicine to East Asia," Shigehisa Kuriyama reminds us, was "one in which the deep incisions and radical excisions of the scalpel challenged the traditional preference for noninvasive herbal cures."[13] Staple classroom surveys of Japanese history, meanwhile, tell readers that surgery "was neglected in East Asia, where invasive techniques ran the risk of doing damage to a body inherited from the ancestors."[14] Physicians such as Hanaoka who sought to apply "deep incisions and radical excisions" therefore required new strategies to maintain the commensurability of medical virtue despite risks of "damage to a body inherited from the ancestors." It is here where translation, in all the multiple senses outlined above, played a pivotal role. As a means by which to posit commensurability between foreign systems, translation functioned as a powerful tool of moral legitimation, bringing surgical techniques into the framework of Neo-Confucian universalism. By manipulating aspects of word, image, register, and narrative between Chinese, Dutch, and Japanese, tests on patients of procedures foreign, untested, and often violent, might be construed as virtuous acts.[15]

The stakes of such legitimation are particularly heightened in the case of cancer, epitome of the "incurable" disease. Thematized already in the Hippocratic corpus, incurability posed a vexing dilemma for physicians' virtue, trapped as they were between the Scylla of condemnation for withholding treatment and the Charybdis of iatrogenic death through uncertain procedures. Notably, premodern discussions often came with a masculine inflection, centered on the question of whether withholding treatment might be considered an act of "bravery" (*andreia*), with all its echoes of martial valor.[16] These gendered stakes grew increasingly complicated as early modern discourses of cancer in both Europe and East Asia came to focus specifically on the female breast

[11] Ren-Zong Qiu, "Medicine—The Art of Humaneness: On Ethics of Traditional Chinese Medicine," *J. Med. & Phil.* 13 (1988): 277–99, 283; Yi-Li Wu, *Reproducing Women: Medicine, Metaphor, and Childbirth in Late Imperial China* (Berkeley: Univ. of California Press, 2010), 54–83; Wu, "A Trauma Doctor's Practice in Nineteenth-Century China: The Medical Cases of Hu Tingguang," *Soc. Hist. Med.* 30 (2016): 299–322.

[12] Daniel Trambaiolo, "Ancient Texts and New Medical Ideas in Eighteenth-Century Japan," in *Antiquarianism, Language, and Medical Philology: From Early Modern to Modern Sino-Japanese Medical Discourses*, ed. Benjamin A. Elman (Leiden: Brill, 2015), 81–104, 82–6.

[13] Shigehisa Kuriyama, "Introduction," in Kawakita et al., *History of Ideas in Surgery* (cit. n. 5), ix–xiv, ix. Recent research has revealed certain exceptions to this characterization; see Wu, "A Trauma Doctor's Practice" (cit. n. 11), 300–1; Wu, "Between the Living and the Dead: Trauma Medicine and Forensic Medicine in the Mid-Qing," *Frontiers of History in China* 10 (2015): 38–73.

[14] Marius B. Jansen, *The Making of Modern Japan* (Cambridge, MA: Harvard Univ. Press, 2000), 211.

[15] On "testing," see Elaine Leong, Alisha Rankin, and Michael McVaugh, eds., "Testing Drugs and Trying Cures," special issue, *Bull. Hist. Med.* 91 (2017).

[16] Heinrich von Staden, "Incurability and Hopelessness: The *Hippocratic Corpus*," in *La Maladie des maladies dans le corpus hippocratique*, ed. Paul Potter, Gilles Maloney, and Jacques Desautels (Québec: Les éditions du Sphinx, 1990), 75–112; H. F. J. Horstmanshoff and R. M. Rosen, "The *Andreia* of the Hippocratic Physician and the Problem of Incurables," in *Andreia: Studies in Manliness and Courage in Classical Antiquity*, ed. Rosen and I. Sluiter (Leiden: Brill, 2003), 95–114.

as the disease's prime locus.[17] Symbolically fraught with fertility, motherhood, and eventually nationhood, the breast under threat of cancer embodied a double-edged imperative: to search far and abroad, on the one hand, for a cure to its ailments, and, on the other, to shield it from cures that might further damage it.

In what follows, I show how the multiple forms of translation at work in Hanaoka's account of Kan functioned to maintain the physician's virtue in the face of this double-edged imperative. After a brief introduction of Heister's *Chirurgie*, the inspiration for Hanaoka's attempt at a surgical cure for breast cancer, I demonstrate how the linguistic translation of the *Chirurgie* into Japanese ran into difficulty regarding its compatibility with the virtue of "humaneness" (*jin*)—a virtue whose gendering in Japanese medical discourse became all the more complicated in the case of breast cancer surgery. Continuing this thread of gender analysis, I turn to the *Nyūgan chiken roku*. There, I argue, Hanaoka employed novel verbal strategies and visual representations to legitimate the testing of dangerous techniques, crafting a new type of ideal patient characterized by *her* bravery and willingness to act as test subject for the male physician. I conclude finally with broader suggestions as to how a focus on the circulation of cures for the "incurable," given the demands incurability places on the limits of corporeal and ethical commensurability, might serve as a privileged site for understanding medicine in cross-cultural movement.

"ONE OF THE BOOKS OF THE RED HAIRS"

Alongside the Dutch translation of Johann Adam Kulmus's *Anatomische Tabellen* (first ed. 1722), Lorenz Heister's *Chirugie* occupies a foundational position in the formative mythos of Dutch medicine in Japan. The former is perhaps better known: Kulmus's *Tabellen* was partially translated into Japanese as *Kaitai shinsho* (A new book of anatomy). The publication of the latter in 1774 is routinely cited as the inaugural moment both of Japan's Dutch Learning movement, and the turn toward anatomical knowledge among Japanese physicians.

Well before 1774, however, Heister's *Chirurgie* had already gained traction in Japan. And unlike Kulmus's *Tabellen*, its story demonstrates how European medical knowledge, rather than precipitating a rupture, remained subject to continued integration within long-standing Sino-Japanese frameworks. Travel back to 1768, and we find the Nagasaki interpreter Yoshio Kōgyū (1724–1800) lending a copy of the Dutch translation of the *Chirurgie*, "recently acquired for the price of twenty barrels of the best *saké*," to none other than Sugita Genpaku (1733–1817), the animating translator behind *Kaitai shinsho*.[18] Although he was at the time unable to read Dutch, Sugita described being so entranced by the "exquisite precision" of the *Chirurgie*'s illustrations that he copied them over in their entirety.[19] Travel back again to 1764, and we find the physician

[17] Marjo Kaartinen, *Breast Cancer in the Eighteenth Century* (London: Routledge, 2013); Yi-Li Wu, "Body, Gender, and Disease: The Female Breast in Late Imperial Chinese Medicine," *Late Imperial China* 32 (2011): 83–128.

[18] Quoted in T. M. van Gulik, "Dutch Surgery in Japan," in *Red-Hair Medicine: Dutch-Japanese Medical Relations*, ed. H. Beukers, A. M. Luyendijk-Elshout, M. E. van Opstall, and F. Vos (Amsterdam: Rodopi, 1991), 37–50, 48.

[19] Quoted in Shigehisa Kuriyama, "Between Mind and Eye: Japanese Anatomy in the Eighteenth Century," in *Paths to Asian Medical Knowledge*, ed. Charles M. Leslie and Allan Young (Berkeley: Univ. of California Press, 1992), 21–43, 29.

Nagatomi Dokushōan (1732–66) describing an encounter with a copy of the *Chirurgie* while in Nagasaki. "Its words have substance," Nagatomi remarked, "and though I haven't tried out its techniques . . . I await future generations [that will do so]."[20]

Late eighteenth-century Japan was not alone in experiencing a Heister boom. First placed on the market at the end of 1718, the *Chirurgie* had been composed by Heister specifically in the German vernacular with the goal of providing barber-surgeons with a systematic outline of professional ethics, and instilling in them the "science and understanding" of prevailing medical theory.[21] Surgery, Heister insisted, was not simply *Kunst*—an art or technique. It was, rather, part of *Wissenschaft*—an ordered and systematic body of knowledge; a science. Surgery's full potential might thus be reached were it to be incorporated into an anatomically rigorous comprehensive system of medicine (*Gesamtmedizin*).

The *Chirurgie* met with quick success; three more revised and expanded editions followed in 1724, 1731, and 1752. During this time, learned physicians from across the continent, too, began to express an interest. Heister set to work on a Latin translation, enlarging the text by nearly two hundred pages and twelve additional figures; in 1739, the *Institutiones chirurgicae* appeared in Amsterdam, with a second edition in 1750.[22] Thereafter came a series of loops and lateral transfers. Upcycled to Latinate audiences, the *Chirurgie* was also downcycled beginning in 1747 in the form of the *Kleine Chirurgie*, which reduced the page count to half that of the 1718 original, and the cost to approximately one-fifth. This abridged edition not only enjoyed its own long republication history but was also translated into Latin as the *Chirurgia parva*. Meanwhile, the original German *Chirurgie* itself witnessed a host of translations into other European vernaculars from 1741 to 1770, including Dutch, English, Spanish, and French.[23] Nor was the *Chirurgie*'s impact restricted to Europe. Heister himself in fact understood his work as global: copies had spread, in his words, "not only in all of Germany, but also the Netherlands, Denmark, Sweden, Prussian, Hungary, and even . . . by means of German ship-surgeons, to the Cape of Good Hope, Batavia, and America—moreover on land to Petersburg and Russia."[24]

One core factor propelling the continued fortunes of both the *Chirurgie* and Heister himself was the surgical treatment of breast cancer. Shortly after the publication of the first edition, Heister was approached by the forty-eight-year-old peasant Anna Bayer, and soon afterward removed from her breast, so he claimed, a tumor weighing twelve pounds using only a scalpel and his bare hands. Heister's account of this observation was quickly incorporated into all later editions of the *Chirurgie*, while a separate set of his case histories prominently advertised his prowess in having removed "the largest among all [tumors] that had even been cut [out] and healed."[25] The feat brought such fame that the *Encyclopédie* chose to feature Heister's operation in detail in its entry on surgery.[26]

[20] Nagatomi Dokushōan, *Man'yū zakki* (Osaka, 1764), 15a.

[21] Lorenz Heister, *Chirurgie*, 2nd ed. (Nürnberg, 1724), 5.

[22] For a bibliographical survey, see Karin Sauer-Haeberlein, "Personalbibliographien der Professoren der Medizin zu Altdorf von 1580–1809" (PhD diss., Erlangen-Nürnberg, 1969), 95.

[23] Marion Maria Ruisinger, *Patientenwege: Die Konsiliarkorrespondenz Lorenz Heisters (1683–1758) in der Trew-Sammlung Erlangen* (Stuttgart: Franz Steiner Verlag, 2008), 50–54.

[24] Quoted in ibid., 53.

[25] Heister, *Chirurgie*, 4th ed. (Nürnberg, 1752), 668; Heister, *Medizinische, Chirurgische und Anatomische Wahrnehmungen* (Rostock, 1753), 980–92.

[26] Antoine Louis and Louis Jaucout, "Chirurgie," in *Encyclopédie, ou dictionnaire raisonné des sciences, des arts et des métiers*, vol. 3 (Paris, 1753), 353; "Chirurgie," *Encyclopédie*, vol. 20 (Paris, 1763), table 28.

The *Nyūgan chiken roku* did not mention the *Chirurgie* by name, but multiple documents by Hanaoka's students explicitly state that through Dutch translations, Heister's undertaking served as the inspiration for Hanaoka Seishū's pursuit of a surgical cure for breast cancer.[27] In the absence of evidence attesting to Hanaoka's Dutch fluency, however, it is safest to assume that this inspiration was indirect, filtered through excerpts, paraphrases, translations, and copied images from other Japanese hands. Of these indirect transmissions, three in particular stand out. The first two have already been alluded to above: the works of Nagatomi Dokushōan and Sugita Genpaku. In his *Manyū zakki* (Travel miscellanea; 1764), Nagatomi emphasized that while Sino-Japanese medicine thought breast cancer "incurable," the *Chirurgie* contained instructions for excising tumors through the aid of a "speedy blade."[28] Meanwhile, Sugita's copies of the *Chirurgie*'s illustrations, which included two figures depicting mastectomies, had begun circulating in manuscript under the title *Yōi taisei* (Compendium of surgical medicine) by 1786 at latest.[29] The third and far more detailed source was the *Yōi shinso* (New book of surgical medicine), a manuscript translation of the *Chirurgie*, the early portions of which had begun to circulate through scribal publication as of 1790.[30]

Although the historiography on Dutch Learning has placed emphasis on *Kaitai shinsho*, it was rather *Yōi shinsho* that marked the single largest translation project for Dutch medicine in Japan, mobilizing three generations of scholars. Initial work was conducted from 1771 to 1774, while Sugita and his collaborator, Maeno Ryōtaku (1723–1803), were at work on the *Kaitai shinsho*. As their source, they employed the 1755 second edition of the Dutch translation of Heister's vernacular German *Chirurgie*, titled *Heelkundige onderwyzingen* (Surgical instructions). This Dutch translation featured Heister's text in the upper two thirds of the page, and a running commentary by the translator, guild surgeon Hendrik Ulhoorn (ca. 1692–1750), in the bottom third. Sugita and Maeno ignored Ulhoorn's commentary, focusing their efforts only on Heister's text. Their hope, reflected in the parallelism of titles, was to issue *Yōi shinsho* as a companion piece to *Kaitai shinsho*.[31]

Unfortunately, *Kaitai shinsho*'s very success put a halt on *Yōi shinsho*. In the wake of the former, Sugita and Maeno saw a rapid rise in the number of their patients, siphoning time away from the task of translation. In 1778, however, Sugita spied a new opportunity in his new and ambitious apprentice, Ōtsuki Gentaku (1757–1827). Sugita tasked Ōtsuki to act as middleman for the completion of the unfinished *Yōi shinsho*. Ōtsuki received rough translations of the Dutch from Sugita and brought these drafts to Maeno for consultation and correction. The corrections were then returned to Sugita for the incorporation of changes. Once finished, Sugita again gave the materials over to Ōtsuki, who drew up a fair copy.[32]

[27] *Nyūgan zufu Hanaoka-ke* (1849), 1a, manuscript, Tokyo, Ken'i kai Library.

[28] Nagatomi, *Man'yū zakki* (cit. n. 20), 15a.

[29] Ishida Junrō, *Oranda ni okeru Rangaku isho no keisei* (Kyoto: Shibunkaku Shuppan, 2007), 119–20.

[30] For an analysis of the different manuscripts and their circulation, see Yoshida Tadashi, "Haisuteru no *Yōi shinsho* no hon'yaku," in *Ōtsuki Gentaku no kenkyū*, ed. Yōgakushi kenkyū kai (Kyoto: Shibunkaku Shuppan, 1991), 45–96.

[31] Ōtori Ransaburō, *Igaku shoshi ronkō* (Kyoto: Shibunkaku Shuppan, 1987), 145.

[32] *Yōi shinsho*, vol. 1 (1790), 7b-9a, manuscript, Kyoto University Library, Kyoto, Fujiwara Collection.

Such a process of shuttling between Sugita, Ōtsuki, and Maeno continued until 1785, when Ōtuski left for Nagasaki. Returning to Edo the following year and finding Sugita ill, Ōtsuki then finished the remaining portions of the translation by himself, resulting in a complete but partially uncorrected manuscript by 1790. Meanwhile, Ōtsuki had established his own academy of Dutch Learning, the Shirandō. Students there edited the uncorrected portions of the manuscript. Bit by bit, these remaining sections were released in manuscript form, until a fully corrected copy was finished in the early 1820s. It was Ōtsuki's hope to have this printed, but his hopes, for reasons unknown, were dashed midway. Three volumes comprising the table of contents and the introduction were printed in 1825, with a promise that more volumes would appear the following year.[33] None ever did. Scribal circulation, however, facilitated the *Yōi shinsho*'s circulation within the medical community, albeit in a piecemeal form under varied titles.

Much attention has focused on the post-1790 segment of this tale, for the collected efforts of Ōtsuki and his students to put out a full corrected version of Heister's *Chirurgie* are central to the history of the Shirandō, one of the most famous academies for Dutch Learning in early modern Japan.[34] Heister, in fact, was such a figure of veneration at the Shirandō that his portrait was prominently displayed in their communal dining hall. Ōtsuki's later student, Katsuragawa Hoken (1797–1845), would use this portrait to paint a likeness of Heister on a hanging scroll, representing Heister as a Confucian sage.[35] Researchers have therefore striven to provide detailed accounts regarding how techniques were translated, which techniques were prized, and what form they took in pedagogy and practice.

For our purposes, however, it is the early period of the *Chirurgie*'s translation that is of interest, not only because these early manuscripts were the only ones to which Hanaoka would have had access, but because the problem that emerged during this formative stage has been erased by histories all too focused on the transmission of surgical techniques. Instead, if we shift our gaze laterally, it becomes clear that techniques were arguably not the greatest dilemma of the translation process at the outset. From 1778 to 1785, Maeno Ryōtaku and Sugita Genpaku, through the intermediary of Ōtsuki Gentaku, labored on the translation of Heister's *Chirurgie*. And during precisely this period, Maeno was at work on another project—a project entitled *Jingen shisetsu* (Personal discourse on the word *jin*).

MEDICAL VIRTUE IN TRANSLATION

Translated alternatively as "benevolence," "humaneness," "humanity," "humanitarianism," and "kindheartedness," among others, the character *jin* 仁 (C. *ren*) had long been established as a central virtue within Confucian discourse.[36] Among the *Analects*' multiple glosses of the term, the most frequently cited is Confucius's terse

[33] See the title page to Sugita Genpaku et al., *Yōi shinsho*, vol. 1 (Tokyo, 1825).
[34] Yoshida Tadashi, "Anatomy in Rangaku," *Journal of the Japan-Netherlands Institute* 1 (1989): 21–38.
[35] Reinier H. Hesselink, "A Dutch New Year at the Shirandō Academy," *Monumenta Nipponica* 50 (1995): 189–234, 199–202.
[36] For the ease of readers, I will hereafter only use the Japanese *jin* instead of the Chinese *ren*, even when referring to literary Sinitic.

response to Fan Chi: *jin*, Confucius instructs us, consists of "caring for one's fellow humans."[37] In the *Mencius*, *jin* is ordered as the first and most essential among all basic virtues, pointing to our innate capacity for "compassion."[38]

Yet to fully grasp the import of the *Jingen shisetsu*, we must situate Maeno's interest in the term *jin* not merely within a general Confucian ethics of sagely or gentlemanly self-cultivation, but within the specific articulation of *jin* as the cardinal virtue of the physician's persona. Such articulation found shape as early as Sun Simiao's (581–682) "On the Absolute Sincerity of Great Physicians" (652), the first systematic statement of Chinese medical ethics. Referencing the *Mencius*, and merging this with Buddhist ethical values, Sun demanded that all "great physicians . . . develop first a marked attitude of pity and compassion."[39] Sun's expression was picked up in vernacular Japanese medical writings at the latest by 1303 in Kajiwara Shōzen's *Ton'ishō* (Jottings of the simple physician). "If the great physician is to heal sickness," Kajiwara writes, "he must first awaken a mindset of pity and compassion, and vow to relieve the suffering of all living things."[40]

The early modern period with which we are concerned, however, possessed a far more explicit formulation: "Medicine is the art of *jin*." A constant refrain in explications of Sino-Japanese medical ethics down to our own time, the phrase first occurred in the preface to the *Yishuo* (Medical stories; 1189) of the Southern Song physician Zhang Gao (1149–1227). It was then taken up again in Xu Chunfu's (1520–96) *Gujin yitong daquan* (Complete compendium of ancient and recent medical systems; 1536), where it served as part of Xu's definition of the "Way of Medicine" (C. *yidao*).[41] Whether Zhao or Xu served as the medium of introduction to Japan is unclear. The *Yishuo* was reprinted in Kyoto in 1658, and then again in 1659 with diacritical marks to aid Japanese readers; the *Gujin yitong daquan* was similarly reprinted with diacritical marks from 1657–60. Much clearer is that the phrase's popularization in Japan owes itself decidedly to the work of Kaibara Ekiken, who made the slogan "Medicine is the art of *jin*" the central motif of his vernacular medical advice book *Yōjōkun* (Rules for nourishing life; 1712).

The precise behavior that comprised *jin* varied. After gesturing to broad generalities of "comforting the old, raising the young, and protecting oneself," the *Yishuo* referenced the "art of *jin*" again in its closing afterword in order to outline a more concrete program of self-cultivation: here, the "art of *jin*" consisted in the determined study of the ancients, the broadening of one's knowledge through the observations of others, and the constant accumulation of one's own firsthand experience through practice.[42] Xu Chunfu, on the other hand, saw the phrase "art of *jin*" as marking a source of fundamental tension in medicine. Xu attacked practitioners for reducing "art" to pure

[37] *The Original Analects*, translated by E. Bruce Brooks and A. Taeko Brooks (New York: Columbia Univ. Press, 1998), 89–90.

[38] *Mencius*, translated by Irene Bloom (New York: Columbia Univ. Press, 2009), 35–6, 92.

[39] *Zhongguo kexue jishu dianji tonghui: Yixue juan*, 7 vols., ed. Yu Yinggao (Zhengzhou: Henan Jiaoyu, 1994), 2:196; Paul U. Unschuld, *Medical Ethics in Imperial China: A Study in Historical Anthropology* (Berkeley: Univ. of California Press, 1979), 24–34.

[40] Quoted in Fuse Shōichi, *Ishi no rekishi: sono Nihonteki tokuchō* (Tokyo: Chūō Kōronsha, 1979), 95.

[41] Yamamoto Tokuko, "Chūgoku ni okeru 'I wa jinjutsu' no kigen," *Nihon ishigaku zasshi* 28 (1982): 205–7.

[42] Ibid., 207.

technique, and neglecting its connection to *jin*. "The Way and the art have come into conflict," Xu lamented, "going against the heavenly principle of living things."[43] The necessary path forward was to harmonize *technē* with *ethos*; to pursue the "art of *jin*" required balancing the pursuit of virtue with the development of one's craft.

A condemnation of profit mongering, central both to Xu Chunfu's remonstrance and the collected "stories" of the *Yishuo*, also carried over distinctly into the medical culture of Tokugawa Japan (1600–1867). Medicine, at least in theory, was not to be used for financial gain. Although never manifested in positive law, the notion of charitable practice seems to have evolved into a quasi-governmental mechanism, with physicians enacting an "allocative responsibility" as de facto welfare providers.[44] One notes, for instance, the frequent invocation of *jin* as a regulative ideal in juridical settings where poorer patients, unable to settle their physicians' bills, had their debts annulled. One notes also the documented reticence of physicians to make explicit statements of their fees in advertisements, billboards, and other signage, preferring to negotiate this instead discreetly on an individual basis.[45]

Yet this importation was no mere repetition. In popularizing for Japanese audiences the phrase "Medicine is the art of *jin*," Kaibara Ekiken also mapped its meaning onto a social world decidedly feudal and decidedly masculine. In particular, he embarked on an extended analogy between good physicians and loyal retainers. A retainer entering into the service of a lord, he explained, "forgets his own self, and does everything for the sake of his lord. In his principles and loyalty, regardless of the amount of his stipend, he is willing to sacrifice even his own life."[46] Physicians, similarly, should "devote their whole energies to saving lives and healing the illnesses of their patients," displaying the same traits of loyalty and selflessness.[47]

Refigured as the heroic self-sacrifice of the loyal retainer, Ekiken's "art of *jin*" offered an early and important legitimizing framework for the proactive adoption, integration, and study of Dutch surgical techniques. While it would be false to conclude, along with the sixteenth-century Jesuit visitor Luis d'Almeida (c. 1525–83), that Japan was "ignorant of surgery," the surgery practiced in Sino-Japanese medicine cast invasive techniques to its margins, if not exterior.[48] Since the early years of the Heian court (794–1185), it was rather the field of "massage" that encompassed bandaging and

[43] Xu Chunfu, *Gujin yitong daquan* (1536), Chinese Text Project, accessed July 12, 2019, https://ctext.org/wiki.pl?if=gb&chapter=354834#p67.

[44] Naoki Ikegami, "Economic Aspects of the Doctor-Patient Relationship in Japan—From the Eighteenth Century until the Emergence of Social Insurance," in *History of the Doctor Patient Relationship: Proceedings of the 14th International Symposium on the Comparative History of Medicine—East and West*, ed. Yosio Kawakita, Shizu Sakai, and Yasuo Otsuka (Tokyo: Ishiyaku Euroameric, 1995), 131–46, 133. On *jin* as an early modern political ideal, see Daniel V. Botsman, *Punishment and Power in the Making of Modern Japan* (Princeton, NJ: Princeton Univ. Press, 2007), 41–58; on the use of *jin* to legitimate colonial medicine, see Shin Sōken, "Nihon kanpō igaku ni okeru jigazō no keisei to tenkai," in *Shōwa zenki no kagaku shisōshi*, ed. Kanamori Osamu (Tokyo: Keisō Shobō, 2011), 325.

[45] Ikegami, "Economic Aspects" (cit. n. 44), 138–9.

[46] Kaibara Ekiken, *Yōjōkun* (1712), in *Kaibara Ekiken Yōjōkun*, ed. Kaibara Shuichi (Fukuoka: Tonshindō, 1943), 158.

[47] Ibid., 158–9.

[48] Quoted in Toshirō Ōmura, "Surgeons in Japan in the 17th and 18th Centuries," in *History of the Professionalization of Medicine: Proceedings of the 3rd International Symposium on the Comparative History of Medicine—East and West*, ed. Ogawa Teizō (Osaka: Division of Medical History, the Taniguchi Foundation, 1987), 177–87, 179. But for exceptions, see Wu, "A Trauma Doctor's Practice" (cit. n. 11); and Wu, "Between the Living and the Dead" (cit. n. 13).

wound treatment, bone setting, and the cauterization of abscesses and abnormal growths. By the twelfth century, surgery had emerged as a distinct field under the category of *geka*, "external medicine," yet only in a secondary relation to internal medicine (*naika*), also termed *hondō*, or the "root Way."[49] Cure by knife was something foreign, something outside tradition—the Japanese word for "scalpel" remains today *mesu*, from the Dutch *mes* (knife).

To this very point, six years before the *Yōjōkun*, Ekiken had already deployed the term "art of *jin*" in a preface authored for Narabayashi Chinzan's (1648–1711) 1706 manuscript, *Kōi geka sōden* (Complete teachings of red barbarian surgery). There, Ekiken argued that the relative paucity of surgical knowledge in both China and Japan had inhibited the full development of their healing capacity. Acquiring knowledge from the Dutch should therefore be seen neither as heterodox nor eccentric, but rather as a necessary extension of a physician's fidelity to the "art of *jin*."[50] True service to a lord did not discriminate between means, so long as they efficaciously served the ends of loyal vassalage.

It is within this context that Ryōtaku's *Jingen shisetsu* must be read. Ekiken had already contended that, just as a retainer served his lord at all costs, the "art of *jin*" entailed seeking cures from all sources, even those of "red-haired barbarians." But doubts remained over whether Dutch "barbarians" possessed *jin* at all. Ryōtaku's *Jingen shisetsu* was a response to this skepticism. Specifically, the *Jingen shisetsu* revolved around the search for an adequate translation for the character *jin* in European languages, primarily Dutch and French. Should this translation exist, then it would seem that European "barbarians," too, embraced a commensurable concept of *jin*.

That Maeno already believed this seems clear; in a manuscript from 1777, he had intimated that Christianity, by "relieving and supporting widows and widowers, orphans, the sick, and the poor," was reminiscent of ethical views proposed in the *Mencius*.[51] The proof now lay in language—in the possibility of translation. As primary candidates for investigation, Maeno selected the following words: *medoogend*, *verdraagzaam*, *genadig*, and *barmhertig*.[52] Each of these was subjected to etymological analysis, often highly speculative, based on information available to him in the Dutch-French dictionaries of Marin and D'Arsy, and the monolingual Dutch dictionary of Meijer.[53] Using these etymologies, Maeno identified corresponding loci in the Four Books and Five Classics. *Verdraagzaam*, for instance, was interpreted by Maeno as containing the element *draag*, for "bearing a load," along with *zaam*, which he read as "together." Such shared load-bearing, in his mind, called forth a passage in the *Zhong yong* [Doctrine of the mean], where the term *zhongshu* (J. chūjo) was glossed as a kind

[49] Frits Vos, "From God to Apostate: Medicine in Japan before the Caspar School," in Beukers et al., *Red-Hair Medicine* (cit. n. 18), 9–26, 14; Ōmura, "Surgeons in Japan" (cit. n. 48), 178.

[50] Kaibara Ekiken, preface to *Kōi geka sōden* (1706), by Narabayashi Chinzan, 1a-1b, manuscript, Tokyo, Kokubungaku kenkyū shiryōkan.

[51] Maeno Ryōtaku, *Kanrei higen* (1777), in *Nihon shisō taikei: Yōgaku-jō* (cit. n. 6), 148.

[52] In what follows, I have retained the archaic orthography for these Dutch terms as used by Maeno, e.g., *medoogend* instead of *meedogend*; *barmhertig* instead of *barmhartig*. The semantic field of these terms overlaps substantially, as Maeno's own work demonstrates, but one might roughly translate them respectively as "compassionate" (in a manner closer to sympathetic), "tolerant," "gracious/clement," and finally also "compassionate" (in a manner closer to merciful).

[53] Pieter Marin, *Dictionnaire françois et hollandois comprenant tous les mots de l'usage avouez de l'Académie françoise* (Amsterdam, 1762); Lodewijk Meijer, *L. Meijers Woordenschat*, 3 vols. (Amsterdam, 1745); Ian Louys D'Arsy, *Het groote woorden-boeck* (Utrecht, 1643).

of golden rule: "What is enacted on oneself and one does not wish, should not be enacted on others."[54] *Verdraagzaam*, then, was correlated with *jin*, but not *jin* itself.

After parallel inquiries with other terms, Maeno ultimately settled upon *barmhertig* as his favored translation of *jin*. *Hart* came from the literal organ of the heart—this seemed to him evident. But what of *barm*? His search through dictionaries resulted in three definitions of the noun *barm*—respectively, a levee, a wave, and trash or waste. These proving unhelpful, Maeno invented his own etymology, suggesting instead that the word came from the adjective *arm*, describing those in poverty or hardship, which had then been transformed by the addition of the prefix *be-* into a verb *bearmen*, or *barmen*. This verb, Maeno claimed, originally meant "to save others," especially those in poverty or hardship. To be *barmhertig* was thus to have a heart that cherished the unfortunate and worked for their salvation.[55]

This analysis of *barmhertig* may have satisfied the *Jingen shisetsu*'s aim to demonstrate that the Dutch, too, had a commensurable concept of *jin*. Yet in translating Heister's *Chirurgie*, it posed problems. In his chapter surveying the surgical field, Heister had dedicated a section to elucidating the "necessary qualities of a surgeon" (D. vereischt wordende Hoedanigheden eenes Heelmeesters; G. Eigenschafften oder *Requisitis* eines *Chirurgi*), borrowing heavily from Celsus. These necessities began with physical characteristics. A surgeon should preferably be young, or at least not too old; must have fixed, steady, and agile hands; should be ambidextrous; and must have excellent sight. Following this, however, Heister expounded at length on the final criterion of the surgeon—a criterion regarding mental disposition. The surgeon, in Heister's words, must be "unperturbed, and when necessary, *onbarmhertig* [G. *unbarmherzig*; uncompassionate]." Only by being *onbarmhertig*, Heister informed readers, could the surgeon "prevent himself from being hindered by the screaming of the patient"—a hindrance that would otherwise lead him either to "cut and do less than is necessary," or else "rush too much, and thereby cause harm."[56]

If medicine lay in the "art of *jin*," and to practice *jin* was to be *barmhertig*, then how could a physician maintain *jin* while adopting the *onbarmhertig* disposition requisite for surgery? Confronted with this literal contradiction, the translation cobbled together by Sugita, Ōtsuki, and Maeno in the final published version of the *Yōi shinsho* evaded any direct associations of *ombarmhertig* with the character *jin*. Instead, they displaced the term onto a field of gendered sentiment. The surgeon, in their rendering, "should not arouse womanly and childish sentiments of care and pity, focusing solely on the pain and suffering of the patient."[57] Phrased in this manner, being *onbarmhertig* did

[54] Maeno Ryōtaku, *Jingen shisetsu* (ca. 1783), in *Maeno Ryōtaku shiryōshū*, vol. 2, ed. Ōita Kenritsu Sentetsu Shiryōkan (Ōita: Ōita-ken Kyōiku Iinkai, 2009), 155. Literally a combination of characters for loyalty/integrity and reciprocity, *zhongshu* has been glossed in English variously as "to be true to the principles of our nature and the benevolent exercise of them to others" (Legge); "doing one's best and . . . using oneself as a measure to gauge others" (Lau); and "dutifulness tempered by understanding" (Slingerland). See Confucius, *Analects*, in James Legge, *The Chinese Classics*, vol. 1 (Oxford, 1893), 4.15; Confucius, *The Analects*, translated by D.C. Lau (London: Penguin, 1979), 4.15; Confucius, *Analects*, translated by Edward Slingerland (Indianapolis, Ind.: Hackett Publishing, 2003), 4.15.

[55] Maeno, *Jingen shisetsu* (cit. n. 54), 153–4.

[56] Laurens Heister, *Heelkundige Onderwyzingen*, 2 vols., translated by Hendrik Ulhoorn (Amsterdam, 1776), 1:17.

[57] Sugita et al., *Yōi shinsho* (1826) (cit. n. 33), 1:7b-8a.

not come into conflict with the higher virtue of *jin*, but rather represented a lapse or weakness. On the one hand, by including the adverb "solely," the translation explicitly avoided a unilateral exclusion of the patient's suffering. On the other hand, by denoting care and pity as "sentiments" rather than virtues, and portraying these virtues as "womanly and childish," the translation implicitly associated *onbarmhertig* with the distinctly masculine persona of the physician as loyal retainer. The "art of *jin*" could therefore encompass surgical techniques that might otherwise seem brutal or invasive—at least in its linguistic representation. Whether it could in practice, and whether patients themselves could be persuaded so, remained to be seen.

OPERATING ON THE BREAST

Not all surgery need be brutal or invasive, but surgical approaches to breast cancer were decidedly so. Fanny Burney's now-canonical letter describing her mastectomy, in its invocation of "a terror that surpasses all description," and "a scream that lasted unremittingly," is perhaps the most-cited locus.[58] But we may look too at the numerous voices of lesser-known patients, rescued by recent research, which together testify to a widespread reticence on the part of women to elect surgical treatment for breast cancer, seeking their remedies elsewhere and everywhere before yielding to the knife.[59]

Pain was only one factor in this reticence. A broader factor was the perceived stigmatization that came from damage to an organ so overcharged with symbolic meaning. Although female breasts were not stressed as markers of sexual differentiation in the *Huangdi neijing* (Yellow Emperor's Inner Classic), they did occupy a prominent position in East Asian social differentiations of gender.[60] The lactating, nursing breast in particular took on pronounced visibility. In Japan, the strong connection between breast milk and health led to its valuation as a medicine for everything from toothaches to smallpox, while a longstanding iconographic tradition employed the withered breasts of women, unable to nurse children, to represent famine, disaster, and misfortune.[61] In China, breastfeeding was so tied to acts of benevolence and caring that images at times depicted male literati lactating as a visual metaphor of their Confucian virtue.[62] This symbolic matrix was paralleled in early modern Europe, where since the seventeenth century breastfeeding had been posited as a broader "civic responsibility" beyond the sphere of private domesticity—a responsibility that by the latter eighteenth century would help tie maternity firmly to the nation.[63]

[58] Quoted in Roy Porter, "Hospitals and Surgery," in *Cambridge Illustrated History of Medicine*, ed. Porter (Cambridge, UK: Cambridge Univ. Press, 1996), 202–45, 218–9; Kaartinen, *Breast Cancer* (cit. n. 17), 101–7.

[59] Robert A. Aronowitz, *Unnatural History: Breast Cancer and American Society* (Cambridge, UK: Cambridge Univ. Press, 2007), 23–6; Ruisinger, *Patientenwege* (cit. n. 23), 217–36; Kaartinen, *Breast Cancer* (cit. n. 17), 39–47, 55–88; Alanna Skuse, *Constructions of Cancer in Early Modern England: Ravenous Natures* (Basingstoke, Hampshire: Palgrave Macmillan, 2015).

[60] Charlotte Furth, *A Flourishing Yin: Gender in China's Medical History, 960–1665* (Berkeley: Univ. of California Press, 1999), 45; Wu, "Body, Gender, and Disease" (cit. n. 17), 83–128, 95.

[61] Sawayama Mikako, *Edo no nyū to kodomo: Inochi o tsunagu* (Tokyo: Yoshikawa Kōbunkan, 2017), 25–32, 169–70, 179–91.

[62] Furth, *Flourishing Yin* (cit. n. 60), 221.

[63] Marilyn Yalom, *A History of the Breast* (New York: Ballantine, 1997), 5, 105–23; Ludmilla Jordanova, *Sexual Visions: Images of Gender in Science and Medicine between the Eighteenth and Twentieth Centuries* (New York: Harvester Wheatsheaf, 1989), 29.

Practitioners themselves, in East Asia as in Europe, proved conservative in their attitudes toward breast cancer treatment, although European support for radical mastectomies appears to have grown in the final decades of the eighteenth century. Conservatism was in part etiological. While in the European case, emerging proponents of anatomical pathology did increasingly advocate localized surgery, more widespread humoral pathology still understood cancer in terms of the excess accumulation of black bile, whose stasis led to the formation of a scirrhus, which could then mature into a cancer. This, indeed, was the reason why older women—postmenopausal—were thought to be especially susceptible to breast cancer.[64] Most physicians, including Heister, therefore proposed a course of medicines that would promote circulation, purifying bilious accumulations with healthy blood.[65]

Humoral pathology's account of cancer was remarkably compatible with those in premodern China and Japan. To be sure, basic nosological differences existed: there was no separate category of "cancer" as such distinct more generally from lesions (C. *yan*; J. *gan*), and what I have been referring to here as "breast cancer" (C. *ruyan*; J. *nyūgan*) was at the time understood as a kind of deadly lesion endemic to breasts of females older than fifty, identifiable through the shape it took—a "floating lotus"—as it broke through the skin.[66] Still, within this more limited category, most authors considered breast cancer in analogous terms of accumulation and excess—this time of the female vital force of *yin*. In turn, Sino-Japanese texts counseled oral medication and regimens, as well as compresses and ointments, that would aid in dissolving accumulated *yin*.[67]

Compounding the etiological foundations of this conservative approach was also a more general air of resignation. Chen Shigong's (1555–1636) *Waike zhengzong* (1617), cited by Hanaoka at the beginning of this article, emphasized at several different locations in the text that "there has been no account of a cure, now as in antiquity," for breast cancer, and that for those afflicted, "death truly awaits."[68] In the face of certain mortality, many physicians—including those whom Kan visited before Hanaoka—were prone to refusing treatment, citing Sun Simiao's advice that those "on death's door" should not be treated, just as "one should not concern oneself with the politics of a state on the brink of ruin."[69] Meanwhile, for Heister and others of his age, the fact that breast cancer had "no certain cure" supported an attitude of circumspection, since more proactive treatments risked compounding the disease. Even topical treatments such as ointments, Heister remarked, much less venesection, corrosives, and cauterization, might aggravate the situation.[70] Hippocrates's words, Heister conceded, might

[64] David Cantor, "Cancer," in *Companion Encyclopedia of the History of Medicine*, 2 vols., ed. W. F. Bynum and Roy Porter (London: Routledge, 1993), 1:537–61, 541.

[65] In particular, Heister recommended guaiacum officinale, scrophularia nodosa, and the root of cynanchum vincetoxiucum; see Heister, *Chirurgie* (1752) (cit. n. 25), 317. For a survey of pharmacological approaches, see Kaartinen, *Breast Cancer* (cit. n. 17), 23–35.

[66] Chen Shigong, *Waike zhengzong* (1617), 4 vols., 3:17a, in *Zhongguo kexue jishu* (cit. n. 39), 4:476.

[67] Johnston, "Of Doctors" (cit. n. 5), 157–8.

[68] Chen, *Waike zhengzhong* (1617), 1:31b, 3:17a, in *Zhongguo kexue jishu* (cit. n. 39), 4:437, 476.

[69] Quoted in Yasuo Otsuka, "Some Features," in Ogawa, *History of the Professionalization of Medicine* (cit. n. 48), 65.

[70] Lorenz Heister, *Practisches medicinisches Handbuch* (Nürnberg, 1766), 291; also Ruisinger, *Patientenwege* (cit. n. 23), 221.

well be true: "it is better not to treat [breast cancer], for being treated, [the patient] dies suddenly, but not being treated, lives longer."[71]

To operate on breast cancer was thus exceptional, and the onus fittingly fell on those performing operations to legitimate their choice. Heister was acutely aware of this problem of legitimation, for in his efforts to elevate surgery's status, he also chastised surgeons for being all too eager to turn to the knife, resulting in operations that were unnecessary and dangerous, thereby fueling images of surgeons' ignorance and cruelty.[72] Despite his later reputation as the "father of modern surgery in Germany," records of Heister's own practice indicate that he rarely ever counseled aggressive surgical cures: in only eleven of his 213 published case histories did he suggest radical procedures such as mastectomy and trepanation to patients. In many of these cases, patients refused.[73]

What, then, of Heister's storied operation on Anna Bayer, the removal of whose twelve-pound breast tumor helped secure his fame? Here, we might turn our focus to the gendered dynamic through which Heister dramatized the process by which he and his patient arrived at a decision. Heister, in published accounts, depicted himself as averse to surgical treatment, believing that the cancer's advanced stage and the tumor's size had rendered its removal far too dangerous. After he expressed this fear, however, Anna spoke:

> She implored me urgently that I not leave her in this wretchedly painful and frightening condition . . . [She said] she would be satisfied if she, after the extraction [of the tumor], would live even for a few more hours without pain, and free of the pestilence which this large breast caused her, so that she could, under more peaceful circumstances, thank God for the liberation from this greatly burdensome evil. Afterwards, she would no longer fear death, if it should come. Based on this manly explanation [Auf diese so mannhafte Erklärung], I promised to carry out the operation with her soon.[74]

I have deliberately rendered *mannhaft* as "manly," in order to stress the parallels with which Japanese translators handled the term *onbarmhertig*. To recall, the Japanese translation skirted the problems posed by *barmhertig*'s association with *jin* by displacing its negation, *onbarmhertig*, into the realm of "womanly and childish sentiments of care and pity." These in turn contrasted with the masculine persona of the physician as loyal retainer. Heister, in an inverse fashion, displaced the will to surgery onto Anna Bayer's "manly" words. The female patient in his representation was recruited as a complicit masculine voice that pushed the physician himself to overcome his reservations and complete his heroic feat. Put more broadly, re-scripting the gendered relation between patient and practitioner helped defuse the ethical controversy of breast cancer surgery. Recasting a female patient as the source of bravery and courage served not merely as proof of consent, but as a proactive means of enablement. The difference in the end, one might say, was a matter of when one stopped listening. Male surgeons could ignore cries of anguish when pressing knives to female flesh, for they had already heeded women's manly will to be operated upon.

[71] Heister, *Heelkundige Onderwyzingen* (cit. n. 56), 1:327. For the original, see Hippocrates, *Aphorisms*, 6, 38 in *Hippocrates*, vol. 4, ed. W. H. S. Jones (Cambridge, MA: Harvard Univ. Press, 1953), 189.

[72] Heister, *Heelkundige Onderwyzingen* (cit. n. 56), 2:846.

[73] "Heister (Laurent)," in *Grand dictionnaire universel*, vol. 9, ed. Pierre Larousse (Paris, 1873), 142; Ruisinger, *Patientenwege* (cit. n. 23), 73.

[74] Heister, *Wahrnehmungen* (cit. n. 25), 981.

"USE ME AS YOUR TEST!"

No surviving evidence confirms that Hanaoka read Heister's specific dramatization of the Anna Bayer case before his 1804 encounter with Kan. Yet we may note how closely their tropes align: the reticence and reserve of the male physician; the courageous resolve of the female patient, ordering the male not to hesitate. To this Hanaoka added a further imperative at the scene's end. "Use me as your test!" Kan allegedly exclaimed. And it is this attempt to negotiate a new legitimation of testing that marks the *Nyūgan chiken roku*'s novelty.

The character I translate as "test"—the Chinese character *yan* 驗 (J. *ken*)—has long played a key role in analyses of the impact of "Western science" in Japan. Translating the same Chinese character instead as "experience" and "experiment," these analyses claim that Western science contained a more pronounced emphasis on experiential and experimental knowledge, thereby challenging a textually oriented "Neo-Confucian" paradigm of natural inquiry.[75] The statements of actors themselves at the time are then marshalled to prove this claim, for figures such as Sugita, Maeno, and Ōtsuki, among others, did frequently invoke the term *ken* in their discussions of the particularity of Western science. So we read that *Kaitai shinsho*'s introduction of anatomical dissection challenged audiences to observe the body as it really was, and not as it had been depicted in Chinese texts. So we read of Ōtsuki telling us that "techniques in Western medicine are refined, because they obtain these from real experience [*jikken*] and real results [*jikkō*], honing them through practice."[76]

More recent research has rightly pointed out that philological approaches to "text" themselves embraced rigorous empiricism; that the study of nature in Japan was no less experientially oriented than that in Europe; that statements such as Ōtsuki's quoted above are better understood as the polemical utterances of a group of scholars seeking to capture ground from Chinese schools of medicine. They have furthermore shown that rather than a dichotomy between experiential versus textual knowledge, the question at stake was rather one of *how* one experiences, whether through the seeing of the eye or the touching of the hand.[77]

Without contradicting these valuable reorientations, I would suggest instead that physicians such as Sugita, Maeno, and Ōtsuki indeed perceived something distinct in the notion of *ken*, but that this newness, rather than being solely on the order of the epistemological, occurred also at the level of the ethical. Approaching *ken* as a problem of virtue, their concern was just as much about the nature of knowledge derived from *experiment* as with the nature of a moral economy that might allow for more aggressive practices of *testing*. Foreign techniques, based on distant understandings of the body

[75] Takahashi Shin'ichi, *Yōgaku* (Tokyo: Nihon Shuppansha, 1939); Nakayama Shigeru, "Kindai kagaku to yōgaku," in *Nihon shisō taikei: Yōgaku-ka*, ed. Hirose Hideo, Nakayama Shigeru, and Ogawa Teizō (Tokyo: Iwanami Shoten, 1972), 441–78, 458–9; Satō Shōsuke, *Yōgakushi no kenkyū* (Tokyo: Chūō Kōronsha, 1980), 6–8; Satō, "Iwayuru 'Yōgaku ronsō' o megutte," *Yōgaku* 1 (1993): 1–14, 6.

[76] Sugita et al., *Yōi shinsho* (cit. n. 33), 1b.

[77] Kuriyama, "Between Mind and Eye" (cit. n. 19), 21–43; Maki Fukuoka, *The Premise of Fidelity: Science, Visuality, and Representing the Real in Nineteenth-Century Japan* (Stanford, CA: Stanford Univ. Press, 2012); Federico Marcon, *The Knowledge of Nature and the Nature of Knowledge in Early Modern Japan* (Chicago: Univ. of Chicago Press, 2015); Trambaiolo, "Ancient Texts" (cit. n. 12), 81–104; Yulia Frumer, *Making Time: Astronomical Time Measurement in Tokugawa Japan* (Chicago: Univ. of Chicago Press, 2018).

and disease and perceived often as dangerous in their invasiveness, could only be accepted if testing despite the risks could be made virtuous. This had already been the case, even before the Dutch, with more aggressive techniques counseled by certain Chinese texts. Painful forms of scraping and bloodletting found in works such as the *Shazhang yuheng* (Jade standard of sand-rashes and swellings; 1675), for instance, elicited theoretical interest among Japanese physicians of the late seventeenth and early eighteenth century, but met with resistance when it came to use on the actual bodies of patients.[78] Contemporaneous with the case of Kan and Hanaoka were debates over Chinese variolation methods: some Japanese practitioners explicitly warned against these techniques on grounds that the "risks seemed too great and the benefits uncertain."[79] This resistance would carry over to the introduction of the Jenner vaccine to Japan, a protracted process that began in the 1820s and lasted well until century's end, necessitating decades of attempts to "mitigate concerns about the European origins of the practice" before its final acceptance.[80]

In Hanaoka's own time, the domestic contours for broader ethical debate over dangerous or aggressive techniques had been set by the controversy over Yoshimasu Tōdō's (1702–73) *tenmeisetsu*, or "discourse on fate." As put forth in Tōdō's posthumous *Idan* (Medical judgments; 1759), the *tenmeisetsu* argued that physicians should not be concerned with the life or death of the patient, as life and death were determined by the will of the heavens. Instead, a physician's sole concern should be understanding the diseases with which he was confronted and deciphering their origins and cures, regardless of the patient's actual survival.[81] The *tenmeisetsu* came under swift condemnation soon after the *Idan* was published.[82] The distinction between understanding a disease and saving lives, critics asserted, was little more than sophistry. For example, as the physician Kamei Nanmei (1743–1814), a friend of Ōtsuki and former student of Nagatomi Dokushōan, retorted, it seemed fundamentally contradictory to claim that "the operation was a success, but the patient died."[83]

Debates over the *tenmeisetsu* would have been quite familiar to Hanaoka, who had spent three months in his youth studying in Kyoto under Yoshimasu Nangai (1750–1813), the adopted son and disciple of Tōdō. Moreover, the limits of testing in relation to patient harm had occupied a controversial role in Hanaoka's career. If the anecdotal record may be trusted, then, in his attempts to devise the final anesthetic used on Kan, Hanaoka had first tested substances on his own mother and his wife, in the process permanently blinding the latter.[84] Then, there was the matter of Kan herself. Although she awoke from her anesthetic slumber to great relief, the tumor excised

[78] Daniel Trambaiolo, "Epidemics and Epidemiology in Early Modern Japan: Japanese Responses to Chinese Writings on Warm Epidemics and Sand-Rashes," in *Translation at Work: Chinese Medicine and the First Global Age*, ed. Harold J. Cook (Leiden: Brill, 2020), 157–75, 166, 168.

[79] Daniel Trambaiolo, "Vaccination and the Politics of Medical Knowledge in Nineteenth-Century Japan," *Bull. Hist. Med.* 88 (2014): 431–56, 437.

[80] Ibid., 440.

[81] Tateno Masami, "Yoshimasu Tōdō no tenmeisetsu ni tsuite: Chūgoku kodai igaku shisō to no renkan kara," *Nihon ishigaku zasshi* 43 (1997): 459–78; Trambaiolo, "Ancient Texts" (cit. n. 12), 94–5.

[82] Aoyama Kenpei, "*Idan*, *Seki idan*: Tenmeisetsu o chūshin toshite," *Nihon tōyō igaku zasshi* 54 (2003): 287–303.

[83] Quoted in Otsuka, "Some Features" (cit. n. 69), 66.

[84] Kure, *Hanaoka Seishū* (cit. n. 1), 51–3. But cf. Matsuki, *Seishu Hanaoka* (cit. n. 3), 30, 52–3, for a more skeptical account.

from her breast, she returned in the second month of the new year to report renewed pain. Her cancer, it appeared, had metastasized, and shortly thereafter, she succumbed. It was around this time that the *Nyūgan chiken roku* was written. Kamei Nanmei's paradox had never been more apt. With his patient dead, how could Hanaoka claim the operation a success?

So it was that Hanaoka enlisted new representations of the patient herself to construct out of the episode a tale of virtue. Commonly slotted within the genre of "case histories," the *Nyūgan chiken roku* in fact presents itself as a document equally interested in memorializing Kan as a model patient for future generations. After introducing Kan and explaining his course of treatment, Hanaoka concluded with a brief statement of his own:

> I had long desired to make an attempt at treating breast cancer. However, the patients with whom I met, upon hearing of the violent assault of the treatment, all cowered in fear and left. This woman was not like that. She entrusted her body to the treatment, and to me. Happily, I have dispelled my doubts from past years. To show this to those seeking the same, I have simply made these images and written this, as seen above.[85]

The demonstrative pronoun "this" in the phrase "to show this" points ambiguously both to the preceding description of Hanaoka's surgical procedure overall and to the "woman" who, rather than "cowering" in fear, entrusted herself to her physician. The inclusion of Kan's portrait as well as its specific style further this duality, and attest to the uniqueness of the *Nyūgan chiken roku*. Case histories themselves had become common in East Asian medicine during the second half of the Ming period (1368–1644), in response to an expanded landscape of commercial publishing and expanded medical marketplace.[86] Patient portraits were not, however, a common feature of this genre.[87] More generally, the illustrations in case histories could roughly be said to have been "diagrammatic" or "schematic" rather than "representational" or "mimetic."[88] That is, rather than indexing the presence of an "identifiable patient as the bearer of a specific pathology," they depicted the ideal visualization of a disease on a decontextualized body.[89] Consider again figure 1 in contrast with the wholly anonymous and

[85] Hanaoka, *Nyūgan chiken roku* (cit. n. 1), 266–8.

[86] Christopher Cullen, "Yi'an (Case Statements): The Origins of a Genre of Chinese Medical Literature," in *Innovation in Chinese Medicine*, ed. Elisabeth Hsu (Cambridge, UK: Cambridge Univ. Press, 2001), 297–322; Charlotte Furth, Judith T. Zeitlin, and Ping-chen Hsiung, eds., *Thinking with Cases: Specialist Knowledge in Chinese Cultural History* (Honolulu, HI: Univ. of Hawai'i Press, 2007); Asaf Goldschmidt, "Reasoning with Cases: The Transmission of Clinical Medical Knowledge in Twelfth-Century Song China," in *Antiquarianism, Language, and Medical Philology: From Early Modern to Modern Sino-Japanese Medical Discourses*, ed. Benjamin A. Elman (Leiden: Brill, 2015), 19–51, 23–5; Joanna Grant, *A Chinese Physician: Wang Ji and the 'Stone Mountain Medical Case Histories'* (London: Routledge, 2003), 21–50; Gianna Pomata, "The Medical Case Narrative in Pre-Modern Europe and China: Comparative History of an Epistemic Genre," in *A Historical Approach to Casuistry: Norms and Exceptions in a Comparative Perspective*, ed. Carlo Ginzburg and Lucio Biasiori (London: Bloomsbury, 2018), 15–46.

[87] Vivienne Lo, "Imagining Practice: Sense and Sensuality in Early Chinese Medical Illustration," in *Graphics and Text in the Production of Knowledge in China*, ed. Francesca Bray, Vera Dorofeeva-Lichtmann, and Georges Métailié (Leiden: Brill, 2007), 383–422; Lo and Penelope Barrett, eds., *Imagining Chinese Medicine* (Leiden: Brill, 2018); Kuriyama, "Between Mind and Eye" (cit. n. 19); Fukuoka, *Premise of Fidelity* (cit. n. 77), 34–9.

[88] Larissa N. Heinrich, *The Afterlife of Images: Translating the Pathological Body between China and the West* (Durham, NC: Duke Univ. Press, 2008), 56–9, 128.

[89] Sander L. Gilman, "Lam Qua and the Development of a Westernized Medical Iconography in China," *Med. Hist.* 30 (1986): 50–69, 63.

Figure 2. *Breast cancer in the* Yizong jinjian *[Golden mirror of the orthodox lineage of medicine] (early 18th century). Wellcome Collection.*

abstract illustrations of breast cancer and ulceration from contemporaneous Chinese texts (see figs. 2 and 3): Kan is rendered with a detailed sensitivity and intimacy, her name and origin alongside, in a manner that stresses individuality.

Yet even the stakes of this portrait run even deeper. Kan smiles peacefully, as if in no pain from her affliction. And though her left breast, swollen, is bared, its unassuming presence does not compel the viewer to inhabit a "clinical" or "anatomo-pathological" gaze. This tranquility of the image makes better contextual sense once we situate it beyond the realm of medical illustration, and in relation to Japanese conventions of personal portraiture. At its broadest level, portraiture in Japan was bound closely to strict codes of access and consent. To grant an artist permission for a long sitting represented a significant act of intimacy. Persons of higher rank were known to refuse sitting

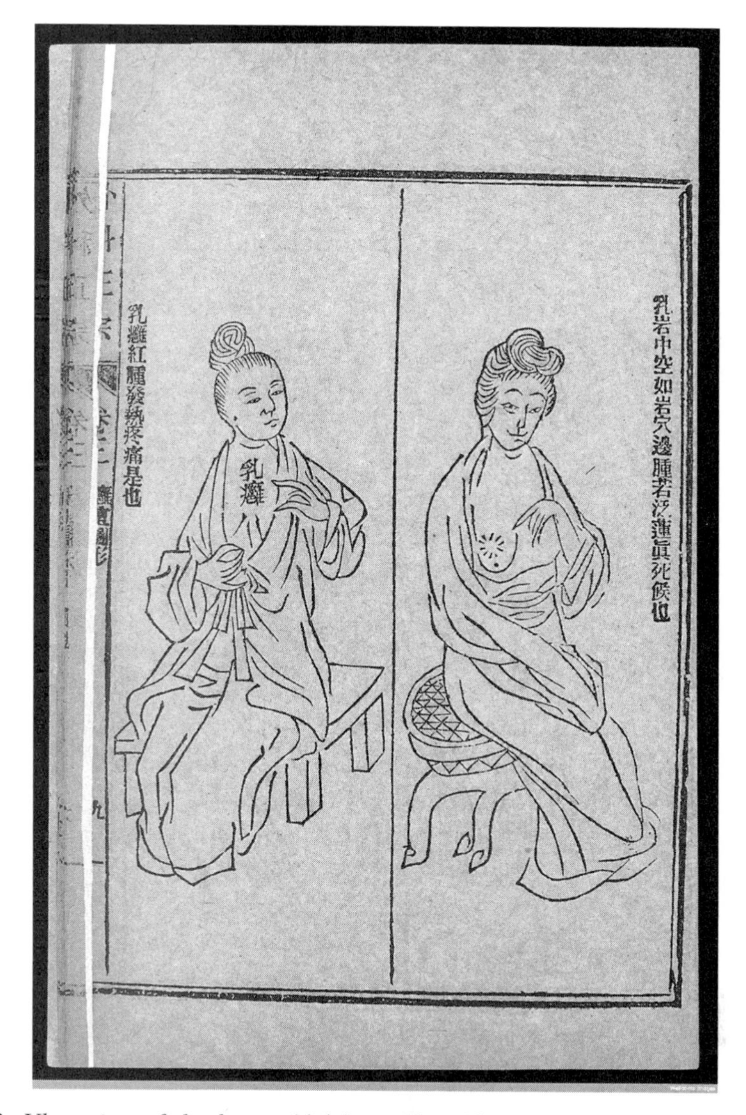

Figure 3. *Ulceration of the breast* (right) *in Chen Shigong's* Waike zhengzong *(1617). Wellcome Collection.*

for portraits, instead offering a list of desired clothing and facial features that were to be synthesized by the artist's imagination into a final image.[90] In this sense, an unperturbed Kan exposing her beast helped further Hanaoka's fashioning of a subject willingly offering herself up to the physician.

Moreover, one primary use of portraits in early modern Japan was as a preparation for impending death. These portraits sought to preserve the ideal essence of that person in their "completed state"—an expression of spirit over matter that would "take over as the human frame collapsed"—and were marked by a specific orientation of

[90] Timon Screech, *Obtaining Images: Art, Production, and Display in Edo Japan* (London: Reaktion, 2012), 172.

the sitter's body and face.[91] Portraits were traditionally placed on the north wall of rooms, facing south. Subjects facing death were therefore depicted looking to their own right—the viewer's left—so that their gaze would fall to the west. In the west lay the Pure Land, the fabled realm where all souls would arrive once they had reached enlightenment, freed from the cycle of earthly reincarnation. To look toward the Pure Land as one died thus expressed the hope that the deceased would at last find her eternal rest; in the cases of eminent personages, this was at times even literalized by the depiction of bodhisattvas in the distance beckoning in welcome.[92] Thus Kan smiles in her westward gaze. She had, Hanaoka suggested, come to terms with her death, facing it bravely, openly, and serenely.

Over the following decades, Kan's portrait circulated through forms of scribal publication bound closely with apprentices of the Hanaoka school. Disciples recopied it, often separately from the actual case history, changing details of hand-coloration along the way. At least one surviving copy of the portrait was rebound as a handscroll, a format that brought with it connotations of a high-end text destined for veneration.[93] More than a case history for spreading new therapies, the *Nyūgan chiken roku* was also a document which, in memorializing Kan, projected the model for a new kind of patient. This new patient was one who, by the courageous exposure of her own body to the unknown, created a space for Japanese physicians to test dangerous foreign procedures while maintaining a commitment to a Confucian norm of *jin*. By demanding to be used as tests, patients legitimated testing as a virtue.

MEDICAL TRANSLATION AND THE COMMENSURABILITY OF VIRTUE

As word of his operation on Kan spread, Hanaoka's reputation grew, giving rise not only to a thriving practice but to the largest private medical school of the era.[94] In response to this expansion, Hanaoka's relationship to his patients came to be increasingly codified, a transformation that would produce the first documented liability waivers in early modern Japan.[95] These waivers were employed almost exclusively for cases of breast cancer surgery. They are all the more astonishing in the face of evidence that no significant malpractice legislation or litigation occurred during the Tokugawa period.[96]

It would be misleading, however, to view the *Nyūgan chiken roku* as motivated by socioeconomic concerns about malpractice. Rather, the text represents a moment in

[91] Ibid., 173, 177; Harold Bolitho, *Bereavement and Consolation: Testimonies from Tokugawa Japan* (New Haven, CT: Yale Univ. Press, 2003), 45.

[92] Michele Marra, *The Aesthetics of Discontent: Politics and Reclusion in Medieval Japanese Literature* (Honolulu, HI: Univ. of Hawai'i Press, 1991), 94; Screech, *Obtaining Images* (cit. n. 90), 174. More generally, monks nearing death were advised to face westward and meditate in anticipation of their afterlife. See Paul Groner, "Icons and Relics in Eison's Religious Activities," in *Living Images: Japanese Buddhist Icons in Context*, ed. Robert H. Sharf and Elizabeth Horton Sharf (Stanford, CA: Stanford Univ. Press, 2001), 114–50, 150; Alan Cole, "Upside down/Right Side up: A Revisionist History of Buddhist Funerals in China," *Hist. Relig.* 35 (1996): 307–38, 324.

[93] *Hanaoka Seishū no jikken zu*, n.d., manuscript, National Diet Library, Tokyo. On the scroll as a high-end format, see Sasaki Takahiro, *Nihon koten shoshigaku ron* (Tokyo: Kasama Shoin, 2016), 9–29.

[94] Over the course of its existence, Hanaoka's school is reported to have trained over 1,300 students. See Otsuka, "Some Features" (cit. n. 69), 70.

[95] Kure, *Hanaoka Seishū* (cit. n. 1), 447.

[96] Johnston, "Of Doctors" (cit. n. 5), 168.

the disruption and resettlement of an ethical worldview—one in which the testing of more radical procedures was emerging as something that might be pursued as an essential part of the virtuous physician's persona. To assert this virtue and render radical testing compatible with existing discourses of *jin*, Hanaoka constructed more than a patient who passively acquiesced in the form of a liability waiver. He fashioned instead a robustly active patient whose very bravery pushed him into new realms of practice.

The translation of Dutch medicine in early modern Japan was therefore always more than a matter of the commensurability of techniques, substances, and bodily imaginations. For these new techniques, substances, and bodily imaginations—yet *un*tested, and to many traditional practitioners dubious, if not dangerous—also demanded a commensurability of virtues to authorize their testing as a humane act. Those invested in importing Dutch medical knowledge to Japan, as we have seen, explicitly sought to affirm the translatability of *jin* as *barmhertig*, and employed strategies of gendering to maintain this translatability in the face of other contextual barriers. In the *Nyūgan chiken roku*, Hanaoka himself drew on their efforts, further complementing them through verbal and visual devices that would depict Kan as an exceptional female patient heroically embracing the risk of death.

Indeed, it is possible that increasingly "global" histories of medicine might profit at large from examining translation as a vehicle not only of epistemological shifts, but of ethical shifts. Here, "incurable" or "hopeless" cases such as breast cancer may prove to be especially useful objects of study. Much as the sociology of scientific knowledge (SSK) privileged situations of controversy as an object of study, arguing that they made visible the "taken-for-granted," so too could we suggest that situations of incurability make visible the insufficiency of presuming everyday pragmatism as an adequate account for the translation of medical knowledge.[97] The presumption of this pragmatism as a mechanism for translation has been evident in studies of global circulation that have of late stressed the maxim of "translating what works."[98] Some have gone so far as to stress efficacy as a "universal concept," shaped by a "*conditio humana* from which nobody can escape," and fundamentally determining the success or failure of medical knowledge in transit.[99]

There are reasons, however, to remain circumspect—to resist this trend wherein notions of efficacy and utility have become "taken for granted." Medical knowledge, as I have argued here, is not translated simply with an aim toward effectiveness. Instead, efficacy and utility are but one set of virtues to be negotiated with others through translation—including, for instance, the problem of how to translate virtues such as benevolence. Even more fundamentally, the relative value and meaning of notions such as "efficacy" and "usefulness" are culturally inflected: what counts as efficacious, what counts as useful, varies significantly by time and place.[100] In this sense, "hopeless" cases

[97] Harry M. Collins, *Changing Order: Replication and Induction in Scientific Practice* (London: Sage Publications, 1985), 5–6.

[98] Harold J. Cook, *Matters of Exchange: Commerce, Medicine, and Science in the Dutch Golden Age* (New Haven, CT: Yale Univ. Press, 2007), 339–77.

[99] Beatriz Puente-Ballesteros, "Chocolate in China: Interweaving Cultural Histories of an Imperfectly Connected World," in Cook, *Translation at Work* (cit. n. 78), 58–107, 65, 100.

[100] Nathan Sivin, "The Question of Efficacy," *Asian Med.* 10 (2015): 9–35, esp. 12–13. See more generally François Jullien, *Traité d'efficacité* (Paris: Grasset, 1997). On the need to understand how "usefulness can mean different things in different times and places," see Eric Moses Gurevitch, "The Uses of Useful Knowledge and the Languages of Vernacular Science: Perspectives from Southwest India," *Hist. of Sci.* (2020): 1–31, 9–10.

like Kan's reveal how, even in the direst of circumstances, at the limits of the body and its suffering, medical translation was always still concerned with something far more than the raw techniques that would effect a cure. To operate on Kan's body required the reorganization of a social and moral life-world. To translate was to produce that reorganized life-world: one in which new and desired cures, their efficacy untested, might be morally permissible.

Notes on Contributors

Tara Alberts is Senior Lecturer in Early Modern History at the University of York. Her research examines encounters and exchanges between Europe and Asia between ca.1500 and ca.1700. She is the author of *Conflict and Conversion: Catholicism in Southeast Asia, 1500–1700* (Oxford University Press, 2013), and the co-editor (with D. R. M. Irving) of *Intercultural Exchange in Southeast Asia: History and Society in the Early Modern World* (I. B. Tauris, 2013). Her current projects examine circulations and translations of materials, ideas, and practices relating to health around Southeast Asia, and the intersection of the spiritual and material in beliefs about healing in the early modern world.

Benjamin Breen is Associate Professor of History at the University of California, Santa Cruz, and the author of *The Age of Intoxication: Origins of the Global Drug Trade* (Pennsylvania, 2019), which won the American Association for the History of Medicine's William H. Welch Medal in 2021.

Montserrat Cabré is Professor of the History of Science at the University of Cantabria, Spain. Her research interests include medieval women's healthcare practices, medical and natural philosophical conceptions of the gendered body, and the history of women's knowledge; she has published widely on these topics in several languages. Currently, she is working on a team project on the hermeneutics of the visible body in late medieval Latin medicine, as well as on a monograph on the cultural history of women's healing practices in the Crown of Aragon between 1300 and 1500.

Sietske Fransen is Max Planck Research Group Leader of the group "Visualizing Science in Media Revolutions" at the Bibliotheca Hertziana–Max Planck Institute for Art History in Rome. She has co-edited (with Niall Hodson and Karl A. E. Enenkel) *Translating Science in Early Modern Europe* (Brill, 2017), and she is co-editor (with Katherine M. Reinhart) of "The Practice of Copying in Making Knowledge in Early Modern Europe," special issue, *Word & Image* 35, no. 3 (2019). Her primary research interests are in visual and verbal communication of early modern scientific and medical practices.

Pablo F. Gómez is Associate Professor in the Department of Medical History and Bioethics, and the Department of History at the University of Wisconsin, Madison. His first book, *The Experiential Caribbean*, won the 2019 William H. Welch Medal and the 2018 Albert J. Raboteau Book Prize. He is interested in histories of knowledge making, and health and corporeality in the early modern world, with a particular focus on the histories of Latin America, the Caribbean, and the African diaspora. His latest book, *The Gray Zones of Medicine* (an edited volume), explores the history of health practices in Latin America from the sixteenth to the twenty-first century through the biographies of unlicensed healers. Gómez is currently working on a history of the quantifiable body, and medical arithmetics, emerging in the world of slave trading in the sixteenth- and seventeenth-century Mediterranean and Atlantic worlds. He is also collaborating on several projects largely related to global histories of science and medicine and histories of enslavement, health, and bodies in the Caribbean Latin America.

Shireen Hamza is a historian of science and medicine in the medieval Islamic world. Her dissertation is about the ways transregional medicine (*ṭibb*) was made local by Muslims across the Indian Ocean world. She has also published research on the history of sexuality and the body. She is a doctoral candidate in the Department of the History of Science at Harvard University with a field in Critical Media Practice, and a managing editor of the Ottoman History Podcast.

Hansun Hsiung is Assistant Professor in Modern Languages and Cultures at Durham University. He works at the intersection of the global history of science and media history in the nineteenth century. His book manuscript, *Learn Anything!: Cheap Print and the Diffusion of Western Knowledge*, traces networks of reprinting and translation across South, Southeast, and East Asia that formed the groundwork for "Western knowledge" in Japan. His publications have appeared in *Isis*, *Contemporary Japan*, *PMLA*, and the *Harvard Journal of Asiatic Studies*.

Elaine Leong is Lecturer in History at University College London specializing in histories of everyday science, medicine, and technologies in early modern Britain. Her first book, *Recipes and Everyday Knowledge: Medicine, Science, and the Household in Early Modern England* (Chicago, 2018), won the 2019 Margaret W. Rossiter Prize. She has co-edited a number of volumes of essays and journal special issues, including (with Alisha Rankin) *Secrets and Knowledge in Medicine and Science, 1500–1800* (Ashgate, 2011), (with Carla Bittel and Christine von Oertzen) *Working with Paper: Gendered Practices and the History of Knowledge*

(Pittsburgh, 2019), and (with Claudia Stein) *The Cultural History of Medicine in the Renaissance* (Bloomsbury, 2021).

Projit Bihari Mukharji is Associate Professor at the University of Pennsylvania. His research focuses on the history of science and medicine in modern South Asia. Mukharji is the author of three monographs: *Nationalizing the Body: The Medical Market, Print and Daktari Medicine* (London, 2009), *Doctoring Traditions: Ayurveda, Small Technologies, and Braided Sciences* (Chicago, 2016), and, most recently, *Brown Skins, White Coats: Race Science in India, 1920–66* (Chicago, 2022).

Ahmed Ragab is a historian, physician, and documentary filmmaker. He is Associate Professor of the History of Medicine at Johns Hopkins School of Medicine, and the founding director of the independent Center for Black, Brown, and Queer Studies. Ragab's research focuses on the history of medicine in the premodern Middle East and the Islamic world as well as on colonial and postcolonial medicine, science, and technology. He also studies and publishes on gender and sexuality in the medieval and early modern Middle East; postcolonial studies of medicine, science, and religion; and other questions in the history of medicine, science, and religion. He is the author of *The Medieval Islamic Hospital: Medicine, Religion, and Charity* (Cambridge, 2015), *Piety and Patienthood in Medieval Islam* (Routledge, 2018), and *Medicine and Religion in the Life of an Ottoman Sheikh* (Routledge, 2019).

Alisha Rankin is Associate Professor of History at Tufts University and co-editor of the *Bulletin of the History of Medicine*. Her first book, *Panaceia's Daughters: Noblewomen as Healers in Early Modern Germany* (Chicago, 2013), won the 2014 Gerald Strauss Prize for Reformation History. She co-edited (with Elaine Leong) a collection of essays titled *Secrets and Knowledge in Medicine and Science, 1500–1800* (Ashgate, 2011). Her most recent book, *The Poison Trials: Wonder Drugs, Experiment, and the Battle for Authority in Renaissance Science* (Chicago, 2021), examines poison antidotes, medical testing, and ideas of proof. She has published widely on a variety of topics in the history of science and medicine, and she is working on a new project on witchcraft, magic, and medicine.

Daniel Trambaiolo is Assistant Professor of Japanese Studies at the University of Hong Kong. He has published articles and book chapters on the history of vaccination, ways of conceptualizing epidemics, and other aspects of early modern Japanese medical history. He is currently working on a book manuscript titled *Ancient Texts and New Cures: Transformations of Medical Knowledge in Early Modern Japan*.

Dror Weil is Assistant Professor of History at the Faculty of History, University of Cambridge. His teaching and research interests focus on scientific and other textual exchanges between the Islamicate world and China during the late medieval and early modern periods. He has published articles on China's participation in the early modern Islamicate book culture, the fourteenth-century transformation in China's reception of Arabo-Persian astronomy, the role of Chinese-Muslims as agents of scientific knowledge, and the study of Arabic and Persian texts in late imperial China.

Index

L

M

N

O

P

Q

R

X

Y

Z

SUGGESTIONS FOR CONTRIBUTORS TO OSIRIS

OSIRIS is devoted to thematic issues, conceived and compiled by guest editors who submit volume proposals for review by the OSIRIS Editorial Board in advance of the annual meeting of the History of Science Society. For information on proposal submission, please write to the Editors at osiris@bbqplus.org.

1. Manuscripts should be submitted electronically in Rich Text Format using Times New Roman font, 12 point, and double-spaced throughout, including quotations and notes. Notes should be in the form of footnotes, also in 12 point and double-spaced. The manuscript style should follow *The Chicago Manual of Style*, 17th ed.

2. Bibliographic information should be given in the footnotes (not parenthetically in the text), numbered using Arabic numerals. The footnote number should appear as superscript. "Pp." and "p." are not used for page references.

a. References to books should include the author's full name; complete title of book in *italics*; place of publication; date of publication, including the original date when a reprint is being cited; and, if required, number of the particular page cited (if a direct quote is used, the word "on" should precede the page number). *Example*:

[1] Mary Lindemann, *Medicine and Society in Early Modern Europe* (Cambridge, UK: Cambridge Univ. Press, 1999), 119.

b. References to articles in periodicals or edited volumes should include the author's name; title of article in quotes; title of periodical or volume in *italics*; volume number in Arabic numerals; year in parentheses; page numbers of article; and, if required, number of the particular page cited. Journal titles are abbreviated according to the journal abbreviations listed in *Isis Current Bibliography. Example*:

[2] Lynn K. Nyhart, "Civic and Economic Zoology in Nineteenth-Century Germany: The 'Living Communities' of Karl Möbius," *Isis* 89 (1999): 605–30, on 611.

c. All citations are given in full in the first reference. For succeeding citations, use an abbreviated version of the title with the author's last name. *Example*:

[3] Nyhart, "Civic and Economic Zoology" (cit. n. 2), 612.

3. Special characters and mathematical and scientific symbols should be entered electronically.

4. A small number of illustrations, including graphs and tables, may be used in each volume. Hard copies should accompany electronic images. Images must meet the specifications of The University of Chicago Press "Artwork General Guidelines" available from the Editor.

5. Manuscripts are submitted to OSIRIS with the understanding that upon publication copyright will be transferred to the History of Science Society. That understanding precludes consideration of material that has been previously published or submitted or accepted for publication elsewhere, in whole or in part. OSIRIS is a journal of first publication.

OSIRIS is published once a year.

ISSN: 0369-7827 | E-ISSN: 1933-8287

Paperback ISBN: 978-0-226-82156-6

eISBN: 978-0-226-82512-0

Single copies are $35.00.

Address subscriptions, single issue orders, claims for missing issues, and advertising inquiries to *Osiris*, The University of Chicago Press, Journals Division, 1427 E. 60th Street, Chicago, IL 60637-2902.

Postmaster: Send address changes to *Osiris*, The University of Chicago Press Subscription Fulfillment, 1427 E. 60th Street, Chicago, IL 60637-2902.

OSIRIS is indexed in major scientific and historical indexing services, including *Biological Abstracts, Current Contexts, Historical Abstracts*, and *America: History and Life*.

Copyright © 2022 by the History of Science Society, Inc. All rights reserved. The paper in this publication meets the requirements of ANSI standard Z39.48-1984 (Permanence of Paper). ♾

Osiris

A RESEARCH JOURNAL DEVOTED
TO THE HISTORY OF SCIENCE
AND ITS CULTURAL INFLUENCES

A PUBLICATION OF THE
HISTORY OF SCIENCE SOCIETY

CO-EDITORS
W. PATRICK MCCRAY
University of California Santa Barbara

SUMAN SETH
Cornell University

COPY EDITOR
BETH INA
Lincoln, Nebraska

PAST EDITOR
ANDREA RUSNOCK
University of Rhode Island

PROOFREADER
JENNIFER PAXTON
The Catholic University of America

OSIRIS EDITORIAL BOARD

PABLO GÓMEZ
University of Wisconsin–Madison

MATTHEW LAVINE
Mississippi State University
EX OFFICIO

MEGAN RABY
University of Texas, Austin

ALEXANDRA HUI
Mississippi State University
EX OFFICIO

HSS COMMITTEE ON PUBLICATIONS

AILEEN FYFE
University of St. Andrews

JENNIFER RAMPLING
Princeton University

SIGRID SCHMALZER
University of Massachusetts Amherst

W. Patrick McCray
Co-Editor, Osiris
Department of History
University of California, Santa Barbara
Santa Barbara, CA 93106-9410 USA
pmccray@history.ucsb.edu

Suman Seth
Co-Editor, Osiris
Department of Science & Technology Studies
321 Morrill Hall
Cornell University
Ithaca, NY 14853 USA
ss536@cornell.edu